Lecture Notes in Computer Science 6585

Stefano Ceri   Marco Brambilla (Eds.)

Search Computing

Trends and Developments

Volume Editors

Stefano Ceri
Marco Brambilla
Politecnico di Milano
Dipartimento di Elettronica e Informazione
P.za L. Da Vinci, 32, 20133 Milano, Italy
E-mail: {stefano.ceri, marco.brambilla}@polimi.it

ISSN 0302-9743 e-ISSN 1611-3349
ISBN 978-3-642-19667-6 e-ISBN 978-3-642-19668-3
DOI 10.1007/978-3-642-19668-3
Springer Heidelberg Dordrecht London New York

Library of Congress Control Number: 2011922270

CR Subject Classification (1998): H.4, H.3, D.4, C.2.4, F.2, D.1.3

LNCS Sublibrary: SL 3 – Information Systems and Application, incl. Internet/Web and HCI

Typesetting: Camera-ready by author, data conversion by Scientific Publishing Services, Chennai, India

Printed on acid-free paper

Springer is part of Springer Science+Business Media (www.springer.com)

Preface

Searching for information is perhaps the most important application of today's computing systems. In the new century, all the World's citizens have become accustomed to thinking of the Web as the source for answering their information needs, and search engines as their Web interface. Websites reporting a movie's plot, tomorrow's weather in our next destination, the risks of a surgical procedure, the fastest route to a friend's house, the video of the last opera at La Scala can all be found as result of a keyword search. If any Web page in the world stores the answer to our information need, then we expect the search engine to link that page and describe it through a snippet appearing in the first page of the search results.

A few search engine companies are able to meet such expectations, and completely cover the search engine market. However, offering a link to a Web page does not cover all information needs. Many problems cannot be solved by simple keyword-based queries. The notion of "best page" for solving a given problem is typically inadequate when the problem requires solutions spanning over multiple pages. We are indeed accustomed to using a variety of Web resources to solve our problems: while the search engine hints to useful information, the user's brain is the fundamental platform for information integration.

Problems such as "who is the best doctor to cure insomnia in a nearby hospital" can be solved by using the Web multiple times, searching for partial results. Once a hospital's website is located and the listing of its doctors is extracted, one can find out that there is one doctor in the list who has published recent papers on insomnia. While doing so, the user is performing information integration in her brain; specifically, she is applying ranking while extracting hospitals based on proximity and doctors based on their publications on insomnia, then matching on the basis of doctor names. Of course, the best way to build the matching is to use the search engine itself, by entering doctors' names as keywords, extracted from either the hospital search or the literature search, but then the search is less focused and result interpretation is more difficult.

Complex queries are supported in certain domains, such as travels, hotel booking, and book purchasing, by specialized, domain-specific search systems or search engine integrators. It is important to assess how travel assistants solve the problem: they offer a few predefined queries to build the itinerary, then offer additional services (e.g., car rentals, hotels, local events, insurance) so as to complete the plan around the itinerary. Thus, they perform specialized steps of integration by substituting the user's brain, and then let the user enrich the solution incrementally and interactively, with customized interfaces. In other words, they solve complex queries in the context of given domains, which are supported by a substantial business; such specialized search systems dominate over general purpose ones in their domain of expertise, and therefore attract users.

The search computing project (SeCo), funded by the European Research Council as an advanced IDEAS grant, aims at building concepts, algorithms, tools, and technologies to support complex Web queries. The project is now entering the third of a five-year lifespan (November 2008 – November 2013); it proposes a new paradigm for solving complex queries based on combining data extraction from distinct sources and data integration by means of specialized integration engines. Data extraction retrieves data from different sources, ordered based on local rankings, and data integration merges such results into result combinations, with an associated global ranking, such that combinations with the highest ranking are produced as fast as possible; a result combination represents the solution of a complex search problem. Thus, the search computing project has the ambitious goal of lowering the technological barrier required for building complex search applications, thereby enabling the development of many new applications which will cover relevant search needs.

Search computing covers many research directions, which are all required in order to provide an overall solution to a complex search. The core of the project is the technology for search service integration, which requires both theoretical investigation and engineering of efficient technological solutions. The core theory concerns the development of result integration methods that not only denote "top-k optimality," but also the need of dealing with proximity, approximation, and uncertainty. Such a theory is supported by an open, extensible and scalable architecture for computing queries over data services, designed so as to incorporate the project's results by adding new operations, by encoding new join methods, and by injecting new features dealing with incremental evaluation and adaptivity.

A number of further research dimensions complement such core. Formulation of a complex query and browsing over solutions is a complex cognitive task, whose intrinsic difficulty has to be lowered as much as possible so as to meet usability requirements. Therefore, we are investing a consistent effort in the development of user-friendly interfaces which are targeted at assisting users in expressing their needs and then browsing on results. Solving a complex problem requires supporting users in the interactive and incremental design of their queries, thereby assisting search as a long-term process for exploring the solution space; result differences can be better appreciated by visualizing results (e.g., through maps or timelines). The project success also depends on the ability of registering new sources and making them available for solving complex problems; therefore, we have designed abstractions, architectural solutions, and model-driven design tools for service registration and for application development, aiming at assisting service publishing, application design, and query execution tuning. While the current description of Web resources is very simple, so as to enable an equally simple description of Web interactions, we aim at linking the service description to ontological sources, so as to enable high-level expressive interfaces covering the gap from high-level interactions to query expression.

While focusing on technological dimensions, we are also investigating crucial aspects to the project success, such as the business models and user involvement in the design process through user-centered design. We are additionally investigating the use of search computing for scientific applications, such as supporting bio-informatics research by enabling the access to genetic and proteomic data sources.

This book reports the proceedings of the workshop "New Trends in Search Computing," held in Como and Milan during May 25–31, 2010, as the follow-up of the workshop "Search Computing Challenges and Directions," also published by Springer in 2010 (LNCS 5950).

The workshop was divided into eight independent sessions, reflecting the many research directions of the project. It was held during five consecutive days, in Milan and Como, with about 60 participants equally divided between SeCo researchers and international experts. Each workshop session had editors, chosen within the SeCo research team, and external experts, who provided their tangible contribution to the project with feedback, advice, and contributed chapters. Session editors helped us in organizing the book's design, by interacting with the session experts and by shaping up each session and the corresponding book part.

Each part of the book reports the result of a workshop session; it includes one chapter describing the search computing approach to the problem, and one or more additional chapters reflecting the contribution and viewpoints of experts that participated in the workshop, broadening the spectrum of investigations which are currently ongoing in search computing. In some cases, the part is closed by a short chapter reporting different opinions that the workshop participants discussed at panels which closed the corresponding session.

The book is the result of the collective effort of all the project participants and has been reviewed by several experts. We would like to thank all of them for their efforts.

January 2011

Marco Brambilla
Stefano Ceri

Organization

Reviewers

Carlo Batini	Università degli Studi di Milano-Bicocca
Jordi Cabot	INRIA - École des Mines de Nantes
Andrea Calì	University of Oxford
Alex Komoroske	Google, Inc.
Rodrigo Lopez	European Bioinformatics Institute
Ioana Manolescu	INRIA - Université de Paris Sud
Massimo Paolucci	DOCOMO Euro-Labs
Alfonso Valencia	Spanish National Cancer Research Centre
Roberto Verganti	Politecnico di Milano
Gerhard Weikum	Max Planck Institute for Informatics

Part Editors

Part 1: The Search Process	Marco Brambilla and Stefano Ceri
Part 2: Interaction Design	Tiziana Catarci and Maristella Matera
Part 3: Semantic Description	Alessandro Campi and Davide Eynard
Part 4: Rank-Join	Davide Martinenghi and Marco Tagliasacchi
Part 5: Query Processing	Daniele Braga and Michael Grossniklaus
Part 6: Tools and Mashups	Marco Brambilla and Alessandro Bozzon
Part 7: BioSeco	Marco Masseroli
Part 8: Sustainable Exploitation	Emanuele Della Valle

Sponsoring Institutions

The Search Computing (Seco) Project is funded by the European Research Council (ERC), responding to the 2008 Call for "IDEAS Advanced Grants," a program dedicated to the support of investigation-driven frontier research. SeCo started on November 1, 2008 and will last until October 31, 2013.

Table of Contents

Part 1: The Search Process

Part 2: Interaction Design

Part 3: Semantic Description

Part 4: Rank Join

Part 5: Query Processing

Part 6: Tools and Mashups

Part 7: Bio-SeCo

Part 8: Towards a Sustainable Exploitation

Part 1
The Search Process

Search is a complex activity that can be modeled as a multi-step process, leading users from generic information exploration purposes to identifying precise information needs and finally to make access to the information they need. A complex search process is typically characterized by multiple steps, spanning multiple sources of information during long-term sessions; the process is characterized by a continuous refinement of the search goal. During this activity, users combine information finding with source exploration, they aggregate and compose results; occasionally they alternate online interaction with manual note-taking. Such complex information seeking behaviors challenge the search engine interfaces, because they require to support all the stages of information acquisition, from the initial formulation of the area of interest, to the discovery of the most relevant and authoritative sources, to the establishment of relationships among the relevant information elements.

The first chapter of this part presents a vision of how current search engines technology is challenged by understanding the user's needs, either implicitly, trying to guess unexpressed user's intents as if the engine was reading the user's mind, or explicitly, by offering interactive tools that provide hints about the user's intent.

The second chapter describes the exploratory capabilities offered by search computing. It presents interaction options for selecting objects within a resource framework and then shows how search computing users can progressively build complex queries by progressively assembling them, driven by object properties and inter-object connections.

Finally, the last chapter collects the research visions of the workshop's participants and focuses on issues such as helping the user, taking advantage of the wisdom of the crowds, contextualization vs. personalization, and raising the level of interaction with search systems.

The New Frontier of Web Search Technology: Seven Challenges

Ricardo Baeza-Yates[1], Andrei Z. Broder[2], and Yoelle Maarek[3]

[1] Yahoo! Research, Barcelona, Spain
[2] Yahoo! Research, Santa-Clara, CA
[3] Yahoo! Research Haifa, Israel

Abstract. The classic Web search experience, consisting of returning "ten blue links" in response to a short user query, is powered today by a mature technology where progress has become incremental and expensive. Furthermore, the "ten blue links" represent only a fractional part of the total Web search experience: today, what users expect and receive in response to a "web query" is a plethora of multi-media information extracted and synthesized from numerous sources on and off the Web. In consequence, we argue that the major technical challenges in Web search are now driven by the quest to satisfy the implicit and explicit needs of users, continuing a long evolutionary trend in commercial Web search engines going back more than fifteen years, moving from relevant document selection towards satisfactory task completion. We identify seven of these challenges and discuss them in some detail.

1 The Evolution of Commercial Search Engines

Commercial Web search engines have gone through a significant evolution in the last 16 years. We have identified three major stages in their evolution according to the type of data on which the technology has focused at the time:

- **On-Page Data:** First generation engines, like Excite, Lycos, or AltaVista in the middle of the 90's, used standard information retrieval models on crawled pages and used the *on-page* textual data almost exclusively, thus supporting only a syntactic match between queries and documents. The key paradigm was to parse HTML pages, possibly assigning more weights to important sections such as titles and abstracts, or author supplied keywords, and use usual *tf x idf* methods to compute relevance of pages within the Web corpus seen as a flat collection. The main challenge was scale and speed with relevance taking a second seat.
- **Web Graph Data:** Around 1998 a new approach, made popular by Google and eventually adopted by all engines, started to exploit *off-page* Web specific data. Three types of off-page data were leveraged: (1) link (or connectivity) data, initially simply in the form of in-degree (the number of links to a page), later through an analysis of the entire Web graph. The use of this data was based on the idea that "people vote with their links", [13, 4]; (2) anchor text data, that is, how people refer to a page on *other pages*, as a form of surrogated text abstraction of the target page. Here the idea was that "people vote with their labels"[10]; and finally (3) "click-through" data, that is, what results users clicked on as a form of implicit voting, the

S. Ceri and M. Brambilla (Eds.): Search Computing II, LNCS 6585, pp. 3–9, 2011.

idea being that "people vote with their clicks" [12], something that DirectHit used as early as 1997.

Simultaneously an infrastructure revolution took place based on distributed processing over a huge numbers of commodity computers. This novel infrastructure and its dedicated software enabled a "scale revolution" and allowed to crawl, index, store and serve more data than ever before, while respecting response time, latency and freshness constraints.

- **Usage Data:** Finally the third stage, which we are still experiencing, attempts to answer "the need" behind the query, that is, the unexpressed intent that drove the user to make a particular query in the first place. This phenomenon is expressed in multiple manners: First, the engine tries to guess what type of information best answers the intent and to this end, multiple sources of data are integrated in the result page, for example images, maps, videos, stock quotes, weather reports, current prices, tweets, news, etc. Second, additional query assistance tools, deployed both before and after the query is processed are becoming prevalent: for instance, query spell correction à la *did you mean* or query completion after the user entered only a few characters, are deployed on all major search engines. Similarly, results exploration tools (*e.g.*, narrowing search results by type, source, date, translation, etc.) are becoming common place. This is achieved by focusing on yet another type of data: usage data. While previously usage data had been aggregated at page level, the attention has now moved from individual pages to individual users. Users' behavior at every stage of their interaction with the search engine, and even their post-interaction behavior (via browser toolbars, beacons, etc.) can be captured in very detailed logs for an enormous number of queries. By studying users' behavior *after* they make a particular query (*e.g.*, query reformulations, clicks, browsing time), one gains an understanding of users' intent. In practice, efficient statistical methods are and can be used to create adequate pre- and post-search assistance tools.

By monitoring users' activities and gathering usage data at a very large scale, search engines serve users at two levels: *implicitly* by trying to guess unexpressed intent, as if the engine was reading the user's mind and *explicitly* by offering interactive tools, which not only make the search experience more attractive but also provide additional hints regarding the user's intent. We believe these two directions represent the new frontier of Web search and are associated with a number of technical challenges. We list below a few of these challenges and areas where progress is being made and more innovation is to be expected. For readers interested in a detailed coverage of Web retrieval we suggest the chapter by Baeza-Yates and Maarek in [3].

2 Ongoing and New Challenges

As discussed earlier, monitoring users' activities on a very large scale allows to better answer implicit and explicit information needs [5, 19] and more specifically query intent. We have identified seven challenges that we believe represent opportunities for further research and innovation, some already seeing incremental progress happening on a regular basis and others demanding drastic departure from previous art. We have ranked them below by their order of (possibly future) appearance in the Web.

2.1 Query Assistance

Query assistance tools first appeared on the search engine results page, offering alternate query forms in case the user was not satisfied with the returned results. These alternate queries took two possible forms: first related queries, typically derived from the results themselves, and soon after query spelling suggestion that leverages usage data [6]. The novelty in the now famous "did you mean" feature for instance consisted in its learning from usage data rather than using a fixed dictionary. Multiple techniques are used today such as counting most frequent queries at a small edit distance of the original query[1] or looking at query reformulation as users tend to correct themselves if their original query was misspelled [16]. Leveraging usage data via query-log analysis at a large scale gave and is still giving excellent results. It is also one of the best examples of the "wisdom of crowds" [21]. It took a longer time however for the attention to move to the core search box and try helping users formulate their queries even before the query is issued. The first major query assistance tool appeared in 2004 when Google offered " Suggest"[2], as an experimental feature on Google Labs. Thanks to the scale revolution, it was suddenly possible to use as completion dictionary a large query log, which gave the impression of an uncontrolled vocabulary experience. At that stage the corpus was static, but progress in infrastructure allowed Yahoo! Search to launch "Search Assist" in 2007 and the following year Google to launch Suggest on google.com and youtube.com. Nowadays, most engines offer this feature and keep improving it, with better freshness, coverage, locality, instant previews, etc. This represented a critical stage in helping users express their intent, based on the conjecture that the odds are good that a previous user with similar intent found a good query to express it.

2.2 Contextualization

Another trend in recent years has been the *contextualization* of the answer. We use the term contextualization in a generic manner to cover (1) localization (geographical and/or language contextualization), (2) personalization (user contextualization), (3) socialization (that is, take in account the social context), and (4) query intention (intent contextualization), among others. Considering that most users interact little and do not explicitly authorize individual identification (by signing-in for instance), personalizing is a difficult task, which impacts only a small percentage of users. In contrast, intent contextualization relies on analyzing the usage data originating from users conducting the same task. It is both more pragmatic (as it does not require signing-in or explicit authorization), and more effective as it can applied to larger populations of users and thus some limited form of the previous mentioned "wisdom of crowds" intuitions can be applied. Most of all, privacy infringement risks are significantly reduced as no single user is isolated, and techniques are applied to groups of people (small crowds). The effectiveness of these methods come from the fact that while users are all different in their heterogeneous needs or facets, on each facet they are not that different from other users

[1] A well known Google example shows that the correct spelling for the query "Britney Spears" is more frequent by an order of magnitude than its immediate follower, see http://www.google.com/jobs/britney.html

[2] Now called autocomplete [22].

and perform similar tasks, their uniqueness comes from the combination of these facets and on when, how long and how well they conduct those tasks. The challenge here is then to better detect query intent and better contextualize intention. Contextualizing the results affects search results display and the overall user experience (*e.g.*, geographical contextualization may require displaying an interactive map within the search results page) and hence triggers the next challenge: how to present different types of results. Regarding the social aspect, for some search needs the social context is clearly relevant as the Web is a communication media that is owned in a large extent by its users through the Web 2.0 [18].

2.3 Universal Search

Another significant step in guessing the intent of the query is not to require from the user to specify what type (*e.g.*, image, video, map, etc.) or source of data (*e.g.*, news, blogs, encyclopedia, etc.) s/he is more interested in but simply guess for a given query what types and sources should be shown. The goal here is to integrate rich and complex data source in a semi-transparent manner. This concept was coined "universal search" by Marissa Mayer in [14] and continues seeing a great deal of progress. It presents multiple challenges, as it requires "comparing apples and oranges", and more specifically deciding what sources should be probed, how many results from relevant sources should be shown, where in the ranked list these results should be slotted if at all, etc. This area becomes even more intriguing with real-time feeds such as tweets for which relevance needs to be estimated in almost real-time. One open problem is the screen layout for the different types of results if you want to move away from the classical sequential list of ten results. This research area is now called "aggregated search" and naturally leads to the next challenge.

2.4 Web of Objects

A more recent challenge consists of departing from the usual result triplet (title,snippet, link) as a surrogate of a given Web page and returning instead the object that really satisfied her needs. A typical example when searching for an artist is to get as a result not an heterogeneous list of links to his official site, images, videos, fan club, wikipedia entry, lyrics pages, etc. but rather a composite object that integrates all the possible facets that should be relevant to the user. The same goes for famous athletes (for whom the user would like to get team information, recent stats, photos, etc.), restaurants (address, map, opening hours, reviews,etc.), travel destinations (slideshows, weather, hotels, places to see, etc.). The key idea behind the concept of Web of Objects is that the individual pages that typically form the Web are exploded into individual objects that can be recomposed into a synthetic page, which is then shown to users as a search result. Thus, in our previous athlete example, a user searching today for the Viking quarterback "Brett Favre" will see on a Yahoo! Search, as top result, a synthetic concise "mini page" consisting of various objects such as 2010-2011 stats displayed as a dashboard (with QB rating, number of touch downs etc.), extracted from a given Web site), a profile generated from another Web site, links to news, games logs, Scores and Schedules, etc., all extracted from various sites as needed. More generally, integrating Web derived knowledge, well beyond entity extraction, towards building and representing interrelationships between known entities, will enable users to search not only the"Web of Pages" but the "Web of Objects" as detailed in [2, 7].

2.5 Post-Search Experience

As we focus on intent, we must address needs that go beyond simple information discovery. We are still doing very little with results so far. We can "star" them [9], translate the associated page via a single link on Google, or share video results in a social network or via messenger or email in Bing. In addition, major search engines now offer the ability to narrow search results according to various facets [15]. Some results can be displayed in various microformats [17] using Yahoo! SearchMonkey [20] or Google's rich snippets [11] via common agreement between publishers and search engines. Yahoo! Search Pad [8] goes one step further as it automatically gathers clicked results of a same search session and allow users to annotate/edit/share such pads. Yet, it seems that there is a great deal of opportunities for engines to offer additional tools that would facilitate post-search experience, mostly for better exploration and manipulation (filtering, extracting, etc.) of results so as to better satisfy the underlying query intent. This challenge is also related to "universal search".

2.6 Application Integration

A natural extension of the post-search experience would be to go further with result manipulation and facilitate the integration of third party applications to enable a richer, more diversified, and more satisfying user experience. Underlying intent might involve a series of tasks, such as planning for a holiday, organizing a birthday celebration, etc. where search results represent only raw data that need to be digested and processed to satisfy the intent. This mostly unexplored area presents fascinating technical challenges. A small but significant step in this direction was conducted by Yahoo! Search recently via its "QuickApps" mechanism. Third-party applications are offered on the left trail of the results page by partners such as Netflix, the popular US-based flat-rate movie rental/online video streaming services, or OpenTable the online restaurant reservation service. Such applications can be triggered for specific queries and pre-populated with the needed parameters, once a " movie intent" or "restaurant intent" are identified. This application integration approach focuses on facilitating the task behind the query in order to better satisfy users.

2.7 Implicit Search

Finally, the most intriguing of all challenges is "implicit search", which aims at addressing users' needs without requiring them to express a query. As search engines better and better understand their users by monitoring their activities and building sophisticated users models, and multiple contextual signals are accessible (via sensors, GPS, cell information, etc.) one can envision scenarios in which the need can be identified by a simple click or simply as a side effect of another action. A taste for implicit search is offered today in book selling, like Amazon, or movie rental, such as NetFlix, services with their recommendation services. Another example is the recently launched Priority Inbox in Gmail [1]. These are only preliminary steps in this direction. We believe that with the fast penetration of smart phones and cloud computing where all devices can be associated with a single user, search engines will have at their disposal a plurality of signals that will make the difference in personalization and contextualization. One can envision implicit search mechanisms being triggered within various applications as mentioned above, or related content being pushed to users in appropriate contexts. However, most wild scenarios will remain unrealistic if privacy concerns are not addressed and answered in a satisfying manner.

3 Conclusion

Web search has now become a mature technology with high penetration to the general population in most developed countries. We believe that the focus of innovation should move towards the "before" and "after" stages in web search with endless possibilities. Numerous challenges are involved, including but definitely not limited to the seven we have listed here, with multiple research and technology development opportunities. Overall, all of them are focused in improving the overall search experience of the user. Nevertheless, some of these new results will also improve user experience in general.

References

[1] Aberdeen, D.: Email overload? try priority inbox. The Official Gmail Blog (August 2010), http://gmailblog.blogspot.com/2010/08/email-overload-try-priority-inbox.html

[2] Baeza-Yates, R., Raghavan, P.: Chapter 2: Next generation web search. In: Ceri, S., Brambilla, M. (eds.) Search Computing. LNCS, vol. 5950, pp. 11–23. Springer, Heidelberg (2010)

[3] Baeza-Yates, R., Ribeiro-Neto, B.: Modern Information Retrieval. Addison-Wesley, Harlow (2010)

[4] Brin, S., Page, L.: The anatomy of a large-scale hypertextual web search engine. In: Proceedings of the 7th International Conference on World Wide Web, Brisbane, Australia (1998)

[5] Broder, A.: A taxonomy of web search. SIGIR Forum 36(2) (Fall 2002)

[6] Cucerzan, S., Brill, E.: Spelling correction as an iterative process that exploits the collective knowledge of web users. In: Proceedings of Empirical Methods in Natural Language Processing, Barcelona, Spain (July 2004)

[7] Dalvi, N.N., Kumar, R., Pang, B., Ramakrishnan, R., Tomkins, A., Bohannon, P., Keerthi, S., Merugu, S.: A web of concepts. In: PODS, pp. 1–12 (2009)

[8] Donato, D., Bonchi, F., Chi, T., Maarek, Y.: Do you want to take notes? identifying research missions in yahoo! search pad. In: Proceedings of the 19th International Conference on World Wide Web, Raleigh, North Carolina, USA, pp. 321–330 (2010)

[9] Dupont, C.: Stars make search more personal. The Official Google Blog, March 3 (2010), http://googleblog.blogspot.com/2010/03/stars-make-search-more-personal.html

[10] Eiron, N., McCurley, K.S.: Analysis of anchor text for web search. In: Proceedings of the 26th Annual International ACM SIGIR Conference on Research and Development in Informaion Retrieval, SIGIR 2003, pp. 459–460. ACM, New York (2003)

[11] Goel, K., Guha, R., Hansson, O.: Introducing rich snippets. Google Webmaster Central Blog (May 2009), http://googlewebmastercentral.blogspot.com/2009/05/introducing-rich-snippets.html
[12] Joachims, T.: Optimizing search engines using clickthrough data. In: Proceedings of the Eighth ACM SIGKDD International Conference on Knowledge Discovery and Data Mining, KDD 2002, pp. 133–142. ACM, New York (2002)
[13] Kleinberg, J.: Authoritative sources in a hyperlinked environment. In: Proceedings of the ACM-SIAM Symposium on Discrete Algorithms, pp. 668–677 (1998)
[14] Mayer, M.: Universal search: The best answer is still the best answer.The Official Google Blog (May 2007), http://googleblog.blogspot.com/2007/05/universal-search-best-answer-is-still.html
[15] Mayer, M., Menzel, J.: More search options and other updates from our searchology event. The Official Google Blog (May 2009), http://googleblog.blogspot.com/2009/05/more-search-options-and-other-updates.html
[16] Merrill, D.: http://www.youtube.com/watch?v=syKY8CrHkck#t=22m11s at timestamp 22m11s
[17] Mika, P.: Microsearch: An interface for semantic search. In: Proceedings of the SemSearch 2008 Workshop on Semantic Search at the 5th European Semantic Web Conference, Tenerife, Spain (June 2008), http://sunsite.informatik.rwth-aachen.de/Publications/CEUR-WS/Vol-334/
[18] Ramakrishnan, R., Tomkins, A.: Toward a peopleweb. IEEE Computer 40(8), 63–72 (2007)
[19] Rose, D.E., Levinson, D.: Understanding user goals in web search. In: WWW 2004: Proceedings of the 13th International Conference on World Wide Web, pp. 13–19. ACM, New York (2004)
[20] Searchmonkey, http://developer.yahoo.com/searchmonkey/
[21] Surowiecki, J.: The Wisdom of Crowds: Why the Many Are Smarter Than the Few and How Collective Wisdom Shapes Business, Economies, Societies and Nations. Random House (2004)
[22] Wright, J.: This week in search 10/16/10: Renaming google suggest. The Official Google Blog (October 16, 2010)

Information Exploration in Search Computing

Alessandro Bozzon, Marco Brambilla, Stefano Ceri, and Piero Fraternali

Politecnico di Milano, Dipartimento di Elettronica ed Informazione,
V. Ponzio 34/5, 20133 Milano, Italy
{fistname.lastname}@polimi.it

Abstract. Search computing queries typically address search tasks that go beyond a single interaction. In this paper, we show a query paradigm that supports multi-step, exploratory search over multiple Web data sources. Our paradigm requires users to be aware of searching over "interconnected objects" with given semantics, but each exploration step is simplified as much as possible, by presenting to users at each step simple interfaces, offering some choices that can be supported by the system; choices include moving "forward", by adding new objects to the search, or "backward", by excluding some objects from the search; and the selection and de-selection of displayed results in order to dynamically manipulate the result set. For supporting exploration, we designed a new architectural element, called query orchestrator, which connects the user interface module with the execution engine; the orchestrator maintains the history of the query session and caches query results for reuse at subsequent interactions.

1 Introduction

Search engines, the most popular entry point to the Web, offer user interfaces that are based on keywords. Although advanced search options allow users to combine keywords into complex Boolean expressions, users are not familiar with such query paradigms; therefore, the search engine responds by finding the most authoritative Web pages based on few user keywords, according to page ranking algorithms. Such keyword-based query paradigm is inadequate to express complex search computing queries, especially queries spanning multiple domains of interest.

Therefore, in the Search Computing project we have designed a query language able to express the structure of complex queries, organized as conjunctive expressions over search and exact services [7][8]. The query language yields to a query interface that is a predefined form, where users are asked to enter query constants, so that a parametric query becomes fully instantiated. Query processing consists of optimizing and executing the query, producing results in the form of object combinations, with a tabular representation highlighting their structure (flat schema) and ordering (highest ranked tuples are shown in the top of the table).

To give users some flexibility in exploring the result, we have proposed the *liquid query* paradigm [4][5], which allows users to interact with the search computing result by asking the system to produce "more result combinations", or "more results from a specific service", or "performing an expansion of the result" by adding a sub-query

S. Ceri and M. Brambilla (Eds.): Search Computing II, LNCS 6585, pp. 10–25, 2011.

which was already planned while configuring the query. This result exploration paradigm empowered result browsing in the context of fixed query, and has proven to be fully appropriate in the context of vertical search applications with a structured interaction pattern.

In this chapter we propose a query paradigm that extends *liquid query* by supporting extensibility at query formulation time. By means of such paradigm, the user is supported in expressing fully exploratory queries, starting from an initial status with no predefined query, and enabling a progressive, step-by-step construction of the query itself. The new paradigm consists of exploring a network of connected resources, where each resource corresponds to a clearly identified real-world concept (an "hotel", a "flight", a "hospital", a "doctor"), and the connections have predefined semantics ("hotels" are close to "restaurants", "doctors" care "diseases" and are located at "hospitals"). Such network, called the "semantic resource framework", is built at service registration time, and is described in another chapter of this book [9]. The proposed exploration paradigm exploits query expansion and result tracking, giving the user the possibility to dynamically selecting and deselecting the object instances of interest, and move "forward" (adding one node to the query) or "backward" (deleting one node) in the resource graph. A novel query result presentation paradigm, called *atom view*, supports exploration by visualizing instances of different objects in separate lists, at the same time displaying the combinations they belong to and their global rank. This view allows users to focus at each step on the new results, and therefore is most suitable for a progressive exploration. This presentation paradigm can be easily applied to mobile search interfaces because it displays lists with simpler schema with respect to the tabular view proposed in the original *liquid query* approach (i.e., one object at a time with respect to entire combinations).

Moving from keyword-based or form-based interaction to this new approach is a big shift, which requires user's awareness of the task being performed. We believe that the proposed paradigm can support users in complex search processes, in the spirit of past results of the exploratory search discipline [16][18][19][24][25][26]. The supported search tasks go beyond the typical one-shot, memory-less interaction with a search engine, and span over several steps, with the possibility of suspending and resuming the work (e.g., performing search processes that last several days). Given the long living nature of the search tasks, many subtle problems arise about whether the system should cache data (both at intra-query and inter-query levels) to grant good performances. These issues are also addressed by our solution.

The new approach supports all the stages of information seeking, from the initial formulation of the topic of interest, to the discovery of the most relevant and authoritative sources, to the establishment of relationships among the relevant information elements.

The Chapter is organized as follows: Section 2 presents the general issues related to exploratory search and then focuses on the few existing examples of graph exploration. Section 3 presents the exploratory search paradigm in the context of the semantic resource framework. Section 4 introduces the "atom view" and shows how query formulation and result presentation may alternate on the atom view interface. Section 5 sketches the architecture supporting exploratory search, which is based on the interplay between a query orchestrator (introduced specifically for supporting the exploratory interaction paradigm) and the execution engine (extended for better integration with the orchestrator).

2 Background on Exploratory Search and Resource Graph Browsing

The exploratory search discipline addresses the problem of providing the user with tools and mechanisms for easily moving through an information space until he reaches the information he was looking for. This requires supporting all the stages of information acquisition, from the initial formulation of the area of interest, to the discovery of the most relevant and authoritative sources, to the establishment of relationships among the relevant information elements. All these steps can be performed in an iterative way and should support the navigation of information along semantic connections between data. Some connections could lead to dead ends, and thus require to rollback the navigation history and take other paths towards the information need fulfilment.

A very general view on the tasks and objectives of the user when exploring information is provided by the *information seeking funnel* model proposed in [19], inspired by the *buying funnel* or *sales funnel* in the commercial world, depicting the changing attitude of people at different stages of the buying process, from all those who might possibly be interested in a product or service to those who actually purchase it. An approach similar to the one of potential purchasers appears also in information-seeking users, who are driven to the bottom of the funnel towards information consumption (see Fig. 1).

The first steps of the process include *wandering* (in which the user does not have an information seeking-goal in mind) and *exploring* (in which the user has a general goal but not a plan for how to achieve it). Subsequently, in the *seeking* phase the user clarifies the open-ended information needs that must be satisfied, and finally in the *asking* phase the user identifies an information need that corresponds to a closed-class question. With respect to this model, search computing covers the last three phases: from open-minded exploration (based on the structure of the Semantic Repository Framework) to the request for information based on precise questions.

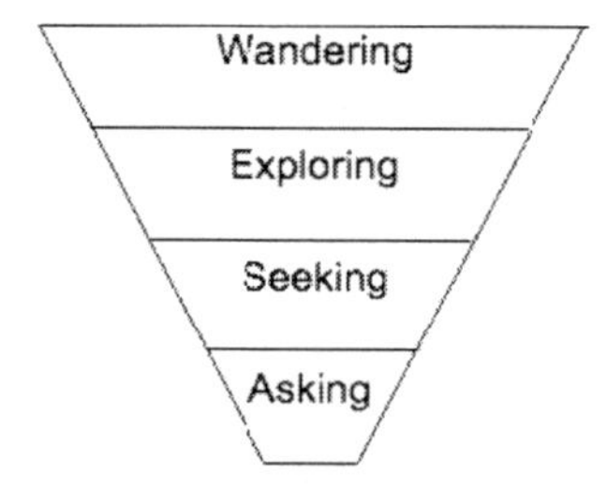

Fig. 1. The Information Funnel

This incremental approach is recognized by basically all the information seeking models: Kuhlthau's Information Search Process depicts information as a process of construction on the part of the individual user through a step-by-step model [14]. Bates' berrypicking model [3] assumes that users jump from source to source and search technique to search technique as a means to build a satisfactory answer to a query. Both models theorize that people search for information by making sense out of multiple search stages. Similarly, orienteering assumes that the user navigates information and gets more and more oriented while getting closer to the result of his search.

A more precise formalization is given by the information foraging theory [18], which assumes that information seekers behave like animals foraging on patches. In this case, patches are information nuggets spread all over the Web and users move around based on patch size and patch transfer effort: they try to intake as much resources as they can, but there is an optimal time limit to be spent on a single patch for maximizing the information ingestion.

Exploratory search can be studied also under the perspective of Interactive Information Retrieval (IIR), which examines the global behaviour of an IR system from the perspective of the user's interaction [19]. In this field, a characterization of the iterative search process has been proposed by Norbert Fuhr [11], who presents a notion of probabilistic result ranking adapted to the IIR context, where the fundamental assumptions of classic probabilistic ranking are not verified, because the relevance of a result seen by the user is influenced by the documents already seen in the same interactive session and result list browsing is not the only and prominent user's task. Fuhr's interaction model considers an IIR session as the traversal of several system states (situations), in which the user has to take a choice for progressing in the resolution of his information need; the cost model for analysing interaction alternatives should considers the situations, choices and expected benefits in the user's IR task. An example of system supporting the interactive information retrieval paradigm is ezDL (Easy Access to Digital Libraries), a front-end application for seeking across different digital libraries, supporting long-lived search sessions though several search tactics, stratagems and strategies, including multiple query perspectives, query history, and result organization tools.

Search computing acts as a facilitator in the exploration process, by increasing the acquisition efficiency at the single patch level and reducing the effort for moving from one patch to the other, thanks to the join paths that link data sources directly related to his initial query formulation and to the search expansion mechanisms that permit the user to reach novel data sources loosely connected to the initial query.

The utility of search computing can be regarded also under another perspective, related to the mind-set and goals with which users afford a search task.

On one side, the user's information need may have a variable degree of precision: a query may look for concrete factual information (e.g., which is the capital of a state) or be rather indeterminate (e.g., getting information on a certain place). On another (complementary) side, the information need can vary quite substantially in its inherent complexity (e.g., finding only generic information about a place, or finding accurate facts targeted to some specific goal, e.g., going to live in a new place).

Queries that are precise and simple are well served by keyword-based search engines, which aim at presenting very quickly the most relevant document. As the complexity and vagueness of the information need increases, the user tends to do more than one query, to refine the input keywords, try alternative formulations, and take note of partial intermediate results, in order to "assemble" the answer.

This is pictorially depicted in Fig. 2 by drawing a kind of borderline (the "note-taking boundary") delimiting a region beyond which one-shot keyword queries are not sufficient to resolve an information need the users resorts to "note-taking" to compose the response to his need.

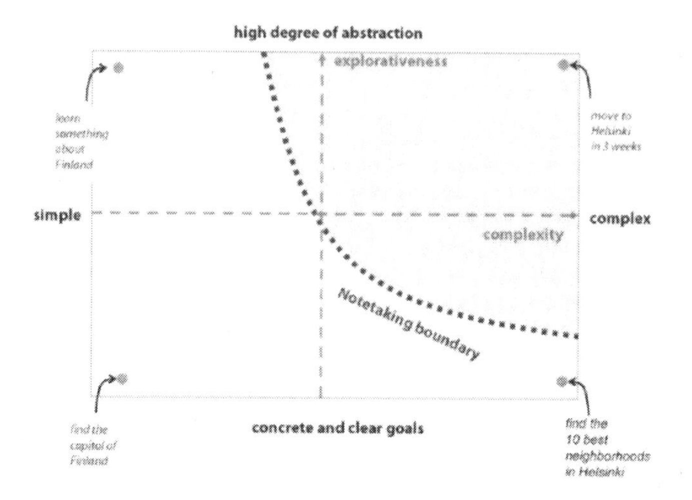

Fig. 2. The note-taking limit [2]

Search computing addresses information seeking tasks that lie just after the boundary of the note-taking limit in exploratory search, where the complexity of the search task and of the associated information is quite high, and the exploration of the options plays an important role in finding the ideal solution.

In the first phases of the interaction, our approach relies on the visual exploration of graphs of resources that can be eventually queried. On this side, we base our work on existing graph exploration techniques, both in terms of general purpose studies and associated with domain specific applications (for instance, [11] proposes visual exploration of biological information).

Some approaches put more emphasis on the aggregation, clustering and dimension reduction of multivariate data [23], so that complex graphs can be reduced in size (both in terms of number of nodes and edges). This technique could help also in our setting, in the event of large resource graphs. Other proposals apply visual transformations to the graphs for giving more emphasis to the items that are deemed more relevant at the moment, e.g., through hyperbolic plan visualization [17]. Similarly, the work in [22] tackles the problem of exploration of large conceptual schemas expressed in the Entity-Relationship notation, by proposing a technique based on graph topology analysis for extracting a meaningful subset of the schema, which is then displayed in the bi-dimensional space using a force-directed placement algorithm.

Finally, some proposals offer domain-specific languages and tools that help generate automatically graph-based visualization and exploration starting from a generic dataset [1].

3 Exploration of the Resource Graph

The search computing paradigm departs quite radically from the traditional keyword-based query interaction typical of search engines. Users are aware of the conceptual structure underlying the Search Computing repository, described in the semantic resource framework (SRF) [9], and sketched in Fig. 3.

The SRF is a high level description of the services that collectively constitute a given "domain of discourse", i.e., a particular area of interest that can be the target of SeCo queries (e.g. tourism information, real estate, scientific publications, etc.). Users access the top-level view of SRF, which is a simple Entity-Relationship model, as shown in Fig. 3 for the tourism domain. Such a view defines the application context, characterized by the presence of named entities (*service marts*, e.g. Hotel, Restaurant, etc.) and relationships (*connection patterns*, e.g. geographical nearness). This top-level SRF view abstracts away from the complexity of mapping service interfaces to data sources and of integrating the different names and formats used by each source to represent its properties, and focuses on a simple, semantic view, which simplifies the exploration of information and the definition of search queries by non-expert users.

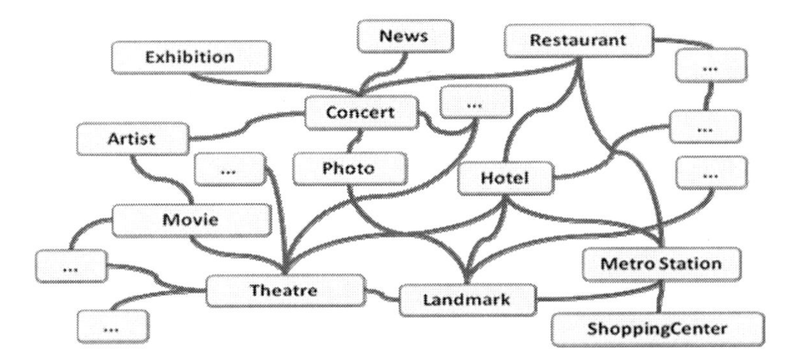

Fig. 3. Objects and connections in the Semantic Resource Framework

Thanks to the SRF, users express their queries directly upon the concepts that are known to the system, such as hotels, restaurants, or movie shows; moreover, users are aware of the connections between the concepts, and therefore they can, e.g., select a show and then relate it to several other concepts: the performing artist, the close-by restaurants, the transportation and parking facilities, other shows being played in the same night in town, and so on. The query is focused (and restricted) to known semantic domains. This can be seen as a major limitation of the query interface, but on the other hand it offers also greater power in organizing the exploration of the search space as a continuous process.

The exploratory query interaction paradigm proceeds as follows. The user starts by selecting one of the available objects, and submits an "object query" to extract a subset of object instances. For example, a user could choose a "*concert*" object, and ask for "jazz" concerts in a "club" in the "Village" area of "New York", or choose a "*restaurant*" object, and ask for "vegetarian" restaurants in the "south end" of "the city" in a given "price range". Object queries are conjunctions of selection predicates: each predicate is represented in the user interface as a widget for inputting the corresponding selection parameters, e.g. a drop-down list for choosing a value among the available restaurant types or city districts, or a slider bar for setting price ranges. Queries results are ranked and the user interface also lets the user specify his ranking preferences, e.g. "distance", "quality", "price range" or a combination of them.

The output is a set of objects that satisfy the query, displayed in rank order; while in this chapter we assume a specific representation of results (by using the *atom view* visualization technique described in the next section), chapter [6] of this book describes visualization mechanisms that take advantage of the properties of the retrieved results, such as the presence of temporal or geo-referenced data types in the object schema. Normally, the system presents a first batch of results, and users browse them; if users are not satisfied, they can ask additional batches, until all objects that satisfy the query are retrieved; showing only the top-ranked object instances is a typical feature of search system.

At this point, the user can select the most relevant object instances and continue with the exploration by choosing the next concept among the various ones that can be connected to the selected objects; after that, (s)he submits another object query, possibly by providing additional selection criteria; the system will then retrieve connected object instances and form a "combination" with the objects retrieved at the preceding steps. Result combinations at any given step of the search process are ranked according to the global ranking criterion based upon the local rank criteria of previously selected objects; if the user wants to preserve previous ranks, he can ask the system not to re-rank object instances or whole combinations selected at previous steps.

Continuing in the above example, after selecting three or four "jazz clubs", the user is offered to add to his "night in the city" plan several additional options: restaurants, exhibitions, movies, music shows, dancing places, and other amenities. If they select a "restaurant", the connection to "concert" maps concerts to the restaurant instances that are close to the concert instances, as shown in Fig. 4. If they further select "metro", they are asked to indicate the starting point of their ride, and the search obtains for every selected object instance at the previous interaction the best metro station and ride, with the detail of trains, line changes, and expected time.

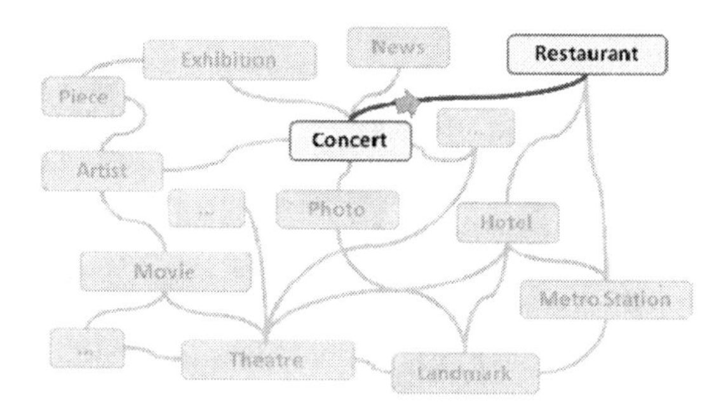

Fig. 4. Selection of two connected objects on the SRF

At any stage, users can "move forwards" in the exploration, by adding a new object to the query, starting from the connections available in the SRF and from the objects that have been previously extracted. Forward path exploration can be applied to all previously extracted objects/combinations or to a subset only, manually "checked" by the user. Users can also "move backwards" (backtrack) in the exploration, by excluding one of the objects from the query, or by "unchecking" some of their previous manual selections of relevant object instances. For example, a user may decide that the bus ride is too inconvenient, prefer to use a car instead, and then explore parking opportunities for the selected restaurants. Fig. 5 shows how nodes of the SRF can be added to and removed from a given query.

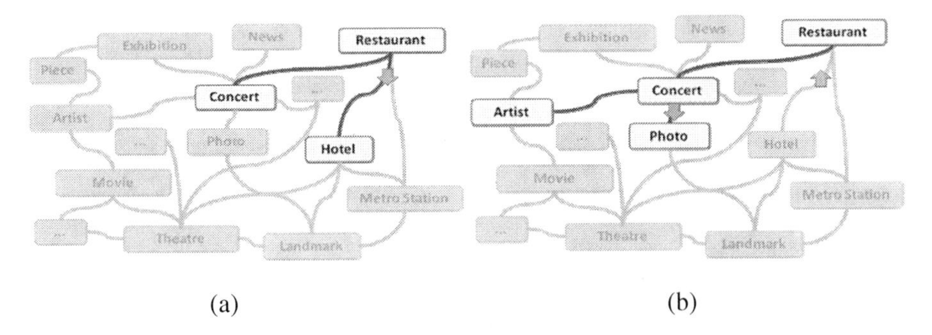

(a) (b)

Fig. 5. Moving "forward" and "backward" on the SRF

At any stage of the exploration, users can store the status of their process, by saving the query that has been formulated so far as well as the results. Such saving, of course, does not guarantee data validity, as obviously data may become obsolete; however, it may save a pattern of process for reuse. Therefore, the same query can be repeated for a different night, saving the exploration effort. Backtracking at the level of individual conditions may help, e.g., in changing the choice of restaurant from "vegetarian" to "Japanese" during reuse. Long-lived processes may be very useful for more complex planning tasks, e.g. exploring the opportunities offered by the real estate market, or planning a summer vacation.

4 Search Computing Information Exploration Interface

To support the exploration of the resource graph as described in Section 3, a set of user interaction mechanisms must be designed. In our work, we investigated various interface options and we put them at work on several case studies.

4.1 Results Visualization

Our first work on exploratory multi-domain search exploited a flat visualization of result combinations in tabular format [4][5], as shown in Fig. 6. This approach proved valuable for introducing the user to the concept of multi-domain result sets: every combination was clearly identified by a corresponding row in the table and one or more table columns represented each composing item. Combinations could be selected and expanded by adding joins with a predefined set of entities. Such entities and their connections were pre-selected at design time by a SeCo expert user, based on their fitness to the specific application scenario.

This approach, though, exhibits some weaknesses in terms of readability and understandability of results: a table tends to quickly become very lengthy, due to the combinatorial nature of the items that match together; moreover, objects of the same type are

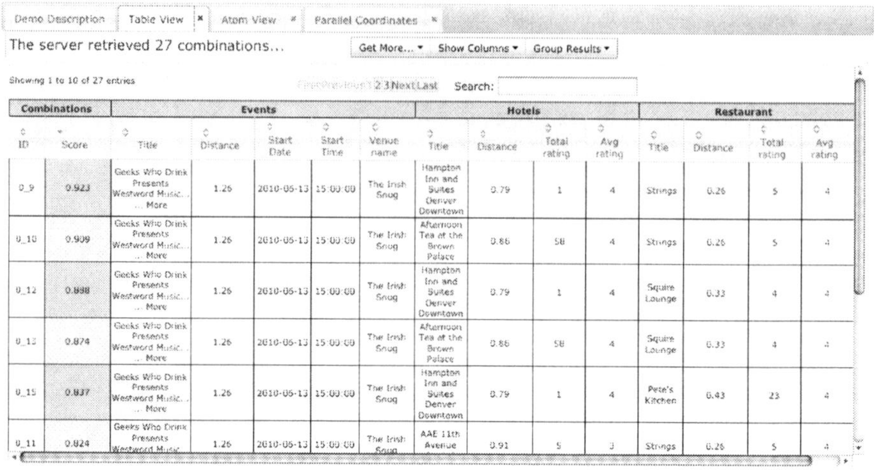

Demo Description | Table View | Atom View | Parallel Coordinates

The server retrieved 27 combinations...

Get More... | Show Columns | Group Results

Showing 1 to 10 of 27 entries

First Previous 1 2 3 Next Last Search:

Combinations		Events					Hotels				Restaurant			
ID	Score	Title	Distance	Start Date	Start Time	Venue name	Title	Distance	Total rating	Avg rating	Title	Distance	Total rating	Avg rating
0_9	0.923	Geeks Who Drink Presents Westword Music... ... More	1.26	2010-05-13	15:00:00	The Irish Snug	Hampton Inn and Suites Denver Downtown	0.79	1	4	Strings	0.26	5	4
0_10	0.909	Geeks Who Drink Presents Westword Music... ... More	1.26	2010-05-13	15:00:00	The Irish Snug	Afternoon Tea at the Brown Palace	0.86	58	4	Strings	0.26	5	4
0_12	0.888	Geeks Who Drink Presents Westword Music... ... More	1.26	2010-05-13	15:00:00	The Irish Snug	Hampton Inn and Suites Denver Downtown	0.79	1	4	Squire Lounge	0.33	4	4
0_13	0.874	Geeks Who Drink Presents Westword Music... ... More	1.26	2010-05-13	15:00:00	The Irish Snug	Afternoon Tea at the Brown Palace	0.86	58	4	Squire Lounge	0.33	4	4
0_15	0.837	Geeks Who Drink Presents Westword Music... ... More	1.26	2010-05-13	15:00:00	The Irish Snug	Hampton Inn and Suites Denver Downtown	0.79	1	4	Pete's Kitchen	0.43	23	4
0_11	0.824	Geeks Who Drink Presents Westword Music	1.26	2010-05-13	15:00:00	The Irish Snug	AAE 11th Avenue	0.91	5	3	Strings	0.26	5	4

Fig. 6. Tabular View Interface: each row represents a combination composed by one object per type

scattered through different combinations, which hampers comparisons within the population of objects of the same type; combinations are also often repetitive (especially when some "good" object dominate others and thus joins with many other in top rank combinations), thus requiring the user to scroll down the table in order to have a comprehensive view of the available objects. These observations prompted for the definition of a novel user interaction interface (described in the next section), but also to various research directions, including visualization of multi-domain contents (illustrated in [6]), diversification of structured results, and so on.

To improve over the tabular representation, we implemented a novel result visualization, called *atom view*, which forms the basis for the exploration of the resource graph and the iterative expansion of the query. When the query is submitted, the search computing engine calculates the result set in terms of combinations of extracted items, as for the tabular interface. However, instead of showing the combinations as table rows, the *atom view* adopts the visualization strategy shown in Fig. 7: items are shown in separate lists, one for each entity involved in the query: in the example of Fig. 7 (central part), the set of hotels is shown side by side with the list of events. The lists are completely independent and show the top-k items retrieved by the query for each entity. In the *atom* view the unified global ranking is displayed in a dedicated widget (Fig. 7, bottom part), where the joined combinations of objects are represented (and ordered) by their global ranking score: relative importance of combinations is represented by the ordering of items, while absolute value of the ranking score is shown through appropriate visual clues (like colour gradient or object size). The visualization of join relationships among the objects in a combination is left to the user interaction: by moving the mouse over the list of combinations, the user can select one and automatically highlight the items in the various object lists that contribute to the combination; vice versa, he can select one element from a list and thus highlight the combinations comprising it. The advantage of this view is the good simultaneous visibility of the items extracted, of their combinations and of the global ranking.

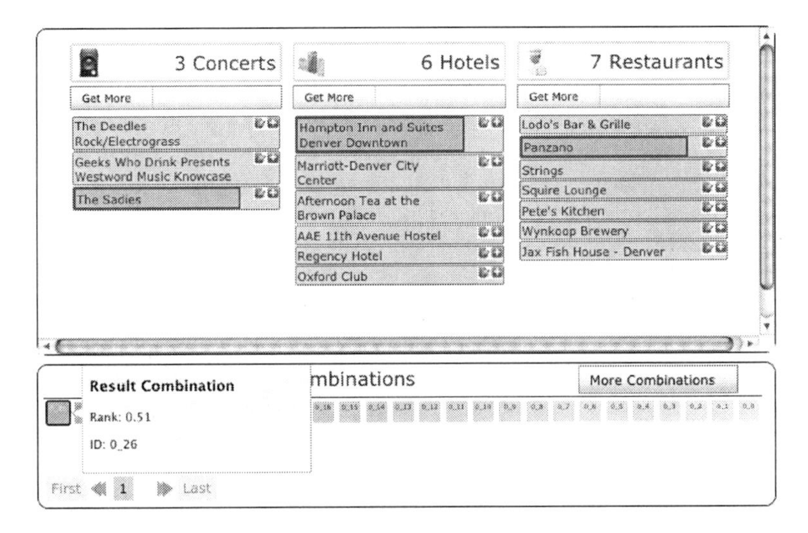

Fig. 7. Screenshot of the *atom view* interface as implemented in the current Search Computing demonstrator (http://www.search-computing.com/demo/UI)

The modular structure of the *atom view* is well-suited to mobile devices (PDA, smartphones, etc.), which demand for simple interfaces, with less information shown on screen and quick links for navigating across pages. Indeed, one can think to an exploratory approach in which information is shown one entity at time, and exploration is performed by selecting a set of entity instances and moving to another set of instances connected to those ones. This would be implemented in a simple interface like the one in Fig. 8, where distinct list of objects (Hotels in the example of Fig. 8.a) are shown in separate screens, and it is up to the user to navigate across such screens to explore the current result set.

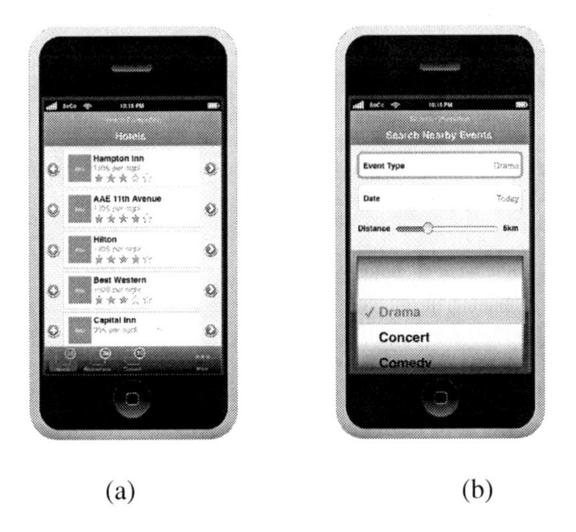

Fig. 8. Example of interaction in a mobile device

4.2 Explorations of Results

The atom view is a good basis for achieving an interface organization capable of supporting the exploration approach better; the envisioned interaction consists of three steps, as shown in Fig. 9: the user submits some initial search criteria referred to a single entity, which drive the retrieval of the instances of that entity only. Among these instances, (s)he selects the ones (s)he is interested in and then proceeds to the selection of the next entity to explore.

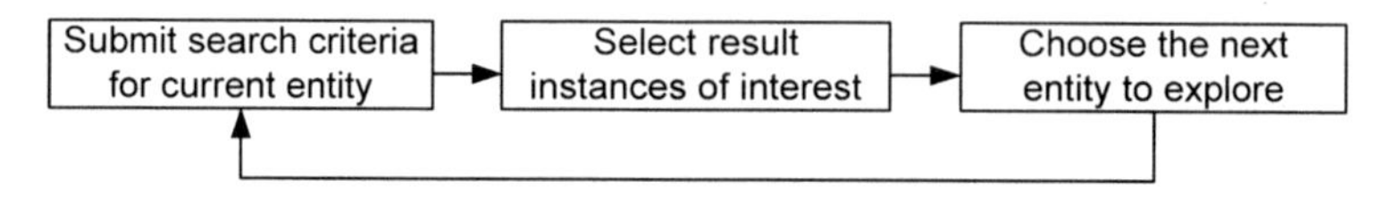

Fig. 9. Iterative approach for entity-by-entity exploration

The selection of the possible exploration directions starts from one of the entities currently part of the result schema. For each entity, the list of possible directions are shown, grouped by the exploration semantics (e.g., space or temporal proximity, match of object names, etc.), as visible in the left part of the interface of Fig. 10: for instance, given a selection of hotels of interest, the user is shown the list of relevant resources grouped by geographical nearness (restaurants, theatres, museums, events), temporal nearness (movie show time, concert), name match (reviews). The user can choose one of the entities and therefore start the exploration loop again by submitting the search criteria (as described in Section 4.1). The exploration path is shown pictorially in the form of breadcrumb links on the top of the screen, forming a derivation tree, thus allowing the user to see the steps explored so far in terms of selected objects. Exploration can be

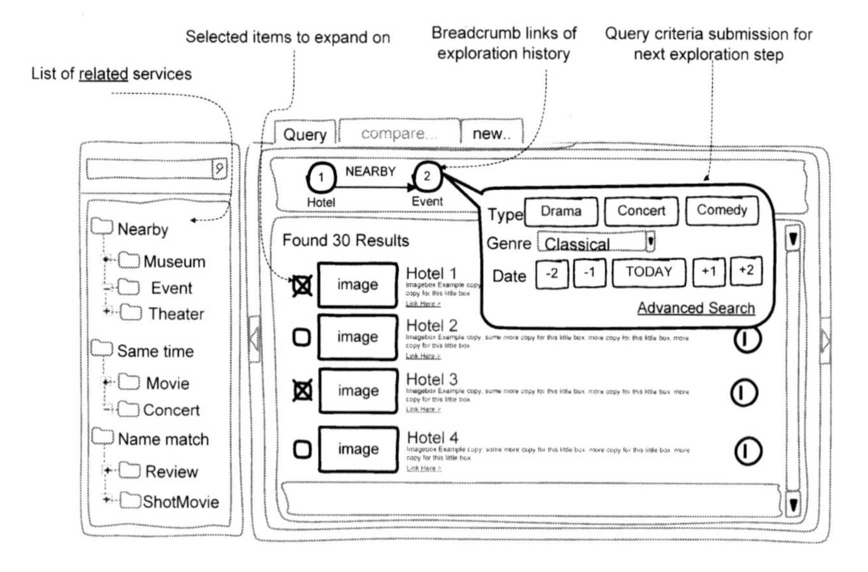

Fig. 10. Selection of next entity to expand towards

backtracked by clicking on the breadcrumb links at the end of the tree (links to the leaves of the tree); this action backtracks the corresponding query execution. Results are not refreshed; therefore backtracking leads the user to a previously retrieved query result. Then, he may choose to change the selection of combinations or to refresh the results before trying a new exploration in a different direction.

The submission of search criteria at every step consists in filling in an input form that asks for a set of parameters based on the input schema of the service invoked to retrieve the entity instances. Such a form can be automatically inferred, and then can be customized by a designer. For instance, Fig. 11 shows a sequence of input forms for searching an Event entity.

Given a selected entity (service mart) in the SRF, the following steps are performed:

- **Concept selection** (Fig. 10): starting from the current exploration status, the user decides to explore a new concept in the resource graph (e.g., an "event" or a "restaurant"), according to the connection paths available from the currently explored information. The connections may be based on geographical distance, temporal closeness, text similarity or other (possibly combined) search criteria.
- **Pattern selection** in some cases, a given object can be selected according to different input parameters and, correspondingly, the query can consist of alternative access patterns, each one with a different set of input attributes. In these cases, the user can select the search modality (e.g., by "genre" or by "date"), as shown in Fig. 11.a.
- **Data source selection** (Fig. 11.b)**:** when multiple physical services implement the same access pattern, the user can pick the one that (s)he wants to use in the query.

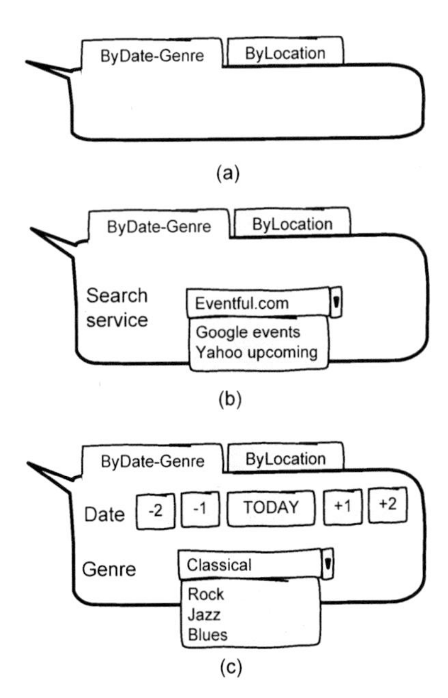

Fig. 11. Query Submission Interface

- **Input of selection criteria** (Fig. 11.c)**:** once the search modality and data source are chosen, suitable values for the set of input attributes of the corresponding access pattern must be provided by users. The interface is kept as intuitive and immediate as possible: all the enumerative attributes are submitted by selection, and also some quantitative ones, when possible, are transformed into pseudo-categorical ones: for instance, the date of the event can be chosen by clicking on a very limited set of items (e.g., yesterday, today, tomorrow) instead of being submitted as a full date. Notice that some input values for the selection criteria inserted by the user may be inconsistent with the search service selected in the previous step, because the service may not cover the subdomain specified by the user (e.g., the user may select a service working only for US locations and then enter a European city as a search parameter). Appropriate input validation will be applied.

5 Infrastructure for Supporting Explorative Queries

According to the exploratory vision presented in the previous sections, the search computing framework must enable the user to perform step by step exploration of the data. Furthermore, the user must be able to easily test and rollback exploration paths, so that various options can be explored with the aim of selecting the best ones. To grant this feature, a search computing system must also provide the possibility of moving back and forward along the navigation history.

To support incremental expansion of search queries, navigation of the information space, and back-forward navigation, a search computing system must manage *queries* and *result-sets* as first class-citizens in the architecture: queries are objects amenable to identification (to be uniquely identified), manipulation (to support forward-expansions and backtracking), storage and inspection (to allow later reuse as predefined explorations); likewise, result-sets need to be identified, inspected and stored to enable their late reuse as search services in other queries.

These requirements have been taken into account for developing the search computing reference architecture as a four-layered framework, comprising *User Interface*, *Query orchestration*, *Execution engine*, and *Semantic resource framework*, plus two auxiliary and shared components that manage the queries and the cached data respectively (see Fig. 12).

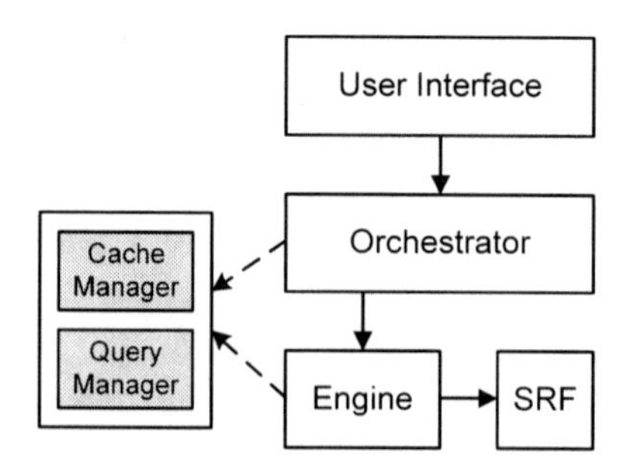

Fig. 12. Overview of SeCo Architecture

The **User Interface (UI)** layer is the front-end of a search computing system and provides to the end-users all the functionalities described in Section 4. Its internal architecture follows the well-known MVC pattern and aims at translating the user interaction events into the appropriate result manipulations or query executions over the search services.

The **Query Orchestrator** is the core component with respect to the management of the exploration features of search computing. In particular, it supports query formulation, evolution and storage. It also offers different ways to manipulate and store search results. Each query is univocally identified in the orchestrator and can be re-executed at any moment. Each result set produced by a query execution is also identifiable and retrievable. These two aspects are the basis for allowing query history navigation: to each step N we associate a query Q_N (comprised with the input parameters needed for its computation), its result set R_N, and the subset of results selected by the user SR_N. Once the user has reached a step N in his exploration, (s)he can move back to steps $N\text{-}1$, ..., $N\text{-}k$, and thus retrieve the old queries (Q_{N-1}, ..., Q_{N-k}, available in the Query Manager component), result sets (R_{N-1}, ..., R_{N-k}), and selections queries (SR_{N-1}, ..., SR_{N-k}) respectively, that (s)he had already explored. If (s)he wants, (s)he can:

- Refresh such results, thus defining a new step M in the navigation, for which the associated query is still Q_N , but the result set is a new one R_M (and the selected results SR_M are obviously reset);
- Submit new search criteria, thus defining a new step P in the navigation, for which the associated query has the same structure of the old one, but with different input parameters, and thus is identified as a different one Q_P , together with the new result set that is produced R_P;
- Start new exploration paths by submitting some expansion query on other search services, thus defining a new set of selected results SR_M and performing a new step S in the navigation, with a completely new query Q_S , which in turn will produce a new resultset R_S;

The *Query Orchestrator* acts as a *controller* that governs the flow of query executions and query/result-set storage and reuse. In such a context, the *cache manager* controls the distributed cache system exploited by the query orchestrator to maintain the current stateful objects for all the active user sessions; the *query manager* manages the current state of interaction for each active user session, keeping track of each user actions' history.

The **Execution Engine (EE)** is in charge of transforming each query into an executable execution plan and of executing it according to the requests of the query orchestrator, eventually providing it with the results. Further details about the EE are reported in [7]. The **Semantic Resource Framework (SRF)** provides abstractions and storage of search services, as described in [9].

6 Conclusions

In this Chapter we have described an evolution of the Liquid Query interface initially implemented for the SeCo system [4][5] that goes in the direction of a more flexible exploration of the resources, modular visualization of the result set and better support for the incremental search process. The implementation of the query expansion facilities described in the Chapter is on-going: the implementation of result displayer with the atom view is completed; preliminary explorations of the SRF with a "forward" version of the orchestrator are supported, while the addition to atom views of the various interactive query features is on-going.

We are experimenting with different allocations of data and functionality between the client and server tier of the architecture. In parallel, research work is being done on result set visualization, as explained in [6] This work will be integrated into the user interface for exploratory search, by complementing the atom view visualization with other mechanisms (maps, timelines, diagrams and charts), while preserving the capability of the system to support the entity-by-entity exploration in addition to one-shot querying and result visualization.

References

[1] Adar, E.: Guess: a language and interface for graph exploration. In: Proceedings of the SIGCHI conference on Human Factors in computing systems (CHI 2006), pp. 791–800. ACM, New York (2006)

[2] Aula, A., Russell, D.M.: Complex and Exploratory Web Search. In: Information Seeking Support Systems Workshop (ISSS 2008), Chapel Hill, NC, USA, June 26-27 (2008)

[3] Bates, M.J.: The design of browsing and berry-picking techniques for online search interface. Online Review 13, 407–424 (1989)

[4] Bozzon, A., Brambilla, M., Ceri, S., Fraternali, P.: Liquid query: multi-domain exploratory search on the web. In: WWW 2010: Proceedings of the 19th International Conference on World Wide Web, pp. 161–170. ACM, New York (2010)

[5] Bozzon, A., Brambilla, M., Ceri, S., Fraternali, P., Manolescu, I.: Liquid Queries and Liquid Results in Search Computing. In: Ceri, S., Brambilla, M. (eds.) Search Computing. LNCS, vol. 5950, pp. 244–267. Springer, Heidelberg (2010)

[6] Bozzon, A., Brambilla, M., Catarci, T., Ceri, S., Fraternali, P., Matera, M.: Visualization of Multi-Domain Ranked Data. In: Ceri, S., Brambilla, M. (eds.) Search Computing II. LNCS, vol. 6585, pp. 53–69. Springer, Heidelberg (2011)

[7] Braga, D., Corcoglioniti, F., Grossniklaus, M., Vadacca, S.: Efficient Computation of Search Computing Queries. In: Ceri, S., Brambilla, M. (eds.) Search Computing II. LNCS, vol. 6585, pp. 141–155. Springer, Heidelberg (2011)

[8] Brambilla, M., Tettamanti, L.: Search Computing Tools and Processes. In: Ceri, S., Brambilla, M. (eds.) Search Computing II. LNCS, vol. 6585, pp. 169–181. Springer, Heidelberg (2011)

[9] Brambilla, M., Campi, A., Ceri, S., Eynard, D., Ronchi, S.: Semantic Resource Framework. In: Ceri, S., Brambilla, M. (eds.) Search Computing II. LNCS, vol. 6585, pp. 73–84. Springer, Heidelberg (2011)

[10] Broder, A.: A taxonomy of web search. SIGIR Forum 36(2), 3–10 (2002)

[11] ezDL, Easy Access to Digital Libraries, `http://www.ezdl.de/project`

[12] Fuhr, N.: A probability ranking principle for interactive information retrieval. Inf. Retr. 11(3), 251–265 (2008)

[13] Huttenhower, C., Mehmood, S.O., Troyanskaya, O.G.: Graphle: Interactive exploration of large, dense graphs. BMC Bioinformatics 10(417) (2009), doi:10.1186/1471-2105-10-417

[14] Kuhlthau, C.C.: Kuhlthau's information search process. In: Fisher, K., Erdelez, S., Lynne, E.F., McKechnie (eds.) Theories of information behavior, pp. 230–234. Information Today, Medford (2005)
[15] Kumar, R., Tomkins, A.: A Characterization of Online Search Behaviour. Data Engineering Bullettin 32(2) (June 2009)
[16] Marchionini, G.: Exploratory search: from finding to understanding. Commun. ACM 49(4), 41–46 (2006)
[17] Munzner, T.: Exploring large graphs in 3D hyperbolic space. IEEE Computer Graphics and Applications 18(4), 18–23 (1998)
[18] Pirolli, P., Stuart, K.C.: Information Foraging. Psychological Review 106(4), 643–675 (1999)
[19] Robins, D.: Interactive Information Retrieval: Context and Basic Notions. Informing Science Journal 3(2), 57–62 (2000)
[20] Rose, D.: The information-seeking funnel. In: Marchionini, G., White, R. (eds.) National Science Foundation workshop on Information-Seeking Support Systems (ISSS), Chapel Hill, NC, June 26-27 (2008)
[21] Rose, D.E., Levinson, D.: Understanding user goals in Web search. In: WWW 2004, Proceedings of the 13th International Conference on World Wide Web, New York, NY, USA, pp. 13–19 (2004)
[22] Tzitzikas, Y., Hainaut, J.-L.: How to tame a very large ER diagram (Using link analysis and force-directed drawing algorithms). In: Delcambre, L.M.L., Kop, C., Mayr, H.C., Mylopoulos, J., Pastor, Ó. (eds.) ER 2005. LNCS, vol. 3716, pp. 144–159. Springer, Heidelberg (2005)
[23] Wattenberg, M.: Visual exploration of multivariate graphs. In: Proceedings of the SIGCHI conference on Human Factors in computing systems (CHI 2006), pp. 811–819. ACM, New York (2006)
[24] White, R.W., Muresan, G., Marchionini, G.: ACM SIGIR Workshop on Evaluating Exploratory Search Systems, Seattle (2006)
[25] White, R.W., Drucker, S.M.: Investigating behavioural variability in web search. In: 16th WWW Conf., Banff, Canada, pp. 21–30 (2007)
[26] White, R.W., Roth, R.A.: Exploratory Search. In: Marchionini, G. (ed.) Beyond the Query–Response Paradigm. Synthesis Lectures on Information Concepts, Retrieval, and Services Series, vol. 3. Morgan & Claypool, San Francisco (2009)

Trends in Search Interaction

Ricardo Baeza-Yates[1], Paolo Boldi[2], Alessandro Bozzon[3], Marco Brambilla[3], Stefano Ceri[3], and Gabriella Pasi[4]

[1] Yahoo! Research
Avinguda Diagonal 177, 08018 Barcelona, Spain
rbaeza@acm.org
[2] Università degli Studi di Milano, Dipartimento di Scienze dell'Informazione
Via Comelico 39/41, I-20135 Milano, Italy
boldi@dsi.unimi.it
[3] Politecnico di Milano, Dipartimento di Elettronica e Informazione
Piazza Leonardo da Vinci 32, I-20133 Milano, Italy
{firstname.lastname}@polimi.it
[4] Università degli Studi di Milano Bicocca, DISCO
Viale Sarca 336, I-20126 Milano, Italy
pasi@disco.unimib.it

Abstract. This paper reports the main findings of a panel about trends in search engine interaction, focused upon the use of search engines for performing complex processes[1]. The discussion focuses on the different evolutionary path followed by search engines with respect to other Web and information management solutions, making end users acquainted with the simplistic and never changing keyword-based query paradigm. The analysis delves into the pros and cons of personalization, contextualization, and exploration of Web information, with special attention to the presentation and user interaction aspects. In the end, we also wonder if the keyword-based query paradigm will ever change.

1 Introduction

The technology of search engines and the amount and quality of services they offer have radically evolved in the last twenty years, in parallel with their ever-increasing market value: not only they largely remain the main entry point to the Web for the vast majority of users, but they are constantly trying to develop new features that are more loosely related to search and should serve to enhance the Web-user experience; such functionalities (integrated Webmail, Web-editing of documents, personal calendars, image and video hosting facilities, personalized information delivery, to cite only a few) are actually changing our way of thinking about computations and Web services in general. As often observed, they are in a sense pointing back to the idea of centralized computation, making browsers similar to old-days' terminals.

Notwithstanding this evolution, the basic service provided by search engines has remained the same, and it is surprising to observe how the basic textual keyword-based

[1] Panel Session, Workshop on "Search as a Process", Politecnico di Milano, May 26 2010.

S. Ceri and M. Brambilla (Eds.): Search Computing II, LNCS 6585, pp. 26–32, 2011.

search interface has not substantially evolved or changed over time: alternative proposals (using portals, employing other more sophisticated tools for information hunting, etc.) did not (yet) take off, except possibly in some specialized niches.

Such a long-lasting success is most uncommon in computer science and should suggest that the very concept of keyword-based search is extremely robust although it not always matches users' expectations and satisfaction.

Of course, behind the scenes, search engines changed a lot. Companies such as Google, Yahoo! and Microsoft have tremendously improved the search engine technology, in particular in scalable distributed systems and sophisticated query processing and ranking techniques. While those innovations did not affect the way in which users approach search through simple keyword-based interfaces, they are changing the way in which results are presented, by trying to guess and satisfy better the user's needs. Let us make an attempt to understand more thoroughly in which directions they are moving to reach this goal.

2 Helping the User

A general motto that is repeated constantly by engineers working in search is that one "should help the user find what (s)he's looking for". But not all users are the same, and not all searches are the same, so helping the user really does not mean the same thing for all searches and for all users. The "one size fits all" paradigm commonly applied by search engines does not adhere to the complex and dynamic characteristics of typical search tasks. In a complex search, the user knows what (s)he wants, and this is the result of a complex task (s)he has in mind, while in exploratory search the user does not have a precise idea of what (s)he wants [20], although the result may be fairly simple. Hence, these two dimensions are somehow orthogonal. In the former case, the system should understand that such a search mission is ongoing, and react by offering diverse suggestions and help, possibly keeping track of what the user has collected so far and trying to make sense of it. In the latter case, a more liberal (and prudent) attitude should be adopted, avoiding to interfere too much with the user's search, and act in a way that may sometimes be felt as unwanted or even intrusive.

Many recent works [3][7][9][10] focus on how one may understand whether a complex search mission is underway, for example, by storing and using the amount of knowledge collected from what other users did in the past (an obviously precious source of information that is collected in search engine query logs). It is important to observe that a complex search mission is probably a long-lasting one. Tasks are "long lived" and occur across sessions – people start to plan their trip to Greece one day and they continue the next day - therefore it would be useful for a search engine to recall/reconstruct previous interactions from the same user and to build personalized histories. A personalized and collaborative search approach can be of help in this context. A further difficulty is that users often "intertwine missions", i.e., they switch from a task to another (e.g., alternate work with leisure) - therefore systems should be able to capture interest shifts (especially temporary ones): this ability has proven to be difficult to achieve, although in many cases intertwined missions actually boil down to simple diversions.

A subtle point to be considered here is that complex searches often present multiple facets and may grow in different directions: therefore, reconstructing the state of a search process may include detecting branches in that process and then associating each branch with different results, as diversity and richness of results is not less important than appropriateness and completeness [16].

3 Wisdom of Crowds

An aspect that we already touched upon is the central role of collective past behavior in understanding what a user is doing and in making an effort to be of help. Albeit such activity is worth in general, using collective behavior to help understanding what is going on is particularly important when the user is engaged in a complex search activity. Gathering data about the decisions taken by a large group of individuals in order to assist the user is one of the fundamental ideas behind the so-called wisdom of crowds [18]: if the group is large enough, and if its members are sufficiently independent from one another, their collective behavior contains at the same time sufficient homogeneity and enough diversity to allow one to extract clear trends to cover most of the possible user's intents and needs. In the Web, wisdom of crowds is an alternative name of Web data mining, in particular Web usage mining.

Using collective information in Information Retrieval is not something new. The basic ranking schemes used in the 1960's use the collective wisdom of people writings (e.g. TF-IDF). Later, in the Web, links were used (e.g. in PageRank), that is the collective wisdom of webmasters (that is not true today as any person can add links thanks to the Web 2.0). Today, we can also use the collective wisdom of users, reflected in weblogs as clicked pages or in query logs as queries and clicked results. Hence, the wisdom of crowds is crucial to rank many kinds of objects (e.g. documents) or to find subgroups of experts in social networks.

This form of large-group knowledge is also at the very heart of Web 2.0 [15], and appears in many different forms (collaborative tagging, blog harvesting, query-log mining etc.). Albeit important, though, one should always have a clear understanding of the limits and possible pitfalls behind this idea; not only the information collected may be biased due to the presence of spam [6], but also (and more simply) because often, rephrasing Gresham's law, "bad information drives out good". Even under optimistic assumptions, mediocrity tends to prevail, thus producing suggestions of low or limited quality; and, in some cases, outdated or blatantly wrong data may outweigh good ones.

Some measure of trust or reliability of users may alleviate this problem; yet, even in the presence of a perfectly dependable source of information, a further point cannot be ignored: employing the wisdom of crowds to bias the results of a query is only worth if the user shares the crowd's values, else it may be more harmful than beneficial. This is the case of the collaborative search approach [17], where like-minded people actions are considered to leverage the search undertaken by an individual of the considered community. Collaborative search is an example of context-aware search, where the considered context is the user's social context. An important aspect of the problem is to understand user's trusted fellows, because adding such information to the result may be very relevant in convincing that result's additions and corrections are acceptable. Making a user satisfied requires some kind of sentiment analysis [14] to determine what the user really wants; otherwise, departing from the "neutral" result of a query (the one natively produced by the search engine algorithm) could cause angriness or disaffection.

We remark that our observations should not be overstated; in most cases, using defensive algorithms and leveraging on large unbiased datasets is enough to obtain useful signals. Our main point was that helping users is quite difficult and may be risky, and that in some cases helps may have gone too far without really improving their precision and recall relative to the user's query. Finally, introducing a bias in the results based on a "machine driven" and too user independent interpretation of the user's preferences yields an ethical question: should the search engine present the answer a user wants, or should it present the "bare truth"?

4 Contextualization and Personalization

Contextualization and personalization are an essential ingredient of modern search, as suggested also in the previous section. The notion of context in the search task may be referred to several components (search context, document context etc.), among which a central role is played by user's context.

As previously outlined, making a search outcome tailored to a specific user (or group of users) is a difficult although challenging task: how to collect and represent users' preferences is one of the most important aspects, which has been extensively investigated in the last years [11][13]. Personalization has been recently addressed not only as a pool of techniques aimed to improve search by taking advantage of a single user's preferences, but also by exploiting the user social context, as previously outlined with reference to collaborative search [17][19].

One of the main issues in personalization is the construction of users' profiles: their definition is based either on the implicit monitoring of the user's behavior (Web logs, past queries, Web pages copied on the user's desktop, etc.) or on explicit user indications or on both. There are only a few things in the user's profile that don't change, e.g., her/his birth date, being a fan of a given soccer team (or of a kind of music), but the rest may change, e.g., the preferences for restaurants may change depending, for example, on the user's current location. For this reason profiles should be defined as dynamic and adaptive pieces of knowledge, and possibly represented to reflect the multi-dimensional and faceted nature of the multiple user's interests. In fact, as each of us shows multiple personalities (a.k.a. *personas*) when interacting with a search engine, the user profile should be able to represent multiple user preferences, according to the context in which the search task is being executed.

User profiles do represent long term topical interests: it is therefore hard to associate a specific topical preference with random sessions, much in the same way it may be difficult to guess one's preferences based on limited human interactions with short sentences. This is the reason why collecting and representing user's actions should be based on several techniques. Moreover, to better represent users' preferences, the user's cognitive (i.e. topical) context should be enriched with the knowledge of her/his geographic and social context. There is in fact a great convergence on the importance of contextualization in space and time: for instance, if one is searching on his/her mobile, most likely (s)he is not attaching a complex task (e.g., next summer's trip to Greece), but rather (s)he is trying to localize a close-by service. Then the search answer (i.e. the user-tailored search process) should take the geographic context into account.

The consideration of multiple preference dimensions gives rise to the interesting problem of aggregation: how to aggregate the various contextual components to define the final ranking of the retrieved information items? Also the aggregation process could be driven by user preferences [4].

Personalization and contextualization in fact typically concern not only the results to be presented but also (and more importantly) their presentation and their order. However, the general approach to ranking sees everyone agreeing on the fact that total order is misleading, as users are only concerned by partial orders. Regarding multi-dimensional search, for many users could be easier to work on one dimension at a time; it could be then appropriate to build search protocols that suggest at each stage the results which optimize one dimension; of course, in such scenarios the ordering of dimensions chosen by the protocol becomes essential, but it could also be user-driven.

Another important issue is to go beyond the ranked list presentation of search results; a first step was mixing other results in the standard page result (e.g. universal search in Google), but the problem of combining different types of results in a single screen is mostly unsolved. Some recent and interesting approaches are aimed to help users to visually identify bad and good results through two or three-dimensional presentations of search results, by also taking into account the user's preferences [1].

An important topic is the impact of personalizing the result. To do personalization well we need enough data from the user as well as explicit consent, due to privacy issues (e.g. login authentication that implicitly approves the terms and conditions of the interaction); the overall percentage of users satisfying these two restrictions will be low. This problem is usually faced when developing client-side applications [12]. Despite of these privacy issues, Google personalized search offers an example of centralized solution to the "one size fits all" approach.

On the other hand, contextualizing results (e.g. ranking and displaying the results) according to the intent and context of the query without personalizing at an individual level is another direction that may help groups of people doing the same task regardless of their identity. As log analysis shows, as users perform similar tasks in the Web, this kind of contextualization could have a larger impact. In addition, as we do not need to know the single user, we remove the privacy issues related to personalization [2].

5 Raising the Complexity of the User Interface?

The final question concerns the interaction paradigm: will it always be bound to keyword-based search, or should we expect that the interaction paradigm complexity would rise, at least for complex search, so as to be more expressive? Query interfaces should anyway be kept very simple, featuring interactions where users implicitly choose dimensions as they become available, and where each dimension corresponds to a real-world object, and objects are connected by a graph. This interaction seems to be coherent with the emerging model of "objects-based search" presented by Yahoo as an evolution of page-based search [2][5]. Interaction is obviously more complicated, although the user does not need to know that (s)he is traversing a graph, (s)he is just presented the valid options at each step. This form of interaction is already partially implemented in the form of query suggestion or as the "find-similar" feature of existing search engines.

Still, this seems a rather radical step for search engines at their current stage of development. There is consensus that, while niche users will enjoy more expressive languages - optimistically one could think of a 10% of sophisticated users – the standard user interface of search engines will not change. It is not in the search company's interest to train people in doing "better search": with millions and millions of new users per year, the average user skill is actually reducing, and search companies are most interested in pleasing & capturing the tail of newcomers.

Moreover, recent studies show that the same user may at time act as "dummy" and at time act as a very clever user [8]. Most people are satisfied of results that they get in the "dummy" mode, so they have little incentives in making more efforts, as the interface simplicity pays off. But a complex multi-domain search could still be engaged with a simple keyword-based interface and simple ways of interaction, so there is hope to see search computing queries at work below current interfaces, and not necessarily with a more expressive interaction paradigm.

On a different direction, a rich trend of research is trying to understand natural-language queries (maybe combined with speech recognition tools, and possibly with a verbal output through a voice synthesizer). This is a promising line of study, especially if you think of its application to mobile-based search, but at the moment it seems to be still immature for large-scale implementation.

Users are so satisfied of the simplicity of search engine interfaces that there must be a huge incentive to convince them to go for a more complex kind of interaction; such a situation may happen in the future, at least in some context, but at the moment it appears to be out of reach. This sets the main challenge of search engine evolution: going much beyond the current answering capability without changing the simple user interface.

References

[1] Ahn, J.-W., Brusilovsky, P.: Adaptive Visualization of Search Results: Bringing User Models to Visual Analytics. Information Visualization 8(3), 167–179 (2009)

[2] Baeza-Yates, R., Raghavan, P.: Next Generation Web Search. In: Ceri, S., Brambilla, M. (eds.) Search Computing. LNCS, vol. 5950, pp. 11–23. Springer, Heidelberg (2010)

[3] Boldi, P., Bonchi, F., Castillo, C., Donato, D., Gionis, A., Vigna, S.: The query-flow graph: model and applications. In: Proceeding of the 17th ACM Conference on Information and Knowledge Management, Napa Valley, California, USA, October 26-30 (2008)

[4] da Costa Pereira, C., Dragoni, M., Pasi, G.: Multidimensional Relevance: A New Aggregation Criterion. In: Boughanem, M., Berrut, C., Mothe, J., Soule-Dupuy, C. (eds.) ECIR 2009. LNCS, vol. 5478, pp. 264–275. Springer, Heidelberg (2009)

[5] Dalvi, N.N., Kumar, R., Pang, B., Ramakrishnan, R., Tomkins, A., Bohannon, P., Keerthi, S., Merugu, S.: A Web of concepts. In: ACM Principles of Database Systems (PODS), pp. 1–12 (2009)

[6] Davison, B., Najork, M., Converse, T.: SIGIR Worksheet Report: Adversarial Information Retrieval on the Web, AIRWeb (2006)
[7] Donato, D., Bonchi, F., Chi, T., Maarek, Y.: Do you want to take notes?: identifying research missions in Yahoo! search pad. In: Proceedings of the 19th International Conference on World Wide Web, Raleigh, North Carolina, USA, April 26-30 (2010)
[8] Goel, S., Broder, A.Z., Gabrilovich, E., Pang, B.: Anatomy of the long tail: ordinary people with extraordinary tastes. In: Proceedings of WSDM 2010, pp. 201–210 (2010)
[9] He, D., Göker, A.: Detecting session boundaries from Web user logs. In: Proceedings of the BCS-IRSG 22nd Annual Colloquium on Information Retrieval Research, Cambridge, UK, pp. 57–66 (2000)
[10] Jones, R., Klinkner, K.L.: Beyond the session timeout: automatic hierarchical segmentation of search topics in query logs. In: Proceeding of the 17th ACM Conference on Information and Knowledge Management, Napa Valley, California, USA, October 26-30 (2008)
[11] Kelly, D., Teevan, J.: Implicit feedback for inferring user preference: A bibliography. SIGIR Forum 37(2), 18–28 (2003)
[12] Kobsa, A.: Privacy-Enhanced Web Personalization. In: Brusilovsky, P., Kobsa, A., Nejdl, W. (eds.) Adaptive Web 2007. LNCS, vol. 4321, pp. 628–670. Springer, Heidelberg (2007)
[13] Micarelli, A., Gasparetti, F., Sciarrone, F., Gauch, S.: Personalized Search on the World Wide Web. In: Brusilovsky, P., Kobsa, A., Nejdl, W. (eds.) Adaptive Web 2007. LNCS, vol. 4321, pp. 195–230. Springer, Heidelberg (2007)
[14] Pang, B., Lee, L.: Opinion Mining and Sentiment Analysis. Found. Trends Inf. Retr. 2, 1–2 (2008)
[15] O'Reilly, T.: What Is Web 2.0: Design Patterns and Business Models for the Next Generation of Software (2010), `http://oreilly.com/web2/archive/what-is-web-20.html`
[16] Rafiei, D., Bharat, K., Shukla, A.: Diversifying Web search results. In: Proceedings of the 19th International Conference on World Wide Web, Raleigh, North Carolina, USA, April 26-30 (2010)
[17] Smyth, B.: A Community-Based Approach to Personalizing Web Search (2007)
[18] Surowiecki, J.: The wisdom of crowds. Knopf Doubleday Publishing Group (2005)
[19] Teevan, J., Morris, M., Bush, S.: Discovering and using groups to improve personalized search. In: Proceedings of the ACM International Conference on Web Search and Data Mining, pp. 15–24 (2009)
[20] White, R.W., Kules, B., Drucker, S.M., Schraefel, M.C.: Supporting Exploratory Search, Introduction. Communications of the ACM 49(4), 36–39 (2006)

Part 2
Interaction Design

One of the key aspects of search computing is to facilitate the user in performing complex search tasks, so as to reduce entry barriers, fasten the process, and increase the probability of success. A careful design of interaction is fundamental for interpreting user's queries, for contextualizing them in the user's setting, and for improving the presentation of results through visual helps which enable grasping the differences between alternative solutions.

Interaction design means, on one hand, to focus on the user's activities so as to add implicit semantics to what the user is explicitly performing, thereby giving meaning to his/her actions. On the other hand, interaction design should reflect the technological setting – be it mobile, desktop, or maybe a huge screen – and the choice and presentation of information should be adapted to the setting. Finally, when complex results include many ranking dimensions, for instance because they result from ranking compositions, visualization tricks should be used to let the most important dimensions emerge and be visualized with the most powerful and immediate representation mechanism. These three dimensions do not exhaust the full spectrum of research in interaction design, but certainly provide a good range of research directions for the current stage of search computing.

The first chapter deals with the importance of inferring usage context from user actions to further improve the user experience. Syntax-driven or context-sensitive mechanisms enable the identification of context embedded within some user actions (e.g., filling forms or bookmarking Web pages). Such contextual information, possibly generalized over many instances of use or associated with a specific user profile, can then be exploited to improve search effectiveness; in turn, search complements human thinking and cognitive processes, and users adapt the focus of their actions while performing search tasks.

The second chapter deals with the diversity of interaction and display tools and hence the diversity of modalities of interaction with devices varying from tiny to huge, henceforth also with the mechanisms to adapt and scale data visualizations to such different requirements and the modalities for input and output (e.g., incorporating papers, using speech and other sounds, gestures, etc.).

The third chapter illustrates visualization design in search computing. The main contribution is a visualization model for search computing results, whose data types are used in the process of determining the best visualization patterns; the model assumes the existence of multiple rankings, and chooses the best visual representation for them. Rules for result visualization depend on the visualization model and specific data types, and are applicable regardless of the semantics of result objects within the application.

Context and Action in Search Interfaces

Alan Dix[1,2]

[1] Talis, Solihull Parkway Birmingham Business Park, Birmingham B37 7YB, UK
[2] Lancaster University, School of Computing and Communications, Lancaster LA1 4WA UK
alan@hcibook.com

Abstract. While the web is often described in terms of access to information, it is also a place where people *do* things from booking hotel rooms, to completing their tax return. This paper outlines the ways in which search can form a part of a more action-based view of web interaction. The simplest is that search can be action that the user is performing to get information. However, search can also be used more computationally within an intelligent system that infers appropriate points to trigger interaction (loci of action) and constructs a model of the users context. The resulting picture is a rich interplay between user action and computation, where each inform and influence the other, and where search can form an intimate part both explicitly for the user and embedded within computation.

Keywords: user interaction, context inference, data detector, intelligent user interface.

1 Overview/Motivation

While the web is often described in terms of access to information, it is also a place where people *do* things from booking hotel rooms, to completing their tax return [1]. Theories of embodiment stress that, as creatures *in the world*, our perception is not an abstract gathering of information for processing, but an integral part of our being acting beings [2]. This is equally true in the digital world of the web: we search for information to do things whether on the web, in the world or both; for example, looking at potential holiday destinations, which will later be booked online but visited physically. Even information seeking is an interactive process as made clear in information foraging theory [3], where information we have allows us to make choices of actions (including performing more searches) which lead to further information. This exactly parallels the ecological understanding of perception, where we may turn our head or move to see things better.

Search is thus an integral part of this acting view of human activity on the web, both informing human actions and being one of those actions. If we regard web interaction as solely instrumental: where the user is the only actor and computational elements are tools or data, then that is the end of the story. However, if we allow the computation to take a more active role as mediator or assistant, then we can see a wider role.

S. Ceri and M. Brambilla (Eds.): Search Computing II, LNCS 6585, pp. 35–45, 2011.

For a human listener a request such as "where is the bus" might be interpreted "when is the bus coming, is it late" if said at a bus stop, or "where is the bus stop" if said in a shopping centre. If you know someone well then when they say "RDF" you will know whether they mean "Resource Description Framework" or "Refuse Derived Fuels". When human helpers hear a query or request they will use *context* in order to work out the likely intention. This context may include the nature of the person as well as the situation in which the request occurs. Similarly, context such as personal profile information or the recent activity of a user can be used to interpret the user's actions and information they have been viewing and use this to better respond to the user's future actions and requests.

Furthermore, if you meet a visitor in your hometown and they mention they like art, you may suggest a gallery or local attraction based on that, even *without being asked*. If the person then showed interest you may then go on to suggest the best way to get there, walking, by bus, again without being explicitly asked. That is the human helper may notice *loci for action*, topics or things in the conversation or previous activity that suggest that information or potential actions may be appropriate. In a computational setting this locus may be a phrase in an email or web page being visited, or something in the output of a previous action, for example the destination town after having booked a rail ticket.

As previously noted, search may be a potential action, which may have been suggested based on a locus for action (e.g. searching for information on the destination town), and that search may be influenced by context (e.g. if the user frequently looks up art-related material). However, search can also be used computationally as part of the algorithmic processes that infer potential loci for action and user context.

In the next section we will examine the cycle of digital interaction, seeing both the role that loci-for action and context can play in this and also the way that search can be an enabler and outcome of the cycle. In Section 3 and Section 4, we look in turn at how loci for action and context can be inferred; in each case seeing how search can be used computationally. Section 5 then looks at search as an action itself, but the way in which the loci and context can influence the search process and in particular the choice of search repository.

2 Cycles of Action and Context

Fig. 1 shows the cycle of user action and the role automated context can play in this cycle. The user is engaged in a series of actions: on the web these would typically be clicking a link, hitting a form button, typing in a url, or entering a search string. The actions each create outputs; for example a page of search result, a web page, email message, PDF or Word document (the term 'document' will be used generically to refer to all these).

Sometimes the user may spontaneously invoke an action (e.g. typing a url), but very often actions arise out of the results of previous actions (e.g. clicking a link, reading an email message, opening a document); that is a loci of action, as introduced in the previous section. In some cases the potential loci will be very explicit (link or button); however, in others there may be something in the output document (e.g. a person's name) that may be the trigger for the user's next action, but which is not obviously encoded as such. The user will typically have many choices for a potential next action, and so there is a cycle where the user performs an action, obtains results and may select some part of this output for the next action.

Both the detection of these more implicit loci for action and the execution of the action may be influenced by automatically inferred context based on the user's previous interaction history both long term (building up a broad profile of the user's preferences) and shorter term (understanding the user's current topic of interest).

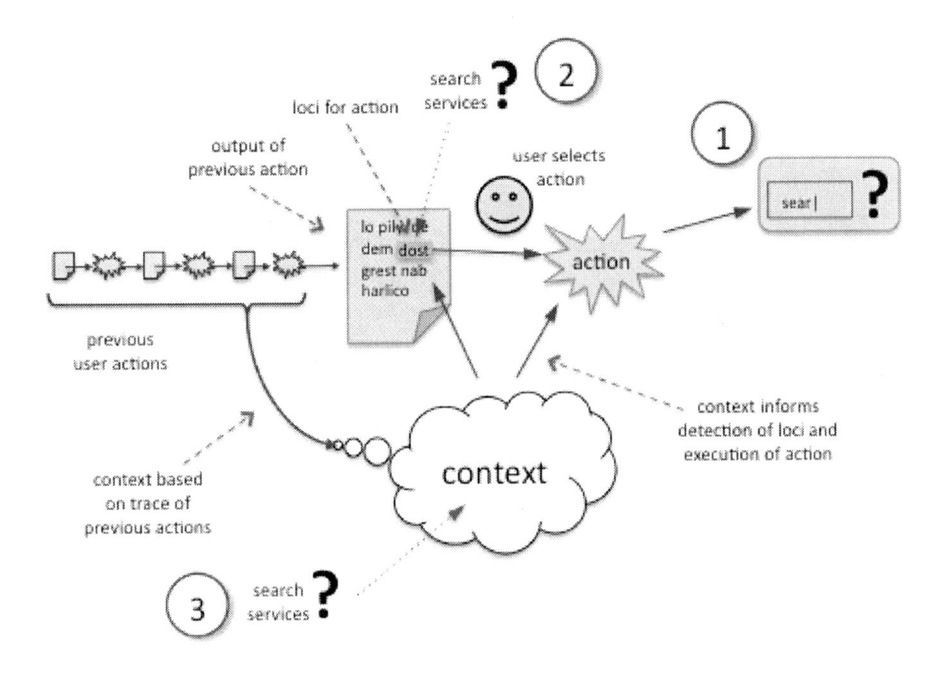

Fig. 1. Cycle of action and context

Search may be involved in different parts of this cycle:

1. the action being invoked may be a search, in which case the loci and/or automatic context may be used to help select appropriate search services, or influence ranking
2. search-based resources can be used together with context in determining loci for action, whether in an web page, email message or document.
3. context inference may use search-based resources in order to enrich trace data

3 Establishing Loci for Action

In some cases the document that the user is viewing may have explicitly encoded loci for further action such as a hyperlink or button. In these cases the application or person that produced the document being viewed determines the next action. However, more interesting are implicit loci, such as names, or places mentioned in text, as these may be used in ways not foreseen by the originator: one user may use a name as input into Google Scholar, another for a Facebook search, and yet another as input into IMDb. If one can determine which parts of the document constitute potential loci and furthermore what kind of data each represents (e.g. person name, place name, telephone number), then it is possible to guide the user towards suitable resources and actions including the choice of appropriate search services.

3.1 Explicit Semantic Markup of Loci

One way in which these loci can be detected is when the document originator has added explicit semantic mark-up. While still not widespread, various forms of markup including micro-formats and RDFa are beginning to be used on the web; for example, Figure 2 shows a fragment of the HTML of a LinkedIn profile page where hCard microformat [4] encoding of vCard [5] is used to mark the words "Alan Dix" as a name where "Alan" is the given name and "Dix" is the family name.

```
<div class="masthead vcard contact">
  <div id="nameplate">
    <h1 id="name">
        <span class="fn n">
            <span class="given-name">Alan</span>
            <span class="family-name">Dix</span>
        </span>
    </h1>
  </div>
```

Fig. 2. hCard microformat on profile page at linkedin.com

Of course, the vast majority of web resources are not marked up in this way, indeed the main reason that the LinkedIn profile can be annotated like this is because it is being generated from a database where the semantics of the name are already explicit. If, for example, the name were mentioned in the middle of an email message from a friend it is unlikely that the writer of the email would explicitly mark-up the content in this way!

3.2 Inferring Loci – Data Detectors

Happily it is possible to automatically detect this type of semantic value within text using a form of local text mining known as data detectors. Data detectors have been used sporadically since the late 1990s when a number of commercial and research systems were developed including work at Intel [6], Apple [7], Georgia Tech (CyberDesk) [8] and aQtive (onCue) [9]. Some of these worked by scanning the complete text available to the user, others some selected portion; for example, onCue worked by looking at the clipboard so that it could react whenever the user copied or cut text. All of these early systems effectively used syntactic methods for recognition, using some variant of BNF, regular expressions, or bespoke code to analyse the text and work out what it represented from patterns of letter use, punctuation etc. For example a UK postcode matches the regular expression below:

```
/([A-Za-z][A-Za-z0-9]{1,3})[ \t]*([0-9][A-Za-z]{2})/
```

Of course, as with any such system, this can lead to both false positives (e.g. "High Court" is classified as a person name because of its length and capitalisation) and false negatives (e.g. the poet name "e.e. cummings" is not recognised because of its unusual capitalisation). Because of this, onCue was deliberately designed using principles of 'appropriate intelligence', embedding the 'intelligent' algorithms within an interaction framework that meant errors in the intelligence did not negatively impact the user experience [9].

3.3 Using Search in Data Detectors

In even earlier work, the Microcosm hypermedia system developed at Southampton University [10] added links dynamically at the server side using a similar form of text mining. However, instead of syntactic rules Microcosm looked up the terms in the source document and matched them against a database linking keywords/terms to particular resources. This is a similar to the way terms link to their topics within Wikipedia by keywords/phrases not full hyperlinks; except that in Microcosm there was no need for *any* explicit mark-up by the page author, not even to say "this a term". Note that this is effectively using *search* in order to deliver 'intelligent' results, in the same way that Google 'suggest' does while typing search terms.

This form of search-based data detection and the more syntactic rules used by most data detectors can be combined. The web-based bookmarking system Snip!t (www.snipit.org) uses technology developed from onCue in order to perform data detection over selected fragments of web pages; Figure 3 shows Snip!t suggesting potential actions after recognising a post code in a selected web page fragment. Snip!t uses some plain syntactic rules like onCue and the other early data detectors, but also performs semantic searches in larger tables of data like Microcosm. These can be used separately for different kinds of data, regular expressions for postcodes, lookup for city names. However, most powerfully then two can be combined.

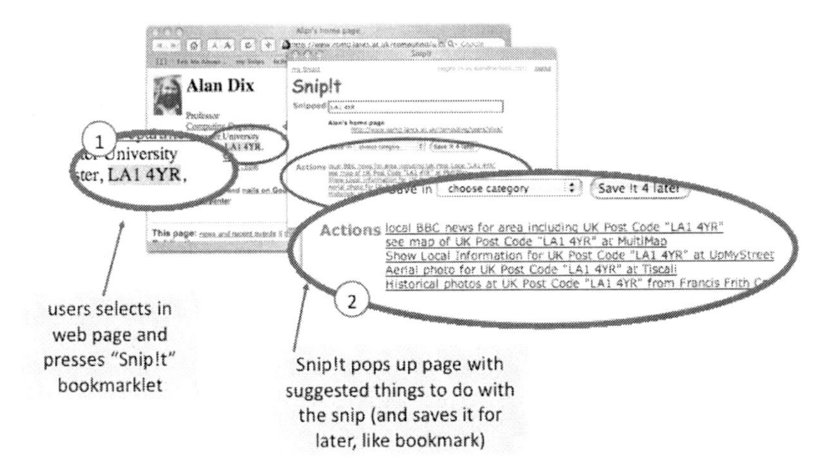

Fig. 3. Snip!t recognises a post code and suggest potential actions

An example of this can be seen in Figure 4, which shows a portion of the XML description file for a person's name. Towards the top it declares "<keys>" and lists "Female First Name" and " Male First Name". These are the names of data types that have already been recognised using a lookup in a large table of male and female first names compiled from US census data. The rest of the XML description then gives syntactic patterns for what could appear before or after this first name in order to form a complete name. While not eliminating false positives and false negatives, the use of lookup/search does increase the accuracy of the recogniser and is also more efficient as the syntactic matching rule is only invoked if a suitable lookup match is found.

```
<beforeafterrecogniser>
  <name>name1Recogniser</name>
  <title>Person Name recogniser</title>
  <keyed>
    <keys>Female First Name, Male First Name</keys>
  </keyed>
  <pattern>
    <pre_context>\W+</pre_context>
    <before>($RE_HONOURIFICS\s*)?</before>
    <key>$RE_CAPSNAME</key>
    <after>\s*$RE_MIDNAMES\s*$RE_LASTNAME</after>
    <post_context>\W+</post_context>
  </pattern>
  <match>
    <type>name</type>
    <description>Person name $$</description>
  </match>
</beforeafterrecogniser>
```

Fig. 4. Snip!t recogniser description file keyed from name lookup

4 Establishing Context

The techniques used to establish loci can also be used as a form of immediate interaction context. If you visit a web page that mentions Milan, it is likely that Milan itself, Italy in general, or maybe research groups within Milan may be your topic of interest. However, when looking at past activity other forms of inference can be used.

Most obviously, the order of past actions can be used to suggest potential future actions, for example, using hidden Markov models or other forms of sequence inference. This can be a powerful technique, but is not strongly related to search computing, so is not discussed further here, but more details can be found in [1].

More pertinently, the input of users can be analysed in a similar way to the outputs of actions. These user inputs are often more likely to refer to data of immediate personal importance and this can be used to yield more precise results. For example, if the term "Tiziana" appears on a web page, it is possible to work out this is a female first name, but it could refer to many Tiziana's. However, if you type

"Tiziana" into an input field, then it is likely that the Tiziana in question is someone you know and local resources, such as an address book, can be searched to determine that this is very likely to be "Tiziana Catarci".

Various projects, including those looking at the 'Semantic Desktop' [11,12] and the TIM project [13,1], consider some form of personal ontology, where personal constructs are linked together, for example explicitly recording that "Tiziana Catarci" 'works in' "University of Roma, La Sapenza", where each of the terms is a semantic entity or relation in the ontology. This data may be mined from existing sources such as address books, calendars and email messages, or entered explicitly using a dedicated interface.

Based on similar knowledge to that embedded in these personal ontologies, an experienced (human) personal assistant would be able to tell that if you had just received an email from Tiziana and then asked for a flight and hotel to be booked, then Rome is a likely potential location, and if so then the hotel needs to be close to "La Sapienza". If a user has a rich digital personal ontology then it becomes possible for a computer to make similar inferences.

In the TIM project and related work we have looked at both rule-based inference of relationships between form fields and also more 'fuzzy' inference using spreading activation. In the case of more precise rules it is possible to infer that if a form has both a name and address field, that the address is likely to be the one associated with the person, and moreover, based on previous form entry, whether it should be the home or business address [1].

Spreading activation is used to suggest potential linkage even where there has been no previous exposure to the precise forms or applications [14]. For example, if an email has been received from Vivi, then the entity representing Vivi is made very 'active' in the ontology; then other entities in the ontology that have close associations with Vivi (maybe a colleague Costas, or her institution 'University of Athens') are also given some activation (see figure 5). In turn entities further and further from Vivi inherit smaller amounts of activation from their neighbours. In this way those entities in the ontology, which are closer to the initially activated entity are most active and this can be used to rank potential candidates to complete future parametrised actions.

This works fine so long as all the relationships are in the personal ontology, but of course a human assistant would know, for example, that Athens is in Greece even if this were not explicitly stated in an address book. Basic spreading activation is limited to the knowledge available, but happily further information is available on the web in both semantic-web linked-data sources [15] and also searchable deep-web search services [16]. It is thus possible to fill the gaps in the locally stored explicit knowledge with this globally stored, often 'common sense' knowledge, for example, knowing (from the personal ontology schema) that 'Athens' is a place, then if Athens is highly activated one can search resources such as the Geonames online database to find information about Athens, which will include that it is the capital of Greece. Initial studies have shown that this web-scale reasoning using spreading activation is feasible and scalable [17].

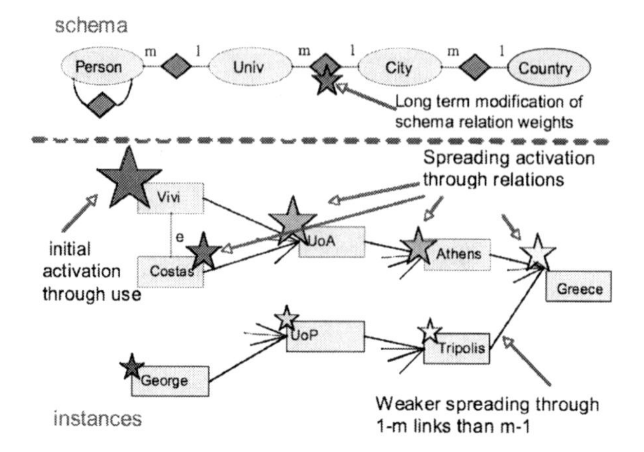

Fig. 5. Spreading activation in a personal ontology

5 Suggesting and Influencing Search

As noted search may be one of the potential actions. If this is the result of a locus for action from a data detector, then the semantics of the identified term can be used to tune the search. For example, Snip!t suggests directory search services such as ZoomInfo if the term is a person name, location-based search services such as recycle-more.co.uk if the term is a post code, and search at acronymfinder.com if the term appears to be an acronym. This already gives a high level of specificity, however, there are some types of term for which there are many possible search services, or the term may be ambiguous. In such case the slightly broader context given by spreading activation may allow greater specificity.

So far this has not been implemented with Snip!t, but the author used somewhat similar techniques during in research at aQtive in the late 1990s. The Open Directory Project was used which has both a hierarchical category system and a large corpus of web material hand-classified in the hierarchy. For any term the relative frequency of the word in pages associated with leaf categories can be used to give an initial 'activation' of the leaf category, which can then be spread both up and down the category hierarchy in order to generate a relevance measure of a term for a category, even if there is no explicit mention of the term in the category. This is a constrained form of spreading activation, as it is limited to a hierarchy, but otherwise very similar to spreading activation on the personal ontology.

The outputs of this process were then used to classify unseen text; for example the term "Chihuahua puppy" would be correctly classified with highest confidence as Chihuahua the dog rather than Chihuahua the place in Mexico even if the term 'puppy' never appeared in pages classified as "pet/dog/chihuahua", because pages in other subcategories of the dog category did mention 'puppy'. This was then used to improve the relevance results of searches of various kinds: web pages, images and online shops. This information could also tune the choice of appropriate search resources (e.g. zoominfo, or IMDb to lookup a name) [18].

6 Discussion

We have seen a number of ways in which search computing interacts with a more action and context focused view of user interaction on the web.

On the one hand search can be used to enhance user interaction by working together with more syntactic methods to allow data detectors to identify loci for actions and more generally text mining to establish potential semantic entities. Similarly when spreading activation is used for context inference, local data stores can be augmented on the fly by external data sources including searches of deep web resources. The appropriate points for augmentation can be identified based on 'chasing' already loaded entities of high activation, and the appropriate deep web resources can be indentified because the personal ontology establishes a clear type for the data.

In the above, search computing was used as a hidden part of the algorithms used to suggest potential future actions, establish user context and based on this aid in the completion of data for actions. However, search computing can also be the final point of such actions; that is the above techniques may yield search as a final action of the user. In these case the fact that loci of action can be identified and given a precise type can help choose or rank appropriate search repositories and moreover this can be tuned further by inferred user context.

More broadly it is interesting to note the way in which user interaction is not only enabled by existing knowledge such as the personal ontology and search data, but is also the source of such knowledge in terms of traces of interaction. It is easy to conceptualise the role of algorithmics in user interaction in terms of the computation that occurs between user actions, that is no more than a variation of a 1960s batch input-output process. However, interactive computation [19] has different properties to plain IO computation, and we need to re-conceptualise with a more *systemic* view where the human activity is integral to the computation, as has been emphasised in recent work on 'human computation' [20, 21].

Knowing that a human will interpret results may change our focus. Rather than seeing algorithms as producing precisely the best the results, we focus more on exposing alternatives and rationale. Similarly knowing that users are acting on outputs of one phase of an algorithm may mean that user actions implicitly feed information back into the automated analysis. For example, faced with the digits "0152410317" in text, data detectors may offer multiple suggestions, not just the 'best' one say as a phone number or a 10 digit ISBN; however if the user selects "phone using Skype" this is a confirmation that it is the phone number not the ISBN.

When we view the computer in this more interactional role, search becomes a natural partner for more traditional algorithmics. In human cognition the relatively slow process of sequential rational thought compliments the massively parallel associative access to past memories. The vast quantities of data available on the web are proving surprisingly powerful [22,23] as a computational resource, and in this paper we have seen examples of how search acting as an associative store compliments more explicit, rule-based algorithms in a similar way to our own human thinking. Search is therefore not only a way to provide information for the human user, but has the potential to allow the computer to act in a more humanly comprehensible way.

References

1. Dix, A., Lepouras, G., Katifori, A., Vassilakis, C., Catarci, T., Poggi, A., Ioannidis, Y., Mora, M., Daradimos, I., Md. Akim, N., Humayoun, S.R., Terella, F.: From the Web of Data to a World of Action. Web Semantics: Science, Services and Agents on the World Wide Web 8(4), 394–408 (2010), doi:10.1016/j.websem.2010.04.007.
2. Clark, A.: Being There: Putting Brain, Body and the World Together Again. MIT Press, Cambridge (1998)
3. Pirolli, P.L.: Information foraging theory: adaptive interaction with information. Oxford University Press, Oxford (2007)
4. Çelik, T., Suda B.: hCard 1.0., `http://microformats.org/wiki/hcard` (accessed December 29, 2010)
5. Dawson, F., Howes, T.: vCard MIME Directory Profile. RFC 2426, IETF (1998), `http://www.ietf.org/rfc/rfc2426.txt`
6. Pandit, M., Kalbag, S.: The selection recognition agent: Instant access to relevant information and operations. In: Proc. of Intelligent User Interfaces (IUI 1997), pp. 47–52. ACM Press, New York (1997)
7. Nardi, B., Miller, J., Wright, D.: Collaborative, Programmable Intelligent Agents. Communications of the ACM 41(3), 96–104 (1998)
8. Wood, A., Dey, A., Abowd, G.: Cyberdesk: Automated Integration of Desktop and Network Services. In: Proc. of the Conference on Human Factors in Computing Systems (CHI 1997), pp. 552–553. ACM Press, New York (1997)
9. Dix, A., Beale, R., Wood, A.: Architectures to make Simple Visualisations using Simple Systems. In: Proceedings of Advanced Visual Interfaces, AVI 2000, pp. 51–60. ACM Press, New York (2000)
10. Hall, W., Davis, H., Hutchings, G.: Rethinking Hypermedia: The Microcosm Approach. Kluwer Academic Publishers, Norwell (1996)
11. Semantic Desktop. semanticweb.org, `http://semanticweb.org/wiki/Semantic_Desktop` (accessed October 2010)
12. Sauermann, L., van Elst, L., Dengel, A.: PIMO - a framework for representing personal information models. In: Pellegrini, T., Schaffert, S. (eds.) Proceedings of I-MEDIA 2007 and I-SEMANTICS 2007 as part of TRIPLE-I 2007, pp. 270–277 (2007)
13. Catarci, T., Dix, A., Katifori, A., Lepouras, G., Poggi, A.: Task-Centred Information Management. In: Thanos, C., Borri, F., Candela, L. (eds.) Digital Libraries: Research and Development. LNCS, vol. 4877, pp. 197–206. Springer, Heidelberg (2007)
14. Katifori, A., Vassilakis, C., Dix, A.: Ontologies and the Brain: Using Spreading Activation through Ontologies to Support Personal Interaction. Cognitive Systems Research 11, 25–41 (2010)
15. Bizer, C., Heath, T., Berners-Lee, T.: Linked Data – The Story So Far. International Journal on Semantic Web and Information Systems, Special Issue on Linked Data 5(3), 1–22 (2009)
16. Sherman, C., Price, G.: The Invisible Web: Finding Hidden Internet Resources Search Engines Can't See. CyberAge Books (2001)
17. Dix, A., Katifori, A., Lepouras, G., Vassilakis, C., Shabir, N.: Spreading Activation Over Ontology-Based Resources: From Personal Context To Web Scale Reasoning. Internatonal Journal of Semantic Computing, Special Issue on Web Scale Reasoning: scalable, tolerant and dynamic 4(1), 59–102 (2010)

18. Dix, A., Karapetis, F., Kefalidou, G.: Concept Classification for Decentralised Search. Working Paper, Lancaster University, November 11 (2005), http://www.hcibook.com/alan/papers/Concept-Classification-for-Decentralised-Search-2005/
19. Wegner, P., Goldin, D.: Computation beyond turing machines. Commun. ACM 46(4), 100–102 (2003), DOI= http://doi.acm.org/10.1145/641205.641235
20. von Ahn, L., Maurer, B., McMillen, C., Abraham, D., Blum, M.: reCAPTCHA: Human-Based Character Recognition via Web Security Measures. Science 321, 1465–1468 (2008)
21. Eagle, N.: txteagle: Mobile Crowdsourcing. In: Aykin, N. (ed.) IDGD 2009. LNCS, vol. 5623, pp. 447–456. Springer, Heidelberg (2009)
22. Halevy, A., Norvig, P., Pereira, F.: The Unreasonable Effectiveness of Data. IEEE Intelligent Systems 24(2), 8–12 (2009)
23. Finin, T., Syed, Z., Mulwad, V., Joshi, A.: Exploiting a Web of Semantic Data for Interpreting Tables. In: Proceedings of the WebSci 2010: Extending the Frontiers of Society On-Line (2010)

Desktop, Tabletop or Mobile?

Moira C. Norrie

Institute for Information Systems, ETH Zurich
CH-8092 Zurich, Switzerland
norrie@inf.ethz.ch

Abstract. We discuss the ways in which search services need to adapt to device and setting to cater for, not only the characteristics of the devices, but also the different types of search typically carried out in different settings. We then discuss the specific challenges of mobile settings before going on to look at opportunities offered by new forms of large interactive surfaces such as digital tabletops.

1 Introduction

Nowadays, users access search services from a range of devices in the workplace, at home and on the move. The capabilities of these devices may vary in terms of the input and output modalities supported as well as physical characteristics such as screen size and resolution. Therefore, the search interface needs to be adapted to the device.

Further, the types of search that users perform may depend on, not only the features of the device, but also the setting. For example, a user at home organising a holiday will typically perform more complex, multi-step searches than a tourist on the move. Consequently, the functionality offered by a search service may be dependent on the setting as well as the device. This means that the adaptation of search services must take into account both functionality requirements and interface design issues.

In this chapter, we will discuss the ways in which search services and their interfaces need to adapt to device and setting, together with the challenges that need to be addressed. We will start by examining the issue of how search requirements differ according to setting, before going on to discuss the specific challenges of mobile settings. Finally, we will look at the opportunities offered by new forms of large interactive devices such as digital tabletops and large multi-touch screens which are now becoming increasingly commonplace in the workplace and also homes.

2 Search and Setting

The types of searches that users perform depends on the setting. To illustrate this, consider the example of tourists. There are three phases that make up a tourist experience–pre-visit, visit and post-visit. During the pre-visit phase, a

S. Ceri and M. Brambilla (Eds.): Search Computing II, LNCS 6585, pp. 46–52, 2011.

tourist may perform various kinds of complex, multi-step searches that involve consulting online travel guides, maps, event calendars, activity sites, restaurant guides and shared photo sites as well as hotel, flight and tour reservation systems.

The searches are often multi-step, not only in the sense that a single search can be an iterative process, but also the fact that the planning process often involves several distinct searches and extends over many sessions. It is not uncommon to hear a colleague say that they spent many evenings over a period of several weeks or even months to do the detailed planning of an extended road trip. Information found in the searches will be compared and aggregated, and possibly later used as input to future searches or stored as part of some general plan.

Searches carried out during the pre-visit phase are therefore likely to be carried out in fixed settings such as the workplace or the home and using desktop computers. We note that the term desktop computer may nowadays refer to a range of devices including laptops, PCs with large or even multiple monitors and multi-touch screens with integrated PCs. Users will often extract information from search results, copying it into other documents or onto paper as a means of recording and aggregating the final or intermediate results of their searches. Sites discovered during the search process that are considered valuable resources may be bookmarked and searches themselves may sometimes be saved by bookmarking the corresponding URLs.

While research projects such as SeCo[1] are carrying out valuable research on supporting complex searches, unfortunately there are limited tools available today to support the kinds of planning activities that involve multiple, complex searches over a period of time. There have been various research projects addressing ways of providing lightweight means of recording and later retrieving useful pieces of information found within Web documents, for example List.it [6], but users still tend to rely on traditional means of copying pieces of information into specially created, and separately managed, paper or digital documents. Further, there is a lack of tools that provide users with simple means of comparing and aggregating that information unless it is well-suited to being managed and manipulated within a spreadsheet.

Once our tourist is on their visit, they may also want to perform various searches such as the classic example of finding a particular category of restaurant close to their current location. Nowadays, many tourists come equipped with a mobile phone capable of supporting such searches either through access to the Web or special mobile applications and downloaded data. A lot of research projects have focussed on the issue of providing tourist guides that are location-aware or more generally context-aware with early examples including Georgia Tech's Cyberguide [9] and the Lancaster GUIDE project [2]. This means that the information returned to the user is filtered so that it is relevant to their current context by automatically refining the search query. General mechanisms for context-awareness can also support other forms of adaptation of, not only content, but also structure and presentation, and provide a basis for adapting Web sites to cater for the characteristics of different devices [3,4].

---

[1] http://www.search-computing.it

It is important to recognise some of the key features of mobile settings that make context such an important factor in search. First of all, searches in mobile settings tend to be much more directed and immediate than those involved in the planning activities described above for the pre-visit phase. Our tourist is hungry and wants to find a restaurant they can go to now for dinner, or they want to find out the direction in which they should walk to reach their required destination. They might be planning possible excursions for tomorrow, but are unlikely to be planning their next full holiday. Searches therefore typically involve a single search process and the results are directly acted upon rather than being stored for later comparison and processing. Second, users are on the move and therefore may be performing the search while watching out for traffic, listening to announcements, talking to friends or even driving a car. They are therefore easily distracted and the information bandwidth is low. This means that, in contrast to desktop settings where users can comfortably browse and read a lot of information returned by a search, users in a mobile setting may be less interested in complete information and more interested in being provided with a simple, short result.

In the post-visit phase, the tourist may still want to perform some searches either to reminisce or to find out more about places visited. The search processes may therefore again tend to be more directed than those during the pre-visit planning phase since they are often based on a particular place, person or event. However, similar to the pre-visit phase, these searches are often carried out in the comfort of fixed settings such as the home and the user may enjoy searching and browsing for extended periods of time. Further, they may wish to save the results of their searches, possibly copying or linking information found into part of an integrated record of their experience.

3 Search in Mobile Settings

Some of the characteristics and challenges of search in mobile settings were outlined above to highlight the fact that the information bandwidth is typically much smaller than in workplace or home settings and searches therefore need to be much more focussed in terms of the results returned. This is an issue both of the content and presentation of results. The use of context, and specifically location, is an important factor in increasing the relevance of the results and reducing the number of possible results to be considered by the user. However, the design of the interface also plays an important role in ensuring that the most relevant information is presented to the user in a simple and direct manner. Context could also be used to eliminate the need for input altogether by performing implicit searches triggered purely by changes in context, although there are still cases where users will want to take over control of the system to perform explicit searches that may be entirely independent of their current context [1]. One approach to simplifying the input task is the use of various forms of gestures on touch screens as proposed for example in [8].

Researchers have experimented with both different devices and modes of interaction in mobile settings to alleviate the problems created by the fact that users are often involved in other activities and may have limited hand use due to the fact that they are driving, carrying items or holding the hands of children. Also, they may either be in very noisy environments or in environments such as concerts where any sound would be frowned upon. Further challenges come from limited sizes of screens and keyboards as well as the difficulty of reading screens in sunlight. While each and every research project seems to have found a solution for one or other problem, the most serious challenge for general search services is the fact that the settings are constantly changing as users are on the move. This means that while a system designed to use audio input and output may offer a solution that leaves users hands-free and may be ideal in some settings, it is problematic in noisy environments and useless in settings that require silence. Other solutions designed to work well in indoor settings, may not work so well in outdoor settings. Although the idea of context-aware applications is that they can adapt to context, we have yet to see solutions that really can adapt to the wide variety of contexts that a tourist might encounter in a typical day.

Practical issues that seem to have received less attention by researchers concerned with search applications are those of cost, power and connectivity. Although these are well known issues in research generally on mobile computing, many researchers at the application level tend to assume total connectivity and ignore issues related to cost and limited battery power. Surprisingly, although many research projects address the tourist domain, they often ignore the fact that tourists frequently travel abroad which, given current business models, tends to mean significant increases in cost for any kind of communication and data transfer. In terms of search, this means for example that saving search queries by bookmarking URLs could have serious implications in terms of cost as revisiting results of a previous search could actually generate repeated network access and data transfer. Also, although the world is becoming increasingly connected, there are still places that we encounter on our travels that have no or limited connectivity. Therefore, in mobile settings, the ability to extract and locally store information from search results could be even more important than in desktop settings. In addition, users should be able to easily transfer information extracted from search results in desktop settings to mobile devices for access on the move.

Power is another limited resource, especially during travels when tourists might have restricted or even no access to power supplies. The assumption therefore that tourists can continuously rely on their mobile phones to guide them on their travels and provide them with all the information that they need, when they need it, is not really valid. One advantage of paper maps and guidebooks is that they require no power and therefore the information is truly persistent [11]. To take advantage of this property of paper, along with others such as ease of annotation, researchers have experimented with systems that can bridge the paper-digital divide by augmenting rather than replacing paper documents with digital search services [10]. One such project, EdFest [12], provided users with a paper map, festival guide and bookmark that could be used in isolation or as an interface to an application running on a mobile device that provided them with additional information about locations and events, navigation services and search services. The system was developed using Anoto[2] technology and users interacted with the document using a digital pen and output was provided through an audio channel using a text-to-speech engine.

EdFest is just one of many projects that have investigated the use of non standard technologies for allowing users to interact with search services in a mobile setting. Again, most of these solutions tend to be domain specific or address only a few of the many challenges of providing users with access to information and services while on the move.

4 Exploiting Large Interactive Surfaces in Search

Large interactive surfaces such as digital tables and multi-touch screens are currently attracting a lot of interest in research communities and products are beginning to emerge in the marketplace. So far relatively little attention has been paid to the opportunities that these offer for search services in terms of increased screen real estate as well as more natural forms of interaction. In fact, most of the research related to these devices has focussed on collaboration, even though they have great potential for providing single users with better support for the sorts of complex information tasks that they are involved in every day.

Consider again our tourist in the pre- and post-visit phases of their experience. The search processes were embedded in complex and long-running activities that involved copying, comparing and aggregating information found during searches. On a physical desktop, a user might surround themselves with information, some of it possibly on a computer screen, and some of it in papers piled on the desk. Documents on the physical desktop can easily be moved around during the processing for purposes of comparison and sorting. These documents remain in the user view either as individual documents or document piles. Large interactive surfaces that support multi-touch interaction or a combination of pen and touch may provide much more natural ways of performing similar actions in the digital world than is currently supported by desktop PCs with mouse and keyboard interaction.

The increased screen real estate offered by large screens and digital tabletops could also be hugely beneficial. The results of complex searches can contain a lot of information to be viewed and compared and limited screen size can be frustrating for users and have a significant effect on performance. At the same time, increased real estate can make it much easier to provide users with an overview of the entire workspace for the task at hand and keep track of different searches as well as information extracted from the results of these searches.

It is interesting to note that although a lot of work has been done on the adaptation of Web sites to mobile phones and other forms of small screen devices, for example [13,5,7], adaptation to large screens and large interactive surfaces has not really been considered. We have recently started to investigate the specific challenges raised by adaptation to large screens in terms of both the technological

[2] http://www.anoto.com

solutions and design guidelines. The questions for the search service community to address are therefore how they could make best use of increased screen size in the interfaces to their search services and also how larger screens and more natural forms of interaction could be used to embed these search services in an environment designed to support the entire user task.

5 Conclusions

The scenario of the tourist, considering all three phases of their experience, helps place search as part of complex information tasks and to see how search requirements can vary according to the particular task and setting. The provision of more advanced search services that are capable of addressing typical user needs is a major step towards supporting the kinds of complex information tasks that normal users carry out in everday activities such as planning their next vacation. However, it is important that this be seen as a first step towards providing users with tools to support the entire planning activity and information workflow, taking into account the different settings involved.

While mobile settings have received a lot of attention, they are also extremely challenging in terms of the variety of issues to be tackled and there are still many open problems. At the same time, the more traditional fixed settings of offices and homes are undergoing a radical transformation as the installation of various forms of large interactive surfaces are becoming more commonplace: This should be viewed as a great opportunity for researchers to consider how they can exploit the increased screen real estate and more natural forms of interaction in complex search tasks.

References

1. Ceri, S., Daniel, F., Facca, F.M., Matera, M.: Model-driven Engineering of Active Context-awareness. World Wide Web Journal 10(4), 387–413 (2007)
2. Cheverst, K., Davies, N., Mitchell, K., Friday, A., Efstratiou, C.: Developing a context-aware electronic tourist guide: some issues and experiences. In: CHI, pp. 17–24 (2000)
3. Grossniklaus, M., Norrie, M.C.: An Object-Oriented Version Model for Context-Aware Data Management. In: Benatallah, B., Casati, F., Georgakopoulos, D., Bartolini, C., Sadiq, W., Godart, C. (eds.) WISE 2007. LNCS, vol. 4831, pp. 398–409. Springer, Heidelberg (2007)
4. Grossniklaus, M., Norrie, M.C.: Supporting Different Patterns of Interaction through Context-Aware Data Management. Journal of Web Engineering 7(3), 200–219 (2008)
5. Hattori, G., Hoashi, K., Matsumoto, K., Sugaya, F.: Robust web page segmentation for mobile terminal using content-distances and page layout information. In: Proceedings of the 16th International Conference on World Wide Web, WWW 2007, pp. 361–370. ACM, New York (2007)
6. Kleek, M.V., Bernstein, M.S., Panovich, K., Vargas, G.G., Karger, D.R., Schraefel, M.M.C.: Note to self: Examining personal information keeping in a lightweight note-taking tool. In: Olsen Jr., D.R., Arthur, R.B., Hinckley, K., Morris, M.R., Hudson, S.E., Greenberg, S. (eds.) CHI, pp. 1477–1480. ACM, New York (2009)

7. Laakko, T., Hiltunen, T.: Adapting Web content to mobile user agents. IEEE Internet Computing 9(2), 46–53 (2005)
8. Li, Y.: Gesture search: a tool for fast mobile data access. In: Proceedings of the 23nd Annual ACM Symposium on User Interface Software and Technology, UIST 2010, pp. 87–96. ACM, New York (2010)
9. Long, S., Kooper, R., Abowd, G.D., Atkeson, C.G.: Rapid prototyping of mobile context-aware applications: the Cyberguide case study. In: Proceedings of the 2nd International Conference on Mobile Computing and Networking, MobiCom 1996, pp. 97–107. ACM, New York (1996)
10. Luff, P., Heath, C., Norrie, M., Signer, B., Herdman, P.: Only Touching the Suface: Creating Affinities Between Digital Content and Paper. In: Proc. Conf. on Computer Supported Cooperative Work (CSCW 2004) (November 2004)
11. Norrie, M.: Paper on the move. In: Baresi, L., Dustdar, S., Gall, H.C., Matera, M. (eds.) UMICS 2004. LNCS, vol. 3272, pp. 1–12. Springer, Heidelberg (2004)
12. Signer, B., Norrie, M.C., Grossniklaus, M., Belotti, R., Decurtins, C., Weibel, N.: Paper-Based Mobile Access to Databases. In: Demo Proceedings of ACM International Conference on Management of Data (SIGMOD 2006) (2006)
13. Zhang, D.: Web content adaptation for mobile handheld devices. Commun. ACM 50, 75–79 (2007)

Visualization of Multi-domain Ranked Data

Alessandro Bozzon[1], Marco Brambilla[1], Tiziana Catarci[2], Stefano Ceri[1], Piero Fraternali[1], and Maristella Matera[1]

[1] Dipartimento di Elettronica e Informazione – Politecnico di Milano
{bozzon,mbrambil,ceri,fraterna,matera}@elet.polimi.it
[2] Dipartimento di Informatica e Sistemistica – Università "La Sapienza", Roma
catarci@dis.uniroma1.it

Abstract. This chapter focuses on the visualization of multi-domain search results. We start by positioning the problem in the recent line of evolution of search engine interfaces, which more and more are capable of mining semantic concepts and associations from text data and presenting them in sophisticated ways that depend on the type of the extracted data. The approach to visualization proposed in search computing extends current practices in several ways: the data to visualize are N-dimensional combinations of objects, with ranking criteria associated both to individual objects and to sets of combinations; object's properties can be classified in several types, for which optimized visualization families are preferred (e.g., timelines for temporal data, maps for geo-located information); combinations may exhibit any number of relevant properties to be displayed, which need to fit to the bi-dimensional presentation space, by emphasizing the most important attributes and de-emphasizing or hiding the less important ones. The visualization problem therefore amounts to deciding the best mapping between the data of the result set and the visualization space.

1 Introduction

Information visualization exploits presentation metaphors and interaction strategies for supporting perceptual inferences [1][15]. It is crucial for the success of any modern information-intensive application, particularly in multi-domain search, which produces articulated results comprising different types of objects extracted from multiple data sources and subject to various ranking criteria. The relevant issues in search computing interfaces are a mix of traditional search interface optimization concerns (e.g., how to ensure simplicity and high performance) and of new problems, such as how to exploit domain- and type-specific knowledge about data to achieve a "natural" ranking and display (e.g., using timelines, maps, or different kinds of charts), how to represent relationships among objects coming from different domains, how to convey the data provenance.

The design of a search interface must take into account the conservative nature of the visualization solutions so far adopted by search engines, which have scarcely evolved over the last decade and tend to exhibit a strong stability in the features offered to users, to preserve ease of use and efficacy for a wide audience of much

S. Ceri and M. Brambilla (Eds.): Search Computing II, LNCS 6585, pp. 53–69, 2011.

differentiated adopters. As the tasks supported by search engines increase in complexity, e.g., by providing concept and topic search [14][25], new features are introduced in the search interface of mainstream search engines cautiously, by progressively expanding the basic paradigm of keyword entry and vertical result list presentation. However, various factors still differentiate multi-domain search systems from traditional search engines.

- **Results are assembled from objects,** not necessarily corresponding to documents. For example, an object may derive from a deep Web record, for which there is no surface Web browsable page. This impacts the way in which feedback can be given to the user, because not always a document summary is the best way to provide provenance clues. For instance, some objects may be best presented showing a subset of their attributes and a link to the original information source, if this is available.
- **Results are not individual documents but combinations of objects.** This challenges the flat vertical list presentation approach, because not only provenance and result ranking must be conveyed, but also the relationships that connect objects in a result combination. Feedback on these relationships should be added to the search interface, e.g., to show why a hotel and a conference venue belong to the same combination.
- **Results are typed.** Unlike documents, which are unstructured, multi domain search retrieves typed objects. Knowing the object type is both a challenge and an opportunity: it is a challenge because exploiting the object type for differentiating the representations can make the interface complex and unstable, breaking the interaction style continuity that search engine users appreciate so much. It is also an opportunity, because type information can suggest optimized ways to present the results. For instance, if the interface recognizes geographical or temporal data it can offer the user the option to display results in a map or timeline.
- **Relationships have semantics.** objects forming a combination may be associated for different reasons: they may be close in space or time; they may share attribute values; they may be related in a hierarchy; they may be linked by a domain-specific relationship. In those cases when the interface is aware of the meaning of a relationship, it should represent it in an intuitive way.
- **The result space can be highly dimensional.** answers to multi-dimension queries may require the display of a high number of data attributes and associations; the interface should allow a compact visualization of the "most relevant" aspects of the result set and the easy appraisal of the features that have been hidden in the initial presentation. Given that the page layout is bi-dimensional, suitable summarization mechanisms are necessary to convey more than two data dimensions.

The initial design of data visualizations proposed for search computing were reported in [3] and focused on exploring the primitives for the manipulation of the result set and the refinement of the initial query, based on a very simple tabular representation of combinations. In this chapter, we instead concentrate on result visualization, and specifically on how to convey the relationships that link the objects in a combination, how to exploit knowledge about the type of an object forming a combination, and how to convey the ranking at both the individual data source and at the object combination level. We will in particular introduce one possible approach that capitalizes on well-known principles for data visualizations, trying to maximize their effectiveness for the specific characteristics of the search computing result sets.

This Chapter is organized as follows. Section 2 surveys a gallery of visualization examples from mainstream search engines and discusses the contributions from data visualization literature that can support the envisioned result presentation approach. Section 3 introduces the Search computing visualization problem, which consists of three parts: the data model of the result set, the existence of "expressive" types and dependencies within such model, and the heuristic approach for mapping the result set onto a visualization layout. Section 4 shows, without attempting an exhaustive approach, some of the visualizations that could be generated by applying heuristics. Section 5 presents the conclusions pointing to our future work.

2 Related Work in Search Engine Interfaces and Visualization Techniques

In this section we overview current search engine interfaces and the visualization paradigms for specific data (including categorical, geographical and temporal data).

2.1 Evolution of Search Engine Result Presentation

In recent years, search results visualization has evolved to include multiple features, which depend on the specific object being visualized, enriching the traditional result sets made of ordered list of objects (document surrogates), as the one shown in Fig. 1.

English poetry - Wikipedia, the free encyclopedia
The history of **English poetry** stretches from the middle of the 7th century to the present day. Over this period, English poets have written some of the most ...
en.wikipedia.org/wiki/**English_poetry** - Cached - Similar

Famous **English** Poets and **Poems**
A List of Famous **English** Poets includes **Poems** and Biographical information of the most Famous **English** Poets. Read and Enjoy **Poetry** by **English** Poets.
famouspoetsand**poems**.com/country/.../**English**_poets.html - Cached - Similar

classic **poetry** - 100% free - World **English**
CLASSIC **ENGLISH POETRY**. One of the best ways of improving your English is to read the language. This part of our site includes complete poems by some of the ...
www.world-**english**.org/**poetry**.htm - Cached - Similar

Fig. 1. Basic search engine result set as an ordered list of documents

The result set can be structured in sub-lists, e.g., to highlight entries of different pages within the same Web site, or result items of different nature (e.g., images related to a given Web documents), as in Fig. 2. The information content of result elements can be enriched if the type of the displayed information is known. For example, Fig. 3 shows the use of a timeline for the visualization of temporal data.

Famous **English Poets** and Poems
A List of Famous **English Poets** includes Poems and Biographical information of the most Famous **English Poets**. Read and Enjoy Poetry by **English Poets**.
famouspoetsandpoems.com/country/.../English_poets.html - Cached - Similar

19th Century **English Poets** and Poems
A list of Famous 19th Century **English Poets**. The authors listed on this page include some of the greatest poets of 19th Century.
famouspoetsandpoems.com/.../19th_century_English_poets.html - Cached - Similar
Show more results from famouspoetsandpoems.com

Famous **English Poets**
Visit this site dedicated to Famous **English Poets**. Enjoy the poems by all of the Famous **English Poets**. Read a great selection of poems - dedicated to Famous ...
www.love-poems.me.uk/a_english_poets.htm - Cached - Similar

Fig. 2. A result set structured into a list of documents and a list of images

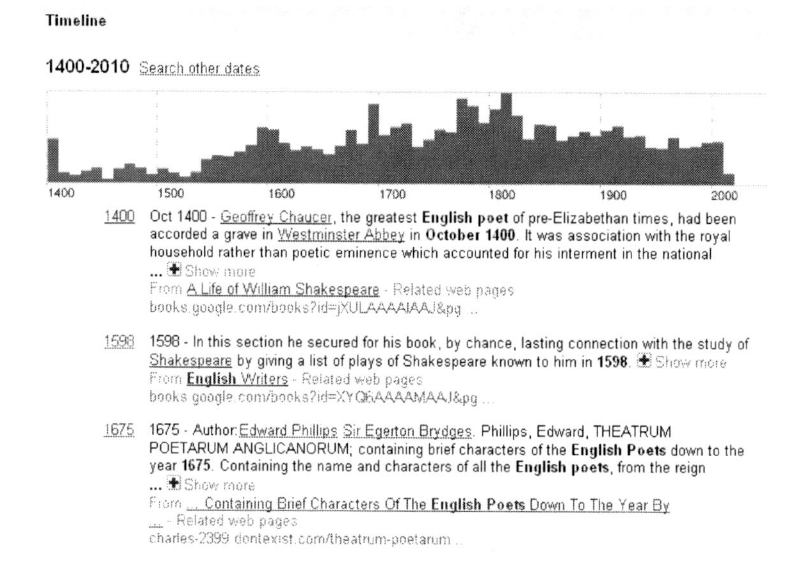

Fig. 3. Timeline visualization for a result set comprising temporally located documents

A special class of typed information is constituted by geo-referential data, which permits the extraction of geographical concepts; a geographical concept is an object associated with geographical coordinates and possibly categorized according to a geographic ontology (e.g., GeoNames ontology classes [13]). Geographical concepts are highlighted in the result set and typically positioned on a map. The identified concept can also be associated with semantically correlated concepts; for example, a town can be related to hotels, restaurants, and places of interest; a touristic place, a hotel or a restaurant can be associated with reviews.

The amount of structure extracted from documental data and the employed visualization primitives depends on the capacity of recognizing domain specific entities. The examples shown so far revolve around the presentation of instances of a single entity. New generation search engines are moving towards to the collection and integration of heterogeneous data sources. Kosmix [25] is an example of general-purpose topic discovery engine, which offers one-page information summaries about a topic. The presented information is retrieved through calls to Web services that extract information from deep Web data sources. The schema of each topic is a complex record type, which not only comprises typed properties of the entity but also associations to other entities representing related topics. For instance, the query "William Shakespeare" returns an instance of the entity *Writer*, as shown in Fig. 4, characterized by information deriving from multiple services (e.g., Biography.com, and Google) and also related to other related writer instances.

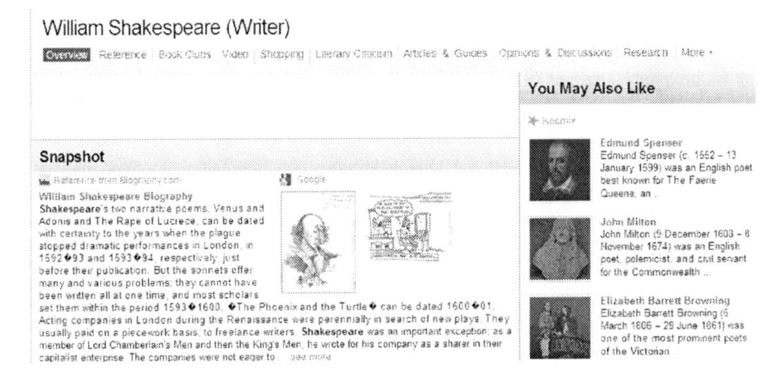

Fig. 4. An instance of the entity *Writer* in the Kosmix topic search engine

2.2 Related Work in Data Visualization

Data visualization has a long-standing tradition, which initially focused on the analysis of alternative visualization techniques for various categories of data [1][4]. Classic works like [27][23] offered guidance in selecting the most appropriate visualization techniques for different types of data (e.g., 1-, 2-, 3-dimensional data, temporal and multi-dimensional data, and tree and network data in [27]). Later works (e.g., [6]) explored the underlying conceptual structure of data-oriented visualization, highlighting a common framework of data visualization strategies (e.g., the data stages/transformation model in [6] and the classification based on data spatialization and visual perception in [26]), giving a deeper rationale to the taxonomies of visualization techniques. For example, in the field of relational data visualization, the work described in [29] proposes three families of 2D graphics that depend on the type of data to be represented on the axes, i.e., *ordinal–ordinal*, when no dependency exists among the different attributes (e.g., a table), *ordinal–quantitative*, where a quantitative variable is dependent on the ordinal variable (e.g., a bar chart), or the ordinal and quantitative data can be independent (e.g., a Gantt chart), and *quantitative-quantitative*, used to represent the distribution of data as a function of one or both quantitative variables, also highlighting causal relationships between the two quantitative variables (e.g., a map). Each family then contains variants depending on the combination of the selected mark type (e.g., rectangle, circle, glyph, text, Gantt bar, line, polygon, and image) and its visual and retinal properties that best suit the characteristic properties of the single objects.

In [31] the author proposes a comprehensive taxonomy of glyph placement strategies with respect to both data types and user's task. The author in particular distinguishes between *data-driven* and *structure-driven* approaches: the former exploit data properties to determine the object location in the visualization space; the latter exploit some (implicit or explicit) relationships between data points, such as a temporal ordering or hierarchical relationships.

Whereas the abovementioned efforts are horizontal, spanning all possible categories of data, other works concentrate on the design, evaluation and comparison of visualization techniques for specific data: categorical (e.g., [19][11]), temporal (e.g., [7][9]), geographical (e.g., [12][20][30]), multidimensional (e.g., [31]), and graph-based data (e.g., [32]) are the most prominent sectors.

A huge amount of literature also concentrates on the presentation generation process. The pioneer approach proposed by Mackinlay [22], and several other successive works (e.g., [15]), exploit data characterization and propose rule-based approaches to map data types to visual elements [5]. In other words, elements of a *data model* are mapped to elements of a *visual model* [15]. The common aim characterizing such works is to automatically derive "adequate" visualizations [5], where adequate means *complete,* i.e., the user perceive from them all the information enclosed within the original data, and *correct,* i.e., no unrequired information is conveyed. By capitalizing on the principles introduced by the above-illustrated works, in the following sections we illustrate how the characterization of the Search computing result set can guide the definition of a visualization process.

3 Visualization Model and Process for Search Computing

3.1 Visualization Model

The result of a Search computing query is made of combinations, where each combination is a record of *N* object instances connected by join predicates and satisfying selection conditions. Each object in a combination has a schema constituted by typed attributes; a subset of the attributes defines the combination *identifier*, which is either system-generated (object identifier) or provided by the real world context (e.g., event's date and time). Objects may have *repeating groups*, where each repeating group is a non-empty set of attributes with multiple values (such as actors of a given movie, with name and gender). Object instances extracted by a query may be locally *ranked*; in such a case one or more object's attributes express the ranking, which is either system-generated (rank position among the object instances) or provided by the usage context (e.g., the rating of a hotel). Combinations have a global rank, expressed as a weighted sum of the local ranks of member objects, normalized in the [0-1] range.

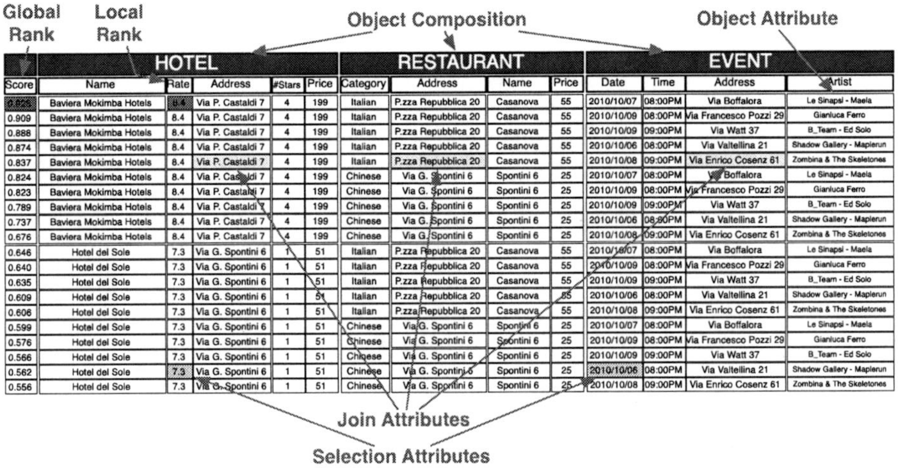

Global Rank | Local Rank | Object Composition | Object Attribute

	HOTEL					RESTAURANT				EVENT			
Score	Name	Rate	Address	#Stars	Price	Category	Address	Name	Price	Date	Time	Address	Artist
0.925	Baviera Mokimba Hotels	8.4	Via P. Castaldi 7	4	199	Italian	P.zza Repubblica 20	Casanova	55	2010/10/07	08:00PM	Via Boffalora	Le Sinapsi - Maela
0.909	Baviera Mokimba Hotels	8.4	Via P. Castaldi 7	4	199	Italian	P.zza Repubblica 20	Casanova	55	2010/10/09	08:00PM	Via Francesco Pozzi 29	Gianluca Ferro
0.888	Baviera Mokimba Hotels	8.4	Via P. Castaldi 7	4	199	Italian	P.zza Repubblica 20	Casanova	55	2010/10/09	09:00PM	Via Watt 37	B_Team - Ed Solo
0.874	Baviera Mokimba Hotels	8.4	Via P. Castaldi 7	4	199	Italian	P.zza Repubblica 20	Casanova	55	2010/10/06	08:00PM	Via Valtellina 21	Shadow Gallery - Maplerun
0.837	Baviera Mokimba Hotels	8.4	Via P. Castaldi 7	4	199	Italian	P.zza Repubblica 20	Casanova	55	2010/10/08	09:00PM	Via Enrico Cosenz 61	Zombina & The Skeletones
0.824	Baviera Mokimba Hotels	8.4	Via P. Castaldi 7	4	199	Chinese	Via G. Spontini 6	Spontini 6	25	2010/10/07	08:00PM	Via Boffalora	Le Sinapsi - Maela
0.823	Baviera Mokimba Hotels	8.4	Via P. Castaldi 7	4	199	Chinese	Via G. Spontini 6	Spontini 6	25	2010/10/09	08:00PM	Via Francesco Pozzi 29	Gianluca Ferro
0.789	Baviera Mokimba Hotels	8.4	Via P. Castaldi 7	4	199	Chinese	Via G. Spontini 6	Spontini 6	25	2010/10/09	09:00PM	Via Watt 37	B_Team - Ed Solo
0.737	Baviera Mokimba Hotels	8.4	Via P. Castaldi 7	4	199	Chinese	Via G. Spontini 6	Spontini 6	25	2010/10/06	08:00PM	Via Valtellina 21	Shadow Gallery - Maplerun
0.676	Baviera Mokimba Hotels	8.4	Via P. Castaldi 7	4	199	Chinese	Via G. Spontini 6	Spontini 6	25	2010/10/08	09:00PM	Via Enrico Cosenz 61	Zombina & The Skeletones
0.646	Hotel del Sole	7.3	Via G. Spontini 6	1	51	Italian	P.zza Repubblica 20	Casanova	55	2010/10/07	08:00PM	Via Boffalora	Le Sinapsi - Maela
0.640	Hotel del Sole	7.3	Via G. Spontini 6	1	51	Italian	P.zza Repubblica 20	Casanova	55	2010/10/09	08:00PM	Via Francesco Pozzi 29	Gianluca Ferro
0.635	Hotel del Sole	7.3	Via G. Spontini 6	1	51	Italian	P.zza Repubblica 20	Casanova	55	2010/10/09	09:00PM	Via Watt 37	B_Team - Ed Solo
0.609	Hotel del Sole	7.3	Via G. Spontini 6	1	51	Italian	P.zza Repubblica 20	Casanova	55	2010/10/06	08:00PM	Via Valtellina 21	Shadow Gallery - Maplerun
0.606	Hotel del Sole	7.3	Via G. Spontini 6	1	51	Italian	P.zza Repubblica 20	Casanova	55	2010/10/08	09:00PM	Via Enrico Cosenz 61	Zombina & The Skeletones
0.599	Hotel del Sole	7.3	Via G. Spontini 6	1	51	Chinese	Via G. Spontini 6	Spontini 6	25	2010/10/07	08:00PM	Via Boffalora	Le Sinapsi - Maela
0.576	Hotel del Sole	7.3	Via G. Spontini 6	1	51	Chinese	Via G. Spontini 6	Spontini 6	25	2010/10/09	08:00PM	Via Francesco Pozzi 29	Gianluca Ferro
0.566	Hotel del Sole	7.3	Via G. Spontini 6	1	51	Chinese	Via G. Spontini 6	Spontini 6	25	2010/10/09	09:00PM	Via Watt 37	B_Team - Ed Solo
0.562	Hotel del Sole	7.3	Via G. Spontini 6	1	51	Chinese	Via G. Spontini 6	Spontini 6	25	2010/10/06	08:00PM	Via Valtellina 21	Shadow Gallery - Maplerun
0.556	Hotel del Sole	7.3	Via G. Spontini 6	1	51	Chinese	Via G. Spontini 6	Spontini 6	25	2010/10/08	09:00PM	Via Enrico Cosenz 61	Zombina & The Skeletones

Join Attributes

Selection Attributes

Fig. 5. Tabular representation of a multi-domain result set

Fig. 5 describes the main elements of the results of the query "*find the best combination of hotels, restaurants and events to spend a nice evening and night in given city* ", which composes of three objects: *events*, *restaurants*, and *hotels*. Fig. 5 can be considered the default way of visualizing a multi-domain result set: the table-based visualization lists all combinations, with their objects and attributes, in descending global rank. However, alternative visualizations are possible, which might highlight few primary visualization dimensions selected within a combination. Such result presentations can take advantage of suitable visualizations, based upon the type of some of the attributes of the result, e.g., maps for geographic locations, timelines for temporal events, or Cartesian spaces for quantitative variables. With this respect, the first step is to identify relevant data types that can guide the selection of suitable primary visualization dimensions.

3.2 Data Type Classification

Visualization of attribute values can be optimized according to their type. As usual, attribute types are classified according to their scale type, as stated in the classic definition of measurement theory [25]:

- **Interval:** quantitative attributes measured relative to an arbitrary interval (e.g., Celsius degrees, latitude and longitude, date, GPA). In this class, two important subclasses are further distinguished for visualization purposes:
  - **Geographic points and addresses:** they admit domain specific operations like the computation of distance, the visualization on maps, the determination of routes, etc.
  - **Time points:** they admit the representation on time scales and calendars, at different domain-specific granularity.
- **Ratio:** quantitative attributes measured as the ratio with a known magnitude unit (e.g., most physical properties).

- **Nominal:** categorical labels without notion of ordering (e.g., music genre). They can be visualized by means of textual labels (for example within tables). In case of a low number of categories, they can be represented through visual clues, for example different shapes or colors.
- **Ordinal:** data values that admit order, but not size comparison (e.g., quality levels).

Nominal data can be also associated with frequency (or relative frequency) values, i.e., the number (percentage) of elements falling into categories. This may imply the use of a combination of visual clues, such as the size of shapes or the opacity of the colors representing categories, or the "construction" of graphics (e.g., bar charts), where categories are labels over the horizontal axis, while frequencies correspond to the vertical axis dimension. Nominal data can also be further distinguished into **taxonomical**, when they admit subset relations (e.g., animal species). Tree based representations (e.g., treeMap) can be adopted in this case [16].

3.3 Visualization Process

The process of generating the visualization aims at producing a representation that maximizes the understandability of the result set, by considering the type and semantics of data and the functions of multi-domain search. The visualization process, schematically illustrated in Fig. 6, maps the result set onto a presentation space by

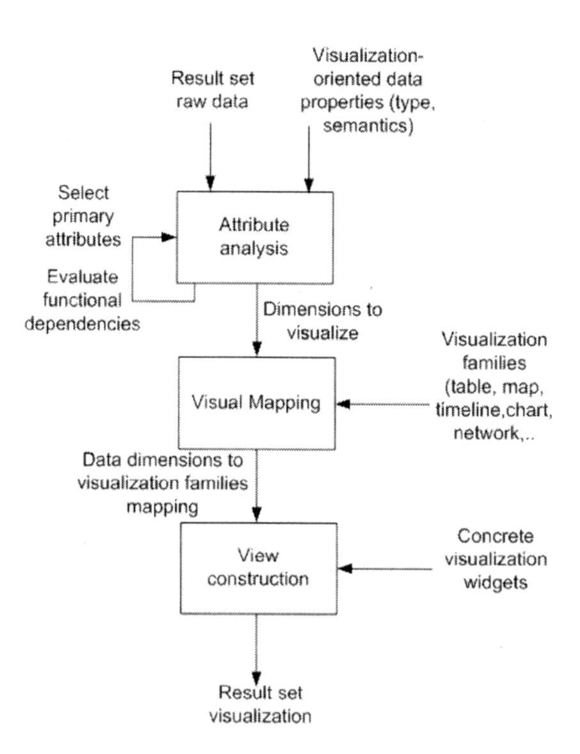

Fig. 6. Main steps in the result set visualization process

considering the types of attributes that describe the properties of combinations, the presentations functions to be implemented, and the visualization families that are best usable for rendering the data dimensions.

The most important visualization dimensions, which need to be highlighted in the visualization and that can guide the identification of the primary visualization space, are determined; to this end, information about object *identifiers*, *join* and *selection conditions*, and dependencies among the different data are exploited.

The output of the visualization process is a page layout (a view, in the terminology of [6]), which assigns data to visualization elements, implementing all the needed visualization functions. The generation process in Fig. 6 can occur at design time (when the query is fixed and non modifiable), at run-time (when the user explores the search space from scratch) and in a mixed mode (when the user starts with a known query and then expands it, by joining in more search services).

In order to identify the primary visualization dimension, one can assume that an ordering of data type exists and that it guides the selection of the primary attributes determining the visualization space. Such ordering depends on the capacity of the data types to "delimit" a visualization space where single objects and combinations included in the result set can be conveniently positioned. For example, interval attributes can be considered the best candidates for primary visualization, since they permit a precise characterization of the position of objects within a bi-dimensional visualization space, followed by ordinal, and nominal attributes, which instead can be adequately represented by means of visual clues. Therefore, if a given object O has a geo-referenced attribute, then its instances can be represented on a map by using that attribute as a primary visualization; similarly, objects with attributes representing temporal events can be placed on a timeline. Representing a combination then amounts to finding suitable representations for the majority of its objects, by highlighting them upon a given visualization space, and then relating together the object instances of a combination through orthogonal visual mechanisms. Once placed on the visualization space, an object is succinctly represented by some of its attributes (e.g., identifiers); typically the local ranking of the object can also be visually represented (e.g., through conventional shapes, or colors). Other attributes are omitted from the visualization, and can be accessed through secondary visualization methods, such as pop-up windows.

In line with the classical approaches for the automatic generation of visualizations [5], this construction can be formalized by heuristic rules, which take in input the characterization of the N dimensions to visualize and emit as output the decision of how to allocate them onto a bi-dimensional representation space, addressing the visualization functions of multi-domain search. A very general scheme is the following:

1. Pick the most relevant dimensions, on the basis of a defined order among the available data types and on the identification of the dimensions that best characterize the majority of objects involved in combinations.
2. Identify the primary visualization space to use;
3. Allocate all dimensions that can be supported by the primary visualization space;
4. Pick the remaining dimensions in order of significance and decide the secondary widgets best suited to represent them;
5. Allocate all remaining dimensions that can be supported ergonomically by secondary widgets.

At step 1, selected data types allow a domain-dependent effective visualization. The order: *geo location→time/date→other types* can be used to exploit geographical information first, then temporal data, and then all the remaining uncharacterized types. At step 2, the most appropriate visualization space is chosen. Guidelines from the literature and best practices in data visualization (e.g., [9] for temporal data, [11] for categorical data, [20] for geographical data) can be used. Also, one can exploit the huge quantity of out-of-shelf components that are more and more offered by visualization projects and providers of software components[1]. At step 4, the dimensions that are not represented by the primary widget need to be considered, and visual clues for their representation selected. Various techniques exist that descend from the classic identification of visual variables by Bertin (position, size, shape, value, color, orientation, and texture), possibly adapted to specific contexts (e.g., thematic cartography [12]).

The primary dimensions for data visualization can be further reduced by exploiting dependencies between objects. For example, if a query includes a 1:M join between two objects O1 and O2, such that a set of instances of O2 is mapped exactly to one instance of O1, then the objects of O2 can be represented by using the primary visualization chosen for O1. For instance, if O1 are hospitals and O2 are doctors, and hospitals have a geo-localized attribute (their address), then it is possible to display doctors on a map by placing them at the same attribute as their hospital.

4 Examples of Visual Representations of Query Results

To highlight the importance of dimension selection, this section presents several examples of visualizations that represent objects, their composition, and their local and global rankings; every example can be generated by suitable applications of generic visualization rules.

4.1 Visualization of Geo-referenced Objects

Fig. 7 presents a visualization example for the query "*find the best combination of hotels, restaurants and events to spend a nice evening and night in given city*". One possible result set is the one illustrated in Fig. 7. The three objects to be displayed have geographic coordinates; therefore, the primary dimension to adopt for their representation is a map. This choice allows us positioning each object instance as a point in the map, which will be conveniently selected so as to include objects in the context (e.g., a portion of a city map). A combination is then a triple of positions on the map; it can be visualized by any representation that puts the three objects together; in the example, we enclose each triple within an area, which is highlighted by means of colors. We then use darkest color for the best combination.

[1] Among the best known examples, the OLIVE library [24] lists data visualization environments clustered according to Shneiderman's classification scheme and the recent effort by IBM [17] offers a community space for publishing visualization widgets and data sets (at the time of writing, 70183 visualizations and 141459 data sets are enlisted).

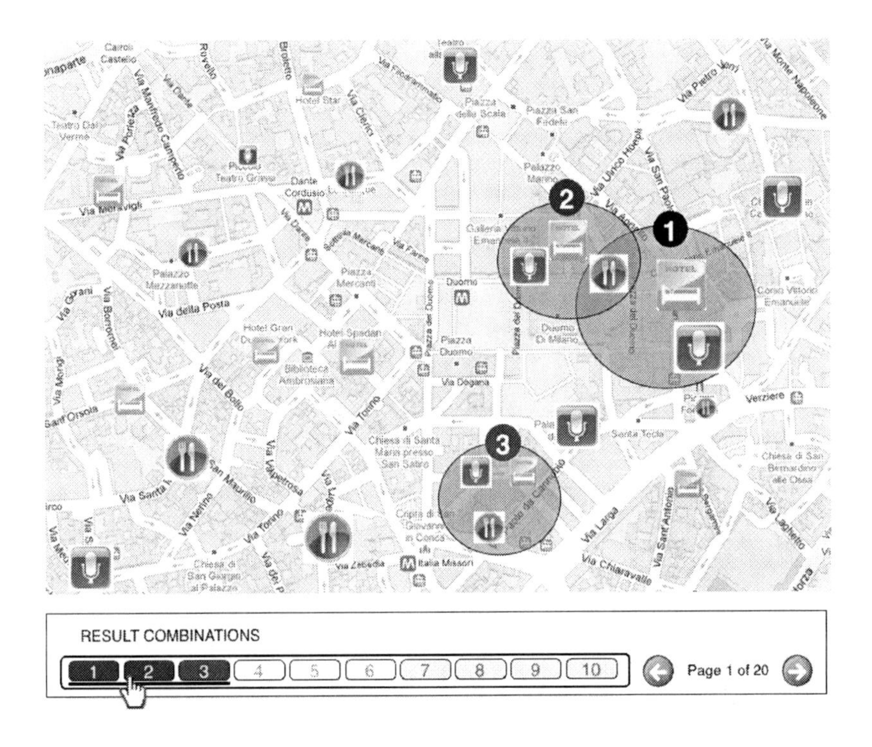

Fig. 7. Visualization of geo-referenced objects

Moreover, each object has a different icon, and the local ranking (representing the relative ranking of the object instance within the selected objects) is represented by the size of the icon. Such representation can show only a few combinations on the same screen, therefore the visual result presents only three combinations (the top-ranked) and a scrolling mechanism allows seeing the following combinations. The other object attributes (e.g., name, stars and price for the hotels) can be displayed in pop up windows, opened by pointing to given object with the mouse.

4.2 Geo-referenced Visualization with Object Dependencies

Object dependencies can be used to associate a visualization dimension to objects that do not have properties with data types that effectively determine a visualization space. Consider a query that searches for *close hospitals to a given location, such that the hospitals have specialists of a given disease (e.g., Parkinson)*; the specialists are ranked, e.g., by the relevance of their published articles on the topic. Doctors have no properties that allow their representation according to a quantitative visualization space, but they are placed at hospitals, which have a specific location. Therefore, doctors can be represented as icons, which are placed on a map at the same location as their hospitals. Their local ranking can be represented through different icon sizes.

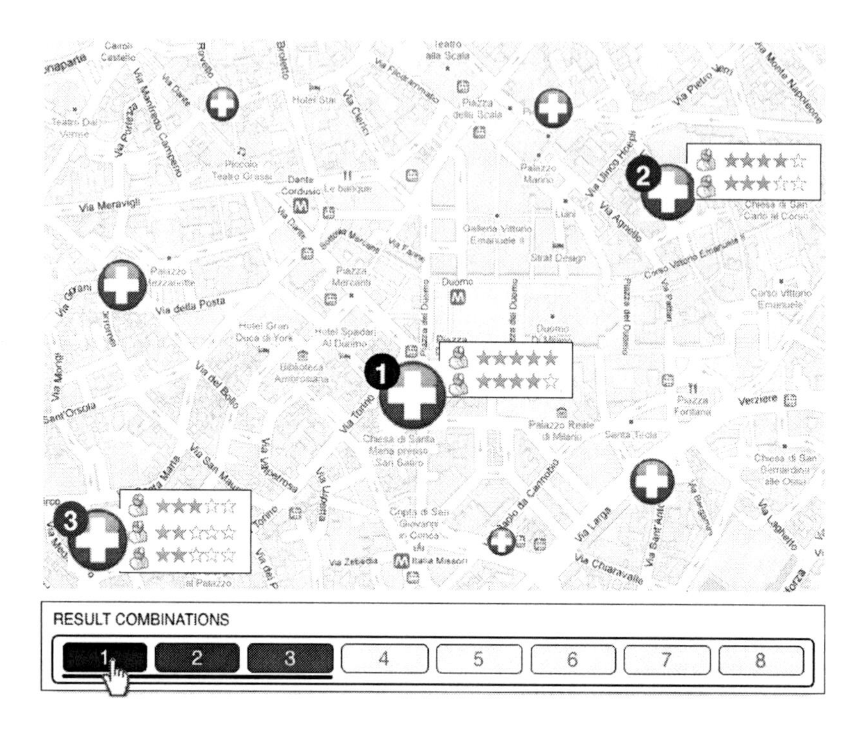

Fig. 8. Geo-visualization with object dependencies

Fig. 8 highlights the dependency of doctors from hospitals; hospitals play the role of aggregators. Being hospitals geo-referenced objects, than the map can be exploited even though addresses do not characterize doctors. The map shows that *hospital 1* has two doctors with high rating; *hospital 2*, instead, has two doctors but with a lower rating, and hospital 3 has three doctors with a lower rating. The hospital index is also a representation of the global ranking, which is in this case globally associated with a hospital, by aggregating the ratings of doctors, and by considering the user's location. Once the different combinations are displayed on the map, then pop-up windows can show more attributes about hospitals and doctors.

4.3 Timeline Visualizations

While geographic maps are very effective for relating objects that are located in space, timelines are effective for relating objects that are located in time. Fig. 9 provides a visualization example for a search about author's productivity, both in terms of publication indexes (e.g., the Hirsch Index) and of yearly production. Authors are locally ranked by their publication index; the yearly productivity of an author is measured by the number of published articles (divided in journal and conferences) and books; the local ranking is a weighted sum of such measures. The global ranking, which takes into account the publication indexes, the yearly production, and how recent is the production (recent production is most highly ranked), presents the "*most productive recent years of high-ranked authors*".

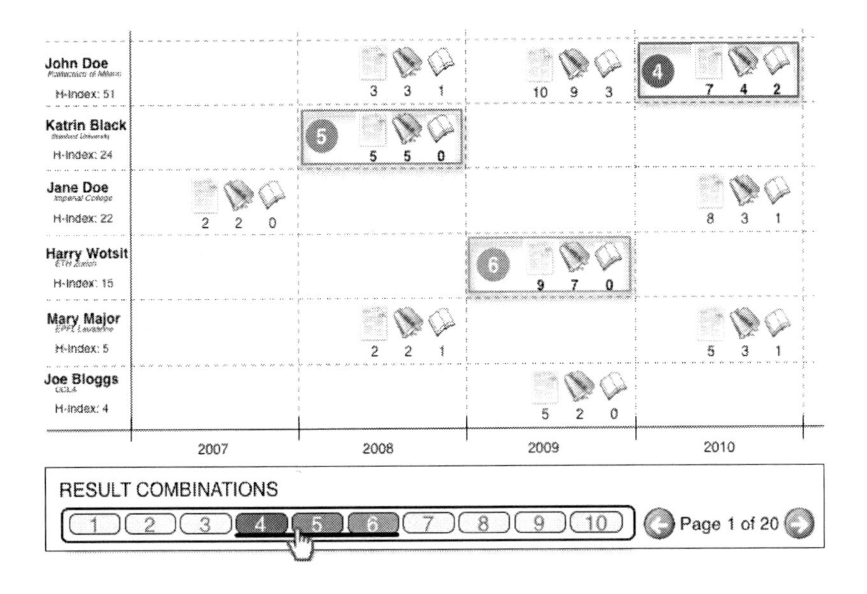

Fig. 9. Timeline representing scientists' productivity over the years

This information is presented on a timeline, having on the time axis the year of production, and presenting every author on the same line. Authors are listed in "order" of their Hirsch ranking, and the visualization highlights the most productive pair author-year, initially centered on that author and year; the author-year pair is highlighted (e.g., being encircled and colored). Author information, reported on the left of the timeline, is the name and H index; yearly production has three icons representing papers, articles and books, with the appropriate number below the icon, placed in the timeline. The three combinations with higher global index are highlighted, and a scroller allows moving to the subsequent combinations; scrolling the combinations has the effect of moving along the timeline and the authors. This representation is inspired by Envision [10].

The automatic generation of such representation uses the assumptions that yearly productivity is related to time intervals (therefore it can be represented by a timeline), and that each yearly productivity item functionally determines an author; therefore, each author can be associated with a single timeline. Then, author's local ranking is the H-index, which is represented, while yearly productivity's local ranking is a combination of the number of publications, which are also represented; and the global ranking is explicitly represented by highlighting. A similar representation could display the severity of hospitalizations of a family of chronically ill patients, with each patient represented as a line in a timeline, hospitalizations reported as intervals on the timeline, and severity of treatment reported as indicators close to the intervals. The presence, type and duration of given treatments could be also reported graphically.

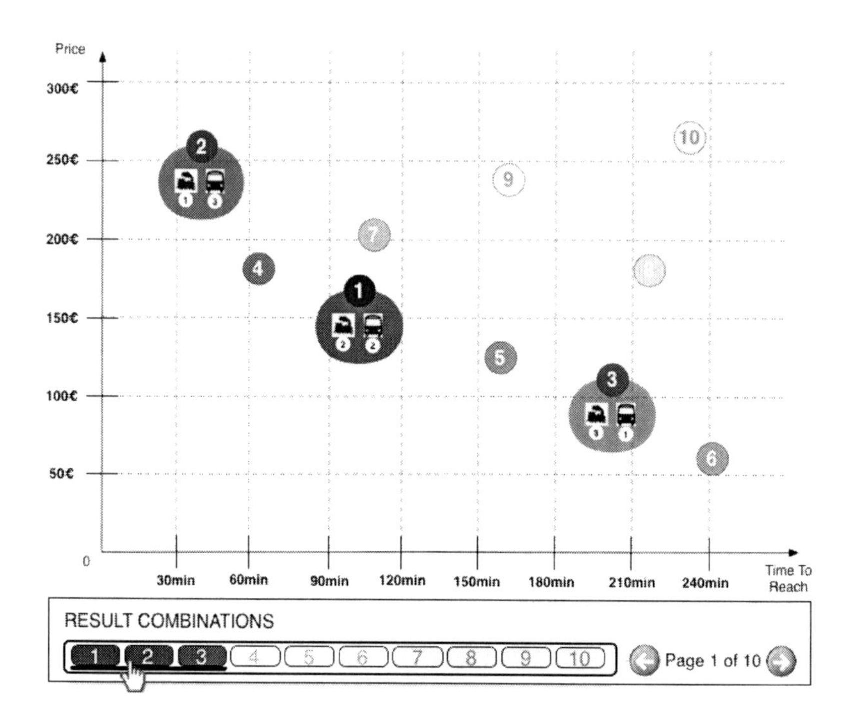

Fig. 10. Representation of combinations ranked by ratio attributes

4.4 Representation in the Lack of Suitable Interval Dimensions

While in the above examples the objects to be displayed include interval dimensions with an associated graphic representation, in the general case objects may lack such a property. Then, their visualization may resort to other object properties, especially when such properties are associated with the object's rankings.

Assume a query about reaching a given location by combining trains and local transportation (taxi or bus). Assume that each such transportation has a cost and duration, and assume that they can be ranked locally and globally by a suitable function of their local and global costs and duration. Note that price and duration both belong to "interval" attribute types according to the categorization of Section 3.2. Then, Price and Time can be used as axes of a Cartesian space.

In the representation of Fig. 10, we highlight the global time and global duration in the space, and then enclose the first, second, and third combinations according to the global ranking (and let other combinations be accessible e.g., through a scroller). The presence of two objects is represented by the fact that each location in the space is associated with two objects, and by clicking on the object one can read the details about the train and local trip, including its duration and cost.

Finally, assume the case when ranking is based on ordinal dimensions. Let's consider an example involving a vacation package where two ordinal dimensions describe the price-range of the package (e.g., in hundreds of Euro) and quality of the hotels being used (from five to three stars), described in Fig. 11.

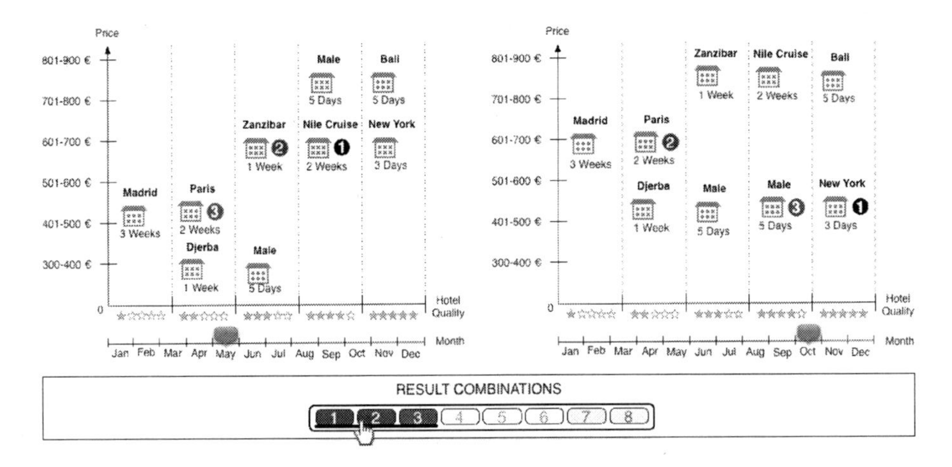

Fig. 11. Example of ordinal-quantitative visualization

Assume then that the global ranking function takes into account both these features and is based on a price-performance ratio. In such case, one of the dimensions (say, the quality of hotels) is used as reference in order to organize the space in five columns, and then the packages are presented as nodes in these columns, named by the principal attraction (e.g., "Nile Cruise"), in price-range order. The global ranking is presented by highlighting the three combinations that present the "best" price-performance according to the global ranking. Fig. 11 also shows the presence of two timelines, selected by the user in the query, and the effect of seasonal changes (e.g., of prices and hence on global ranking) on the choice of top combinations. The system may enable selecting trips (e.g., by country, availability during the year) so as to inspect the vacation packages of interest; once selected, the properties of the packages can be described further by inspecting pop-ups associated with each node.

5 Conclusions and Future Work

This paper has presented the general method that we intend to use in generating visualizations for multi-domain ranked data, and illustrated some representative examples that are currently driving us in building the method. Building a generic visualization tool, able to analyze the visualization data model and produce a suitable representation without being driven by other knowledge about the application domain, is a very challenging task. While the first examples that we have constructed ad-hoc seem promising and yielding to general rules of good applicability, the actual validation of the approach requires a formalization of the data and visualization models, aimed at identifying the most relevant properties that can guide the visualization process. Such formalization will be inspired to some past works that have already identified collections of rules to guide the visualization process for relational structured data, but it will take into account the peculiarity of the Search computing result set.

We will also try to overcome some limitations of past approaches, such as the lack of formal checking of the visualization correctness [4]. An extensive experimentation with users will also allow us to assess the effectiveness of the produced visualizations, and to investigate to which extent totally automatic processes for the visualization generation should be preferred to participatory paradigms where the user is directly involved, e.g., by means of preference expressions, to the construction of the visualization spaces.

References

[1] Bertin, J.: Semiology of Graphics: Diagrams, Networks, Maps (1983)

[2] Bostock, M., Heer, J.: Protovis: A Graphical Toolkit for Visualization. IEEE Trans. Vis. Comput. Graph. 15(6), 1121–1128 (2009)

[3] Bozzon, A., Brambilla, M., Ceri, S., Fraternali, P., Manolescu, I.: Liquid Queries and Liquid Results in Search Computing. In: Ceri, S., Brambilla, M. (eds.) Search Computing. LNCS, vol. 5950, pp. 244–267. Springer, Heidelberg (2010)

[4] Catarci, T., Dix, A.J., Kimani, S., Santucci, G.: User-Centered Data Management. Morgan & Claypool Publishers, San Francisco (2010)

[5] Catarci, T., Santucci, G., Costabile, M.F., Cruz, I.: Foundations of the DARE System for Drawing Adequate Representations. In: Proc. of DANTE 1999 (1999)

[6] Chi, E.H.: A Taxonomy of Visualization Techniques Using the Data State Reference Model. In: IEEE Symp. Information Visualization, Salt Lake City, UT, USA, pp. 69–75. IEEE CS Press, Los Alamitos (2000)

[7] Daassi, C., Nigay, L., Fauvet, M.C.: A Taxonomy of Temporal Data Visualization Techniques. Information-Interaction-Intelligence 5(2), 41–63 (2005)

[8] Elmqvist, N., Stasko, J., Tsigas, P., Meadow, D.: A Visual Canvas for Analysis of Large-Scale Multivariate Data. Information Visualization 7(1), 18–33 (2008)

[9] Fernandes Silva, S., Catarci, T.: Visualization of Linear Time-Oriented Data: A Survey. Journal of Applied System Studies 3(2) (2002)

[10] Fox, E.A., Hix, D., Nowell, L.T., Brueni, D.J., Wake, W.C., Heath, L.S., Rao, D.: Users, User Interfaces, and Objects: Envision, a Digital Library. Journal of the American Society for Information Science 44(8), 480–491 (1993)

[11] Friendly, M.: Visualizing Categorical Data. SAS Publishing (2000)

[12] Garlandini, S., Fabrikant, S.I.: Evaluating the Effectiveness and Efficiency of Visual Variables for Geographic Information Visualization. In: Hornsby, K.S., Claramunt, C., Denis, M., Ligozat, G. (eds.) COSIT 2009. LNCS, vol. 5756, pp. 195–211. Springer, Heidelberg (2009)

[13] GeoNames Ontology, `http://www.geonames.org/ontology/`

[14] Giunchiglia, F., Kharkevich, U., Zaihrayeu, I.: Concept Search. In: Aroyo, L., Traverso, P., Ciravegna, F., Cimiano, P., Heath, T., Hyvönen, E., Mizoguchi, R., Oren, E., Sabou, M., Simperl, E. (eds.) ESWC 2009. LNCS, vol. 5554, pp. 429–444. Springer, Heidelberg (2009)

[15] Haber, E.M., Ioannidis, Y.E., Livny, M.: Foundations of Visual Metaphors for Schema Display. J. Intell. Inf. Syst. 3(3/4), 263–298 (1994)

[16] Hearst, M.: Search User Interfaces. Cambridge University Press, Cambridge (2009), ISBN 9780521113793

[17] IBM. Manyeyes (December 2010), `http://www-958.ibm.com/software/data/cognos/manyeyes/`

[18] Johansson, S.: Visual Exploration of Categorical and Mixed Data Sets. In: Proceedings of the ACM SIGKDD Workshop on Visual Analytics and Knowledge Discovery: Integrating Automated Analysis with Interactive Exploration, Paris, France, July 28-28, pp. 21–29 (2009)
[19] Johnson, B., Shneiderman, B.: Treemaps: A Space-Filling Approach to the Visualization of Hierarchical Information Structures. In: Proceedings of the IEEE Information Visualization 1991, pp. 275–282. IEEE, Los Alamitos (1991)
[20] Maceachren, A.M.: An Evolving Cognitive-Semiotic Approach to Geographic Visualization and Knowledge Construction. Cartography and Geographic Information Systems 19, 197–200 (2001)
[21] Mackinlay, J.D.: Automatic Design of Graphical Presentations, Ph.D. Thesis, Department of Computer Science, Stanford University (1986)
[22] Mackinlay, J.D., Hanrahan, P., Stolte, C.: Show Me: Automatic Presentation for Visual Analysis. IEEE Trans. Vis. Comput. Graph 13(6), 1137–1144 (2007)
[23] North, C.: A Taxonomy of Information Visualization User-Interfaces (1998), http://www.cs.umd.edu/~north/infoviz.html
[24] OLIVE: On-line Library of Information Visualization Environments, University of Maryland College Park (1999), http://otal.umd.edu/Olive/
[25] Rajaraman, A.: Kosmix: High Performance Topic Exploration Using the Deep Web. In: Proceedings of the VLDB Endowment, August 2008, vol. 2(1), pp. 1524–1529 (2009)
[26] Rodrigues Jr., J.F., Traina, A.J.M., de Oliveira, M.C.F., Traina Jr., C.: Reviewing Data Visualization: An Analytical Taxonomical Study. In: Proceedings of the Information Visualization, IV 2006 (2006)
[27] Shneiderman, B.: The Eyes Have It: A Task by Data Type Taxonomy for Information Visualizations. In: IEEE Symp. Visual Languages, pp. 336–343. IEEE CS Press, Los Alamitos (1996)
[28] Stevens, S.S.: On the Theory of Scales of Measurement. Science 103(2684), 677–680 (1946), http://web.duke.edu/philosophy/bio/Papers/Stevens_Measurement.pdf
[29] Stolte, C., Tang, D., Hanrahan, P.: Polaris: A System for Query, Analysis, and Visualization of Multidimensional Databases. Commun. ACM 51(11), 75–84 (2008)
[30] Takatsuka, M., Gahegan, M.: GeoVISTA Studio: A Codeless Visual Programming Environment for Geoscientific Data Analysis and Visualization. Computational Geoscience 28, 1131–1144 (2002)
[31] Ward, M.O.: A Taxonomy of Glyph Placement Strategies for Multidimensional Data Visualization. Journal of Information Visualization 1(3/4)
[32] Wattenberg, M.: Visual Exploration of Multivariate Graphs. In: Proceedings of the SIGCHI Conference on Human Factors in Computing Systems - CHI 2006, pp. 811–819. ACM, New York (2006)

Part 3
Semantic Description

Knowing Web data sources and services in great detail is a fundamental aspect of search computing: the ability of building queries depends on the ability of giving to resources a rich semantic description, so as to characterize the application domain and the search activities that can be performed in that domain. The current conceptual description of Web resources uses a very simple Entity-Relationship model, so as to enable an equally simple description of Web interactions. But we are considering how to link ontological sources to service descriptions, as we expect that richer semantics may cover the gap from informal interactions (such as dialogs or sentences in natural language) to query expression. This part of the book describes the current flow of research in semantic description of Web resources in the SeCo project. It starts by presenting the current search computing model, and then traces its possible evolution, by investigating how descriptions can be linked to ontologies and how the Entity-Relationship model could be replaced by richer ontological models.

The first chapter introduces the Semantic Resource Framework (SRF) as an evolution of the original Service Mart model. It inherits the three-level description of services (conceptual, logical and physical) but it gives to the conceptual level a semantic interpretation, by mapping its concepts to those of the classic Entity-Relationship model: real-world objects or facts are mapped to entities and their connections are mapped to relationships. Connections between entities go beyond pure attribute equality and may represent nearness in space, time, and costs; spatial nearness between addresses or geographic locations is already exploited in many search computing applications.

The second chapter describes a method for automatic schema mapping, taking advantage of semantic annotation and of service normalization. A method for probabilistic lexical annotation relying on WordNet synsets and WSD (Word Sense Disambiguation) techniques finds the probabilistically best lexical relationships between local sources or services and a global ontological schema. The chapter explores how such techniques could be applied in a context of the SRF model to link general ontologies at service registration time.

The third chapter investigates the use of a general ontology, such as Yago, as the conceptual level of the Semantic Resource Framework, substituting for the current Entity-Relationship model. The chapter motivates this approach by showing that it solves practical problems in service registration and query mapping, and draws interesting connections to ANGIE, an ongoing project for mapping ontological queries to Web services.

Semantic Resource Framework

Marco Brambilla, Alessandro Campi, Stefano Ceri, and Silvia Quarteroni

Politecnico di Milano, Dipartimento di Elettronica e Informazione
Piazza L. Da Vinci, 32. 20133 Milano, Italy
{mbrambil,campi,ceri,quarteroni}@elet.polimi.it

Abstract. The Semantic Resource Framework (SRF) is a multi-level description of the data sources for search computing applications. It responds to the need of having a structured representation of search services, amenable to service exploration, selection, and invocation. The SRF aims at extending the Service Mart model used so far in search computing to overcome some of its limitations. The main new features include *external attributes*, which represent the input to be provided by users for accessing objects; *selector attributes*, describing the possibility to map the same access pattern to different services based on some condition; *key attributes* for objects; and a generalized notion of *nearness* between objects. The high-level view presented by SRF is a very simple Entity-Relationship model with objects and binary connections, that can be used for very different query tasks, ranging from custom search applications (i.e. predefined queries) to exploratory search (i.e., exploration of its objects and connections) to natural language interfaces (i.e., query dialogues). Such high-level view should be considered as an initial step in the enrichment of the service repository with additional semantic capabilities.

Keywords: service repository, ontology, semantic annotation, service description, search services.

1 Introduction

Service registration is an essential aspect of the Search computing project; the process is very critical because it must satisfy two conflicting requirements. On one side, services must be described with enough details about their interfaces and deployment so as to support their composition and invocation by means of fully automatic processes. On the other side, the actual mapping of services to real-world objects and facts must be exposed, so as to enable the construction of high-level user interfaces covering the semantic gap between user interaction and service selection.

The service model used for registration must describe not only the object or fact exposed by a service, but also the logic that a specific service performs while accessing an object, so that an interpretation system can select the specific service which best matches the user's requirements expressed in an informal or semi-formal way. Moreover, the model must support processes that aim at recognizing, at service registration time, when services describe the "same" objects or properties through "different" notations (e.g., names or types), so as to support matching. The scope of service registration in SeCo is therefore quite broad, as it must cover aspects ranging

S. Ceri and M. Brambilla (Eds.): Search Computing II, LNCS 6585, pp. 73–84, 2011.

from performance indicators up to the semantic description of services and of their parameters.

The Service Mart model adopted so far in SeCo for service description [6] uses a multi-level modeling approach, consisting of conceptual, logical, and physical layers. The *conceptual level* is a very simple model which characterizes real world entities, called *Service Marts (SM)*, structurally defined by means of attributes, and their relationships. The *logical level* describes the access to the conceptual entities in terms of data retrieval patterns (called *Access Patterns, AP*) described by input and output attributes. Finally, the *physical level* represents the mappings of these patterns to concrete Web *Service Interfaces (SI)*, which incorporate the details about the endpoint and the protocol to be used for the invocation, together with some basic statistics on the service behavior. These in turn may be used for granting agreed levels of quality of service (QoS).

The motivation for a three-layered architecture is due to the following needs: (1) abstracting the conceptual properties of objects from the large amount of services that access them; (2) applying the separation of concerns principle to the service description task, by granting the independence of concept definitions, access methods, and concrete service descriptions.

In particular, the Service Mart model accomplishes the abstraction objective by forcing schema uniformity throughout layers; in this way, it supports a clear definition of how services can be composed and joined. Unfortunately, schema uniformity between layers and service implementations does not always correspond to real-world situations. For instance, the support of object access according to different criteria may require different augmentations of the schema, so as to include additional attributes which describe the access to objects in different context of use. The purpose of this paper is to extend the Service Mart model by making it more expressive and more fitting to the search service description requirements.

The extensions introduced in this paper include: *external attributes*, which represent the input to be provided by users for accessing objects and as such need not to be mapped to object properties; *selector attributes*, describing the possibility to map the same access pattern to different services based on a condition; the definition of *key attributes* for objects; and, for certain domains (such as geographic location), the notion of *nearness* that can substitute for equality in the join of two services.

SRF is a first step in order to add more semantics to service registration; in the next two chapters of the book, further semantic extensions are studied, consisting in the annotation of services with ontological knowledge [3] and in the direct use of an ontological model substituting the SRF [20]. These extensions are under consideration in the project for future extensions.

Section 2 presents a state-of-the-art of the literature in the field. Section 3 briefly summarizes Service Marts from [6] so as to make this paper self-contained. Then, Section 4 introduces innovative aspects of the model (external, selector, and key attributes), and Sections 5 introduces "nearness", i.e. the possibility of accessing objects or of connecting pairs of objects ranked by their "nearness" to other objects, which is defined for given properties (e.g., spatial location, time, money). Section 6 presents a high-level view of SRF that conforms to the Entity-Relationship model, and can be seen as a first step towards adding semantics to service descriptions.

2 Related Work

The work described in this paper is the result of a research stream starting with [19], where the authors propose a Web service management system that enables querying multiple Web services in a transparent and integrated fashion and propose an algorithm for arranging a query's Web service calls into a pipelined execution plan that exploits parallelism among Web services. In this context, SRF is a proposal for increasing the abstraction level and thus facilitating the choice of services and the definition of plans. In the following, we revise the proposals that address similar issues.

Our **conceptual level** description of services through service marts is in line with [8], which describes the idea of the Web of concepts. In this work the term concept refers to things of interest to users of the Web who are either searching for information or trying to accomplish some task. The shift from a Web of pages to a Web of objects is now a recognized trend [2] and several mainstream search engines are following the line by introducing new features in this direction. In [7] the authors propose a conceptual model that describes actors, activities and entities involved in a service-oriented scenario through a glossary of terms.

Also the **logical level** several approaches exist for describing objects on the Web. The most popular ones are based on Google Fusion Tables [11][13][14], a cloud-based service for data management and integration. Fusion Tables enables users to upload tabular data files and provides ways of visualizing the data (e.g., charts, maps, and timelines) and the ability to filter and aggregate the data. It supports the integration of data from multiple sources by performing joins across tables that may belong to different users. There are also several projects related to structured data at Google. Google Public Data[1] is an effort to import public government data and provide high-quality and carefully chosen visualizations of data in response to search queries. The Google Squared Service[2] lets users specify categories of objects and explore attributes of these entity sets. In this case, the data populating the tables is automatically extracted from various sources on the Web, and may not always be accurate.

At the **physical level**, the trend toward the Web of objects is well represented by the Linked Data initiative, which has recently seen an increasing amount of shared information, also thanks to initiatives like the W3C Linked Open Data (LOD) community project[3] and to the dedication of prominent Semantic Web researchers[4]. Outside this initiative, the major search engines are providing facilities for accessing information sources through APIs and query languages. The most known resource is YQL (Yahoo! Query Language)[5] is a language and a platform that lets Web applications query and filter data from different sources across the Internet through SQL-like statements. Similarly, Google Base API[6] also allows one to upload structured data and to query it through the Web. Another relevant aspect of concrete service description is the specification of relevant information for quality of service (QoS) support. QoS has been thoroughly studied

[1] http://www.google.com/publicdata/home
[2] http://www.google.com/squared
[3] http://esw.w3.org/SweoIG/TaskForces/CommunityProjects/LinkingOpenData/
[4] http://www4.wiwiss.fu-berlin.de/bizer/pub/LinkedDataTutorial/
[5] http://developer.yahoo.com/yql/
[6] http://code.google.com/apis/base/

in the past (see e.g. [15], [17]). In line with these approaches, we keep track of service performance for dynamic runtime optimization of execution plans.

Finally, several proposals, such as DAML-S [1], OWL-S [16], COSMO [18] and WSMF [9][10], extend the description of services in a semantic direction, as we plan to do in future work (see the next chapters [3][20] for an overview of the possible research directions).

3 Service Marts

Service Marts are specific data patterns; their regular organization helps structuring Search Computing applications. The most well known data modelling pattern is the so-called "data mart" used in the context of data warehouses; a data mart is a simple schema having one core entity, describing facts, surrounded by multiple entities, describing the dimensions of data analysis [5]. It facilitates the expression of operations for data selection and aggregation (e.g. data cubes, rollup, drilldown). Analogously, a "Web mart" [11] is a pattern introduced in the Web design community to characterize the role played by data items in data-intensive Web applications. Service Marts are instrumental in supporting the notion of "Web of objects" [10] that is gaining popularity as a new way to think of the Web, going beyond the unstructured organization of Web pages.

Figure 1 provides an overview of the approach, by presenting a sample concept Movie, registered as a service mart, together with two associated access patterns and service interfaces, with the respective attribute mappings (notice that mappings are shown only for the first access pattern for clarity). The next subsections describe the semantics and notation of each level.

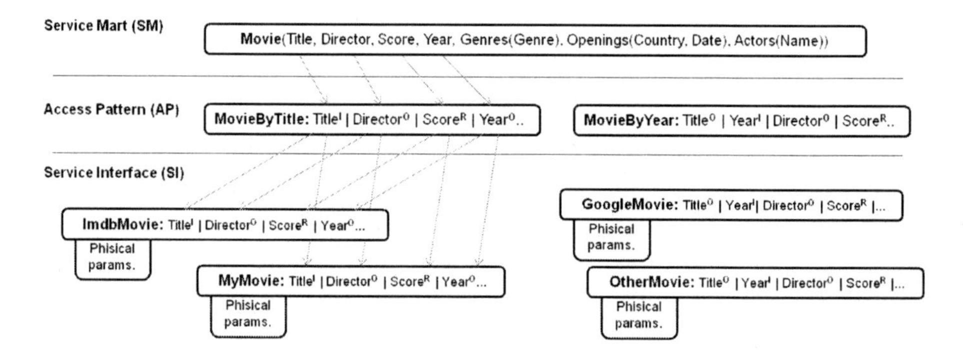

Fig. 1. Example of descriptions for accessing Movie information through the Service Mart, Access Pattern, and Service Inteface layers

3.1 Conceptual Level

At the conceptual level, the definition of a Service Mart includes the object's name and the collection of the object's attributes. All attributes are typed; attributes can be atomic (single valued) or part of a repeating group (multi-valued); each repeating group is a non-empty set of attributes that collectively defines a property of the Service Mart with multiple values (such as "genres" or "actors" of a given movie). The model choice is to support structural complexity with only one level of nesting, rather than arbitrary nesting. The conceptual description of the object "Movie" is:

```
Movie(Title, Director, Year, Score, Language, Genres(Genre), Actors(Name, Sex))
```

3.2 Logical Level

At the logical level, Service Marts are associated with one or more access patterns representing the signatures of service calls. Access patterns contain a subset of the Service Mart's attributes tagged with I (input), O (output), or R (ranking). Ranking attributes may be visible in output. For ease of understanding, if the service call is mapped to a parametric query, input attributes provide query parameters, output attributes provide results, and ranking attributes are used for ordering result instances.

$Movie_1(Title^O, Director^O, Score^{RO}, Year^O, Language^I, Genres.Genre^I, Actors.Name^O)$

$Movie_2(Title^I, Director^O, Language^O, Genres.Genre^O, Actors.Name^O, Actors.Sex^O)$

$Movie_1$ accesses movies by Language and Genre (i.e., "action movies in English"), and results are ranked by Score (a new attribute). $Movie_2$ accesses movies by matching their Title to a string (e.g. "Ben Hur"). We expect few (zero, one, more) un-ranked results.

3.3 Physical Level

At the physical level, each access pattern may be mapped to several service implementations. Each implementation is characterized by a physical URI to be called, a set of physical properties that are specific to the implementation, and a mapping between logical attributes and physical parameters. Services are divided into *exact services* (producing a set of equivalent responses) and *ranked services* (search systems producing a list of results ordered by priority, sometimes explicitly ranked, typically not exhausted by the calling process).

3.4 Connection Patterns

Connection patterns are high-level abstractions of "real world relationships" that provide a simple interface to users and hide implementation details. They are built by means of attributes that share the same domains. At the conceptual level, they are defined by a non-directed edge with a name, e.g.:

```
PlayingMovie(Movie,Theatre)
```

At the logical level they are defined by a (possibly directed) edge with name and join condition:

```
PlayingMovie(Movie,Theatre): (Title=Movie.Title)
```

The above edge represents a join operation between the two access patterns and can either be directed or undirected. In the first case, information is "piped" from one access pattern to another, along connection attributes which are in output in the first service and in input in the second service. As an example, the connection between the following access patterns is directed, following the information flow from $Movie_1.Title^O$ to $Theatre_1.M.Title^I$:

```
Movie₁(Titleᴼ, Directorᴼ, Scoreᴿᴼ, Yearᴼ, Languageᴼ, Genres.Genreᴵ,Actors.Nameᴼ)
Theatre₁(Nameᴼ, Addressᴼ, Movie.Titleᴵ, Movie.StartTimeᴼ)
```

Undirected edges are present when join attributes are both in output from the services, as shown by the following example; note that the Title attribute is labelled "O" in both patterns.

```
Movie₁(Titleᴼ, Directorᴼ, Scoreᴿᴼ, Yearᴼ, Languageᴼ, Genres.Genreᴵ,Actors.Nameᴼ)
Theatre₂(Nameᴼ, Addressᴼ, Movie.Titleᴼ, Movie.StartTimeᴼ)
```

4 Model Extensions

After experimentation, we realized that the basic Service Mart model of the world encountered a number of limitations, and therefore some extensions were needed. We start by extending the model with three orthogonal features: external, selector, and key attributes. Each such feature is described next.

4.1 External Attributes

External attributes are "new" attributes appearing at the logical level, while not being present at the conceptual level; they support object access and ranking. External attributes are appended to access patterns, as illustrated by Figure 2. Attributes for object access support multi-value and similarity access:

- The former case occurs when a query includes one or more values of a multi-valued attribute (e.g., "internet access", "parking", and "wellness area" as required properties of a hotel); in such a case, the service call signature requires providing one or more values in input, which the invoked service will use to select object instances (the specific access predicate being used by the service, e.g. "at least two" required properties for a hotel, is not of our concern).
- The latter occurs when a query finds the instances that most closely match a given input (e.g. the movies whose title most closely matches the word "apocalypse"); in such a case, the service call signature requires two attributes for the same domain, one given in input and one extracted as output, and ranking of selected object instances is based upon best matching.

External attributes for ranking allow services to explicitly describe how their result instances are ranked. They may correspond to specific object attributes (e.g. "stars" or "scores" for hotels) which provide an explicit ranking, or just represent the "position" of the result instance in the result list, used by services returning opaque ranking.

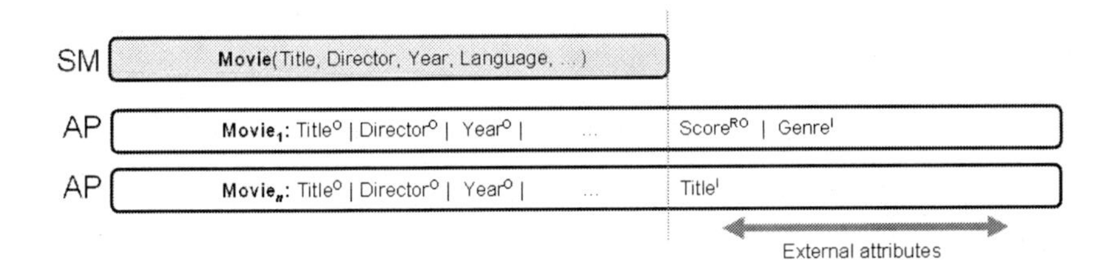

Fig. 2. External attributes for access patterns

4.2 Selector Attributes

Selector attributes support the selection of specific service implementations; they may be present at the conceptual level or be added at the logical level; in the latter case, selector attributes must be labelled "I" (for input). Figure 3 shows "language" as a selector attribute, present in the Service Mart used for accessing *n* movie services (e.g. one for movies in "English" and one for movies in "Italian").

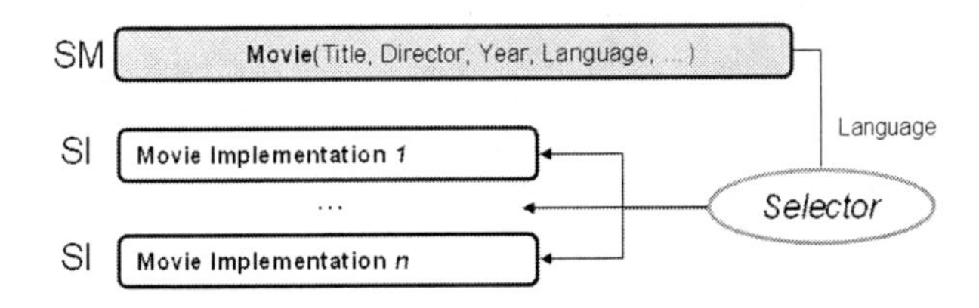

Fig. 3. Selector attributes to choose a specific service implementation

4.3 Key Attributes

In the context of the SeCo project, object identity is used for supporting object caching (so as to avoid repeated queries to the same service) and query sessions (so as to progressively build information relative to the "same" real world object). Both these uses capitalize upon the specific choice of service implementation, performed by the query compiler, and therefore properly belong to the physical level. The latter is also the only level where object identity can be maintained, as it is not realistic that object identity be maintained across different service implementations accessing distinct sets of objects. Thus, we allow each service implementation to be optionally associated with a local definition of "key attributes", corresponding to one or more attributes collectively forming the (primary) identifier of the object instances retrieved by the service implementation. Such key attributes may be present at the conceptual or logical levels or may be added at the physical level. For example, depending on the service interface, hotels may be identified by the pair "city, number" or by an "OID" or "URI" generated by the service.

5 Nearness Support in Accessing Resources

In the Service Mart model, connections of access patterns are based on value equality (equi-joins). However, queries often call for value similarity and/or partial matching; in this case, knowledge of the underlying semantics of attributes supports nearness computation. This aspect is indeed crucial in Web search and intrinsically leads to the concept of ranking based on join conditions. Furthermore, it provides a fundamental rationale for the introduction of external attributes.

In SeCo, the most relevant domain for similarity support is the *spatial domain*, which is used for geographic locations such as the "addresses" of resources. Indeed, the spatial domain is the most common example of application of nearness support as many types of queries are naturally geo-localized, i.e. their results can be visualized upon maps. The trend of offering geo-localized results is common to most search engines, such as Google or Yahoo!

Further domains providing value similarity are the *temporal domain* (describing dates and times) and the *economic domain* (describing costs). Similarity computation between *lexical strings* is also performed to compute term relatedness; this is achieved via morphological operations such as stemming or via external vocabularies such as WordNet [21]. String comparisons may be refined for specific application domains (e.g., the Bio-SeCo application using GeneOntology [12], described in Part 7 of this book). Next, we illustrate similarity support in the spatial domain and focus on providing results close to the current user's location.

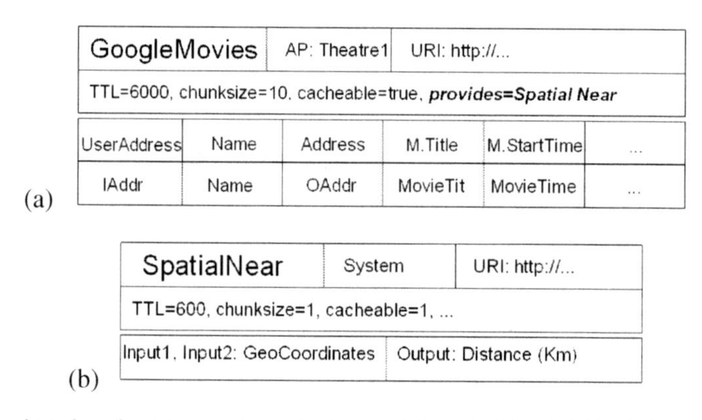

Fig. 4. Service interface for (a) a service natively providing the "Spatial Near" function; (b) an ad-hoc service providing spatial nearness calculation

5.1 Nearness in Single Resources

Ranking by nearness is supported natively by many search services. For instance, the "GoogleMovies" service (shown in Fig. 4(a)) outputs movie shows ranked by distance from the location given as input[7]. In this case, it is sufficient to label the service with the supported similarity semantics, a feature recognized by the SRF.

[7] See e.g. http://www.google.it/movies?near=washington%square%new%york

Note that in Fig. 4(a), "UserAddress" (input location) is modelled as an *external attribute* of the $Theatre_1$ access pattern.

In addition, generic services exist which return the distance between two addresses, provided either as geographic coordinates or as <country, city, street> triples. Such services may be used to rank resources based on their distance from a specific location. Fig. 4(b) represents an ad-hoc service, supported within the SeCo query engine, which takes two coordinates as input and produces their distance as output.

Services for supporting value similarity are engineered starting from services supporting value equality; a number of candidate locations, produced e.g. by location-aware resource selection services, are used to feed the first input parameter of the ad-hoc service, while the second input parameter is set to the user's current location. A sorter is then used to order candidate locations by distance. Caching of triples representing two locations and their distance can be used to reduce calls to the ad-hoc service, and the Search Computing engine supports execution strategies that limit the size of the result set to avoid waiting too long for sorted results (note that sorting is a blocking operation, i.e. an operation that can only be executed when its full input is available).

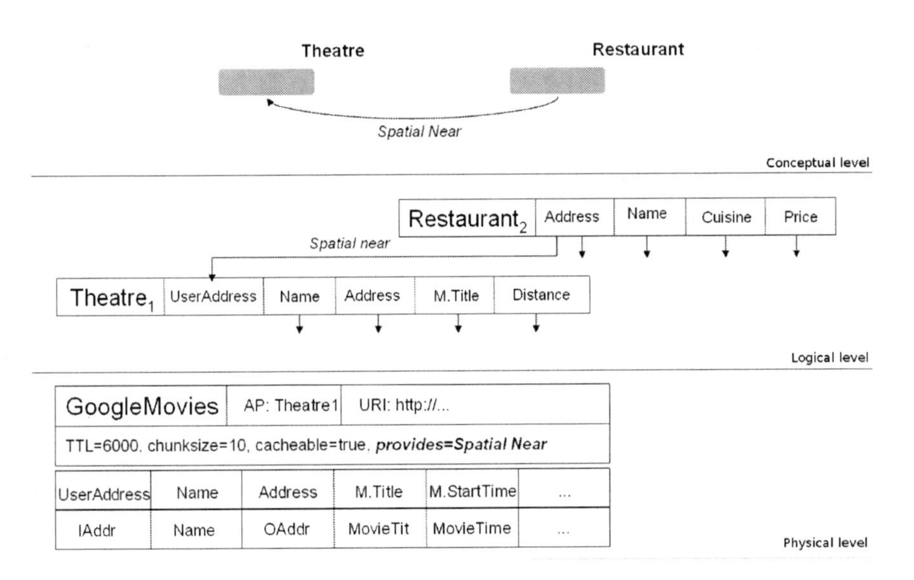

Fig. 5. GoogleMovies service providing spatial nearness: modelling at the physical, logical and conceptual levels

5.2 Accessing Pairs of Resources

The approach illustrated in Section 6.1 can be extended to connection patterns. Fig. 5 shows an example of use of the GoogleMovies service for supporting geographic nearness as a connection pattern. The query searches for theatres (and their movies) close to restaurants, selected (and ranked) in turn according to the user's preference. This is possible thanks to the match between external attributes "Address" and

"UserAddress" in the $Restaurant_2$ and $Theatre_1$ access patterns. Note that the connection at the conceptual level is now directed (because the nearness function is supported by the "theatre" service) and labelled with the name of the *SpatialNear* function.

Alternatively, an ad-hoc service can be used to support geographic nearness, by using the scheme illustrated in Fig. 6. The ad-hoc "SpatialNear" service can only be invoked once the services for "restaurants" and "theatres" have been called, as they use address pairs as input. Note that in this case the connection at the conceptual level, labelled with the name of the *SpatialNear* function, is not directed.

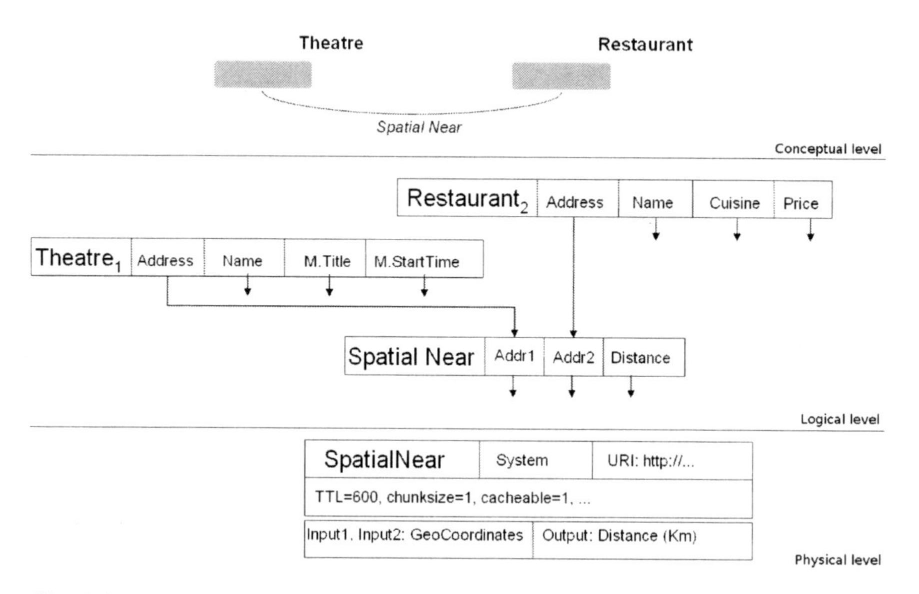

Fig. 6. SpatialNear ad-hoc service: modelling at the physical, logical and conceptual levels

6 Top-Level View of the Semantic Resource Framework

While the service mart model is ideal for registering individual services, as it clusters several service descriptions within a hierarchical multi-level view, its evolution into the SRF highlights fact that services collectively describe a given "domain of discourse", i.e. a particular subset of reality which can be the target of SeCo queries.

The top-level view of SRF is a simple Entity-Relationship model, as described in Figure 7; it defines the application context, characterized by the presence of named entities (service marts) and relationships (connection patterns). Among the possible semantic meanings of relationships, of course we rightfully include nearness as discussed in Section 5.2. This view abstracts away from the complexity of mapping service interfaces to data sources and of integrating the different names and formats used by each source to represent its properties, and focuses on a simple, semantic view.

Thanks to this view, the exploration of information and the definition of search queries is simplified and made more efficient. The "focus" of the exploration can start with a single object and then progressively add more objects, thereby building queries with a fashion, as discussed in the first chapter of this book. Future work for supporting keyword-based or natural language interfaces will also be based on high-level representations of the universe of discourse. Thus, this semantic view, although very simple at this stage, is an important step in the direction of adding semantic power to service descriptions.

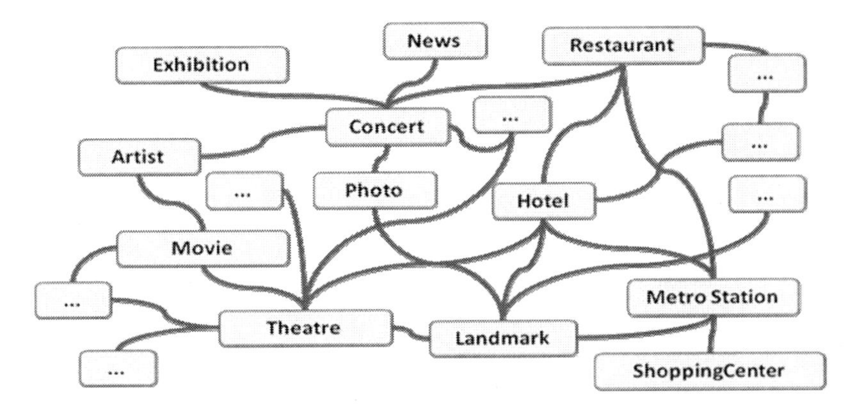

Fig. 7. Example of high-level view offered by SRF

7 Conclusions

Search service management is one of the most critical aspects in search computing. In this chapter, we discussed our current approach to search service conceptualization and registration; the model extends our previous approach in many important directions, such as the support of external, selector, and identifier attributes, and the support of nearness for specific semantic domains, such as distance and time. Registration tools, described in [5], support service registration according to this model and subsequently let query designers specify the search query upon it.

We envision progressive extensions of the SRF for incorporating more semantics; such extensions concern the use of semantic annotations, and the use of ontological knowledge at the conceptual level of the model, thereby adding semantics to the current Entity-Relationship description. These extensions, discussed in the next two chapters [3][20], will further empower designers at registration time, and will facilitate the support of high-level queries.

References

[1] Ankolenkar, A., et al.: DAML-S: Web service description for the Semantic Web, `http://www.daml.org/services/daml-s/2001/10/daml-s.html`

[2] Baeza-Yates, R., Raghavan, P.: Next Generation Web Search. In: Ceri, S., Brambilla, M. (eds.) Search Computing. LNCS, vol. 5950, pp. 11–23. Springer, Heidelberg (2010), doi:10.1007/978-3-642-12310-8_2

[3] Bergamaschi, S., Beneventano, D., Po, L., Sorrentino, S.: Automatic Schema Mapping Through Normalization and Annotation. In: Ceri, S., Brambilla, M. (eds.) Search Computing II. LNCS, vol. 6585, pp. 85–100. Springer, Heidelberg (2011)

[4] Braga, D., Ceri, S., Daniel, F., Martinenghi, D.: Optimization of multi-domain queries on the Web. In: Proc. VLDB, vol. 1(1), pp. 562–573 (August 2008)
[5] Brambilla, M., Tettamanti, L.: Search computing processes and tools. In: Ceri, S., Brambilla, M. (eds.) Search Computing II. LNCS, vol. 6585, pp. 169–181. Springer, Heidelberg (2011)
[6] Campi, A., Ceri, S., Maesani, A., Ronchi, S.: Designing service marts for engineering search computing applications. In: Benatallah, B., Casati, F., Kappel, G., Rossi, G. (eds.) ICWE 2010. LNCS, vol. 6189, pp. 50–65. Springer, Heidelberg (2010)
[7] Chakrabarti, K., Ganti, V., Han, J., Xin, D.: Ranking objects based on relationships. In: Proc. of SIGMOD Int. Conf. on Management of Data, New York, USA, pp. 371–382 (2006)
[8] Dalvi, N., et al.: A Web of Concepts. In: PODS 2009, Providence, Rhode Island, USA, June 29-July 2 (2009)
[9] Fensel, D., Bussler, C.: The Web Service Modeling Framework WSMF. Electronic Commerce Research and Applications 1(2), 113–137
[10] Fensel, D., Musen, M.: Special Issue on Semantic Web Technology. IEEE Intelligent Systems (IEEE IS) 16(2)
[11] Google Fusion Tables, http://tables.googlelabs.com/
[12] Gene Ontology, http://www.geneontology.org/
[13] Gonzalez, H., Halevy, A., Jensen, C., Langen, A., Madhavan, J., Shapley, R., Shen, W., Goldberg-Kidon, J.: Google fusion tables: Web-centered data management and collaboration. In: Proceedings of the 2010 International Conference on Management of Data, SIGMOD 2010, Indianapolis, USA, June 06-10, pp. 175–180 (2010)
[14] Gonzalez, H., Halevy, A., Jensen, C., Langen, A., Madhavan, J., Shapley, R., Shen, W.: Google Fusion Tables: Data Management, Integration, and Collaboration in the Cloud. In: Proceedings of the ACM Symposium on Cloud Computing, SOCC (2010)
[15] Liangzhao, Z., Benatallah, B., Ngu, A.H.H., Dumas, M., Kalagnanam, J., Chang, H.: QoS-aware middleware for Web services composition. IEEE Transactions on Software Engineering 30(5), 311–327 (2004)
[16] Martin, D., Burstein, M., et al.: Bringing Semantics to Web Services with OWL-S. In: World Wide Web, vol. 10(3), pp. 243–277 (September 2007)
[17] Maximilien, E.M., Singh, M.P.: A framework and ontology for dynamic Web services selection. IEEE Internet Computing 8(5), 84–93 (2004)
[18] Quartel, D.S., Steen, M.W., Pokraev, S., Sinderen, M.J.: COSMO:A conceptual framework for service modeling and refinement. Information Systems Frontiers 9(2-3), 225–244 (2007)
[19] Srivastava, U., Munagala, K., Widom, J., Motwani, R.: Query optimization over Web services. In: VLDB 2006, VLDB Endowment, pp. 355–366 (2006)
[20] Suchanek, F., Bozzon, A., Della Valle, E., Campi, A., Ronchi, S.: Towards an Ontological Representation in Search Computing. In: Ceri, S., Brambilla, M. (eds.) Search Computing II. LNCS, vol. 6585, pp. 101–112. Springer, Heidelberg (2011)
[21] WordNet, http://wordnet.princeton.edu/
[22] Xi, W., Fox, E.A., Fan, W., Zhang, B., Chen, Z., Yan, J., Zhuang, D.: SimFusion: measuring similarity using unified relationship matrix. In: Proc. of the 28th Int. ACM SIGIR Conf. on Research and Development in Information Retrieval, New York, USA, pp. 130–137 (2005)
[23] Zaragoza, H., Rode, H., Mika, P., Atserias, J., Ciaramita, M., Attardi, G.: Ranking very many typed entities on wikipedia. In: CIKM 2007: Proc. of the 16th ACM Conf. on Information and Knowledge Management, pp. 118–1018. ACM, New York (2007)

Automatic Normalization and Annotation for Discovering Semantic Mappings

Sonia Bergamaschi, Domenico Beneventano, Laura Po, and Serena Sorrentino

Department of Information Engineering
University of Modena and Reggio Emilia, Italy
name.surname@unimore.it

Abstract. Normalization and lexical annotation methods, developed in the context of matching systems, have proven to be effective for the discovery of lexical relationships among schemata. We will show how these methods are applicable and effective in the context of Semantic Resource Framework to mine the semantics of a web service interface and to discover mappings between them.

Keywords: lexical relationships, probabilistic annotation, word sense disambiguation, label normalization, semantic resource framework.

1 Introduction

This chapter will discuss the applicability of *normalization* and *lexical annotation* methods, developed in the field of schema matching, in the context of web service interfaces. The lexical annotation of a schema element is the explicit assignment of its meanings w.r.t. a lexical resource. Normalization (also called *linguistic normalization* [14]) is the reduction of the label of a schema element to some standardized form that can be easily recognized.

Starting from our previous works in the context of data integration [5,21,26], we propose to apply normalization and annotation methods to mine the semantics of a service, exposed through its interface and to discover connection patterns among web services.

In Natural Language Processing, Word Sense Disambiguation (WSD) is the process of identifying which sense of a word (i.e. meaning) is used in a sentence, when the word has multiple meanings (polysemy). We describe our probabilistic lexical annotation method, which automatically associates one or more meanings to schema elements w.r.t. the lexical resource WordNet (WN) [13], by exploiting a Word Sense Disambiguation (WSD) algorithm, called PWSD (Probabilistic Word Sense Disambiguation) [21]. The accuracy of lexical annotation is affected by labels which are non-dictionary words, such as Compound Nouns (CNs), acronyms and abbreviations which are very frequent on real-world schemata and web service interfaces. We addressed this problem by devising a method to normalize schema labels which is able to semi-automatically expand abbreviations and to *properly* lexically annotate CNs by creating new WN meanings.

Starting from the lexical annotation of schema elements, we can discover lexical relationships between them, on the basis of the relationships defined in WN between

S. Ceri and M. Brambilla (Eds.): Search Computing II, LNCS 6585, pp. 85–100, 2011.

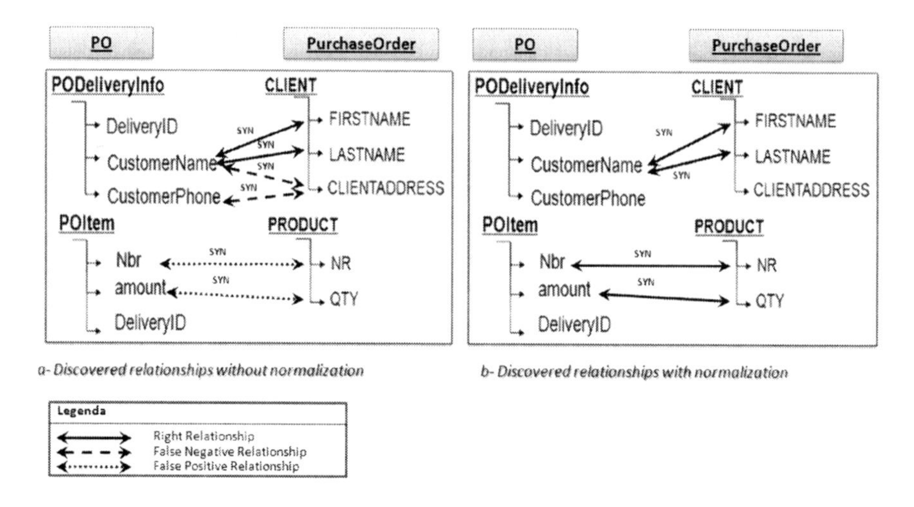

Fig. 1. Example of discovered lexical relationships without (a) and with (b) normalization

their meanings (synsets in WN terminology). Traditional schema matching methods based on string distance techniques [10] do not permit to automatically discover that there exists, for example, a synonym relationship between the two schema elements "amount" and "quantity", as their labels share only few characters. Instead, by using our method, we are able to: (1) automatically annotate these schema elements with the corresponding WN meanings; (2) discover a synonym relationship among them, as they share the same meaning in WN (i.e. the synset "how much there is or how many there are of something that you can quantify").

Moreover, our normalization method improves the quality of semantic mappings by reducing the number of discovered *false positive/false negative relationships*. Figure 1 shows two schemata that need to be mapped/integrated, and compares the relationships discovered with and without normalization. Let us consider, for example, the two schema elements "CustomerName" and "CLIENTADDRESS", respectively, in the source "PurchaceOrder" and "PO", shown in Figure 1(a). If we annotate separately the terms "Customer" and "Name", and "CLIENT" and "ADDRESS", then we might assume a SYN relationship between them, because the terms "Customer" and "CLIENT" share the same WN meaning. In this way, a false positive relationship is discovered because these two CNs represent "semantically distant" schema elements.

Furthermore, if we consider the two corresponding schema labels "amount" and "QTY" (abbreviation for "quantity"), without abbreviation expansion we cannot discover that there exists a SYN relationship between the elements "amount" and "QTY".

In this chapter, we describe the normalization and annotation methods w.r.t. a generic object schema, which may be either a set of data sources or a set of web services (section 2). In Section 3, an example of application of the methods on the Semantic Resource Framework model is shown. Some related works are described in Section 4. Finally, in Section 5, we make some concluding remarks.

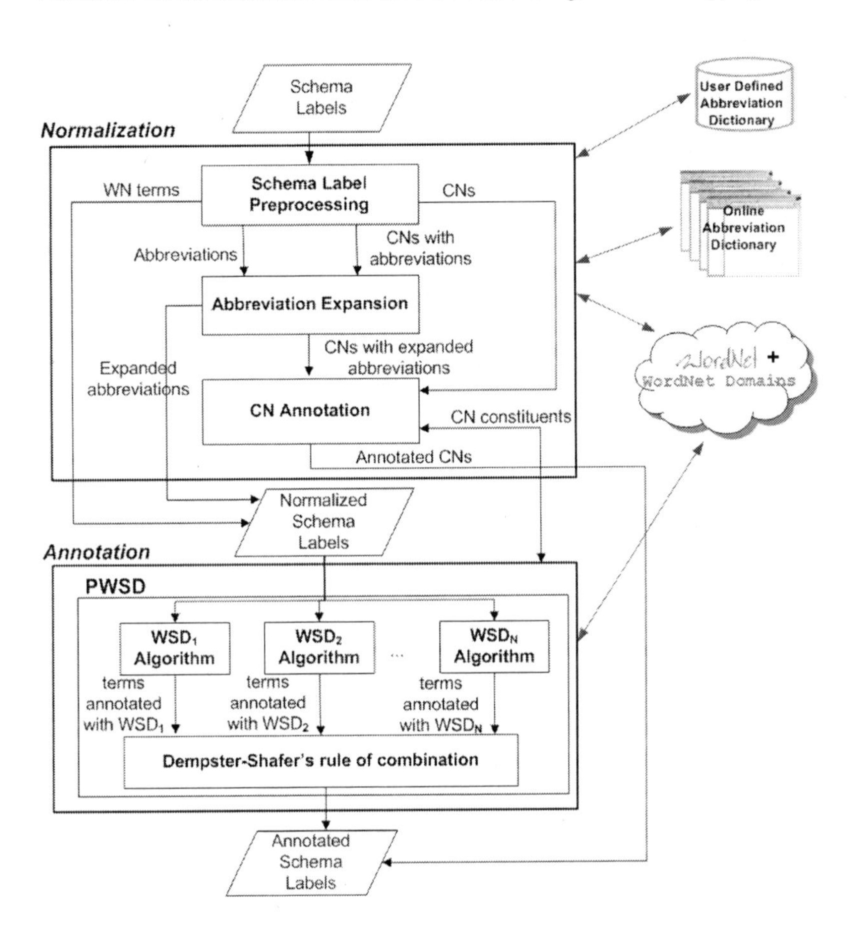

Fig. 2. Overview of the schema label Normalization and Annotation methods

2 Normalization and Annotation of a Conceptual Schema

To describe the normalization and annotation methods, we use a generic Conceptual Modeling Language (CML), which contains common aspects of most semantic data models, UML, ontology languages, such as OWL, and description logics [1].

In the sequel, we use S to denote a *schema* prescribed by the generic CML. Specifically, the language allows the representation of *classes* (unary predicates over individuals), and *attributes*, which can be *simple* (binary predicates relating individuals with values such as integers and strings) or *complex* (binary predicates relating individuals).

Let $S=\{C_1, C_2,...,C_n\}$ be a finite set of classes, where each class, $C=\{A_1, A_2,...,A_n\}$, is described by a finite set of attributes. A simple attribute in the class $Hotel$, representing an hotel reservation web site, may be $name$, while a complex attribute may be $address$; $address$ refers to a class $Address$ in the same schema that combines the information of $street$, $number$, $city$, $zipcode$, and $country$.

Attributes may be subject to constraints such as, cardinality constraints. Such constraints do not influence our annotation and normalization methods and therefore are not taken into account in our generic CML.

Classes are organized in a familiar is-a hierarchy. In the previous example, the class $Hotel$ may be defined as a sub-class of $Service$.

Normalization and annotation methods regard classes and attributes of a schema, referred to as *schema elements* from now on. Each schema element has a name, referred to as *label* from now on.

Definition 1 (Lexical Annotation). *The lexical annotation of a schema element is the explicit assignment of its meanings w.r.t. a lexical resource.*

We define the lexical annotation as the connection of a schema element with its meanings defined in a lexical resource. However from now on, we will also make reference to the annotation of the label of a schema element.

The lexical resource we employ in our methods is WN. WN groups English words into sets of synonyms, called synsets. Each synset represents a distinct concept and is further clarified with a short defining gloss (i.e. a definition and optional example sentences). WN records the various semantic relationships between these synonym sets. These relationships vary based on the type of word; for the syntactical category of nouns, they include:

- hypernyms: Y is a hypernym of X iff every X is a (kind of) Y (person is a hypernym of student, because every student is a member of the larger category of persons);
- hyponyms: Y is a hyponym of X iff every Y is a (kind of) X (student is a hyponym of person);
- holonym: Y is a holonym of X iff X is a part of Y (building is a holonym of window);
- meronym: Y is a meronym of X iff Y is a part of X (window is a meronym of building).

We focus our work on the development of an automatic lexical annotation method able to select the synsets that cover the meaning of a schema element.

Thanks to the WN network of relationships, after the application of the annotation method we can *discover lexical relationships among schema elements.*

Lexical relationships are defined between classes and attributes, and are specified by considering class/attribute labels. Formally, a lexical relationship can be defined as:

Definition 2 (Lexical relationship). *Let t and s be two schema elements and $t_{\#i}$ and $s_{\#j}$ annotations assigned to t and s respectively. A lexical relationship is defined as the triple $< t_i, t_j, R >$ where R defines the type of the relationship between t_i and t_j. The types of lexical relationship are:*

- *SYN (Synonym-of): defined between two elements whose meanings are synonymous (they correspond to a WN synonym set), formally*

$$t \; SYN \; s \; iff \; \exists \; t_{\#i} \equiv s_{\#j}$$

- *BT (Broader Term): defined between two elements where the meaning of the first is more general than the meaning of the second (the opposite of BT is NT, Narrower Term), (it corresponds to a WN hypernymy/hyponymy relationship)), formally*

$$t\ BT\ s\ \textit{iff}\ \exists\ t_{\#i}\ \textit{hypernym of}\ s_{\#j}$$

- *RT (Related Term): defined between two elements whose meanings are related in a meronymy hierarchy (it corresponds to a WN meronymy relationship, i.e. part-of relationship), formally*

$$t\ RT\ s\ \textit{iff}\ \exists\ t_{\#i}\ \textit{meronym of}\ s_{\#j}$$

Let us suppose we want to automatically lexically annotate (annotate in the following) the schema element "address". In WN the noun "address" has eight different meanings, including very similar ones such as "written directions for finding some location; written on letters or packages that are to be delivered to that location" or "a sign in front of a house or business carrying the conventional form by which its location is described". We might generate a single (forced) annotation for each word (as done by many WSD approaches proposed in the literature). However, in many cases, choosing a single annotation would be difficult even for a human annotator: generating more than one probabilistic annotation comes to be a good solution that avoids loosing semantic information.

Uncertainty is an intrinsic feature of automatic and semi-automatic annotation methods and provides a quantitative indication of the quality of the result. In our method, uncertainty is qualified as probability values related to annotations w.r.t. WN.

The strength of a thesaurus like WN is the presence of a wide network of semantic relationships among meanings. Its main weakness is that it does not cover different domains of knowledge with the same level of detail and that many domain dependent terms (called *non-dictionary words*) may not be present in it. Non-dictionary words include CNs, abbreviations and acronyms (from now on, these last two will be referred to simply as "abbreviations"). The result of automatic annotation is strongly affected by the presence of these non-dictionary words in schemata, thus *label normalization* is needed. With label normalization, we mean the process of abbreviation expansion and CN annotation through the creation of new WN meanings.

Definition 3 *(Abbreviation expansion). Let AB be an abbreviation (or short form), abbreviation expansion is the task of finding a relevant expansion (long form) for the given abbreviation AB*[1].

Definition 4 *(CN annotation). Let CN be a non-dictionary compound noun constituted of more words (its constituents). The annotation of a CN is the task of creating a new WN synset starting from the annotations of its constituents.*

[1] The long form that is extracted through abbreviation expansion may not be an entry in WN. This issue remains an open problem. For the moment, we limit ourselves to examine long forms which have an entry in WN (e.g. the long form "Number" for the abbreviation "Nbr") or that correspond to CNs (e.g. the long form "Purchase Order" for the abbreviation "PO").

In the following, we describe normalization, annotation and relationship discovery in detail. In the end of this section, we show some results in term of performance of the methods.

2.1 Normalization

As shown in Figure 2, the schema label normalization method [26] consists of three steps: (1) schema label preprocessing, (2) abbreviation expansion and (3) CN annotation. The input of schema label preprocessing is the set of schema element labels. During this phase, we automatically select the labels to be normalized. The output of this module are the tokenized labels classified into four groups (as shown in Figure 2): *WN terms* (i.e. labels having an entry in WN which do not need normalization, e.g. "Airport") *abbreviations* (e.g. "FLTNO"), *CNs* (e.g. "DepartureAirport"), and *CNs containing abbreviations* (e.g. "ARRAirport").

The abbreviation expansion step is applied on all the schema labels classified as abbreviations or CNs with abbreviations. During this step, each abbreviation is expanded with the most relevant long form by using the knowledge provided by the schema and abbreviation dictionaries. Our method exploits the online abbreviation dictionary Abbreviations.com[2], particularly useful for expanding domain standard abbreviations, and a user-defined dictionary. Since real-world schemata often use application-specific codes (e.g. "X09CCDE") that will not appear in any public dictionary, the designer may enrich the user-defined dictionary with such abbreviations. The user-defined dictionary is initially bootstrapped with schema standard abbreviations and for our example by using the OTA standard[3].

The CN annotation is applied on all the schema labels classified as CNs or CNs with expanded abbreviations. In particular, we focus on a category of CNs called *endocentric*. Endocentric CNs consist of a head (i.e. the categorical part that contains the basic meaning of the whole CN) and one or more modifiers, which restrict the meaning of the head. An endocentric CN exhibits a *modifier-head structure*, where the head noun occurs always after the modifiers. Endocentric CNs are often not included in dictionaries, but they can be interpreted by using the knowledge about their constituents.

This step can be summed up into three sub-steps: (1) CN constituent disambiguation; (2) CN interpretation via semantic relationships; and (3) creation of a new CN synset in WN.

During the first sub-step the constituents of a CN are automatically annotated w.r.t. WN by applying the PWSD algorithm [21] (described in the following) which assigns a set of probabilistic annotations to each constituent. Then, starting from these annotations we perform *CN interpretation*. The interpretation of a CN is the task of determining the semantic relationships holding among its constituents. In particular, we perform automatic CN interpretation by using the set of nine semantic relationships defined by Levi in [16]: CAUSE ("flu virus"), HAVE ("college town"), MAKE ("honey bee"), USE ("water wheel"), BE ("chocolate bar"), IN ("mountain lodge"), FOR ("headache pills"),

[2] `http:\www.abbreviations.com`

[3] OpenTravel Alliance XML schema for the travel industry. Available online at `http://www.opentravel.org/`

FROM ("bacon grease"), and ABOUT ("adventure story"). At the end, we establish a new WN synset for the CN: first, we derive the gloss starting from the discovered Levi relationship and by exploiting the glosses of the CN constituents; then, the new meaning for the CN is inserted in the WN thesaurus by automatically creating a hypernym (and the opposite hyponym) relationship between the new synset and the synset of the CN head, and a generic (*Related term*) (which corresponds to the WN relationships *member meronym, part meronym, substance meronym*) between the new synset and the synset of the modifier.

However, the insertion of these two relationships is not sufficient; it is also necessary to discover the relationships of the new inserted meaning w.r.t. the other WN synsets. To this end, we use the WNEditor tool [4] to create/manage the new synset and to set relationships between it and the existing WN ones. As the final goal of our method is to produce a set of probabilistic annotations for each schema elements, we compute the probability associated to the new synset as the product of the probability values of the individual constituent annotations[4].

2.2 Probabilistic Lexical Annotation

As shown in Figure 2, the output of the normalization method will be the input of the annotation method. Lexical annotation is performed by PWSD, an automatic algorithm that combines several WSD algorithms. In this way, the process is not affected by the effectiveness of a single WSD algorithm in a particular context or application domain. PWSD satisfies three important constraints: (1) it is an automatic technique (only a few configuration settings are required), (2) it is flexible (i.e., it can combine any set of WSD algorithms[5]), and (3) the output of the method does not commit to an exact synset for a term under consideration, but to a set of possible senses that represent the term. We use the Dempster-Shafer theory of evidence [24,20] to combine annotation outputs obtained by the WSD algorithms. By using the Dempster-Shafer's theory of evidence, PWSD associates a probability value to each sense selected to disambiguate a term; this value shows the uncertainty of the disambiguation process.

Given a schema element t, the PWSD algorithm associates a set of probabilistic annotations to t:

$$PM(t) = \{< t_{\#i}, P(t_{\#i}) >, ..., < t_{\#n}, P(t_{\#n}) >\}$$

Eventually, to minimize the introduction of errors, probabilistic annotations with a probability value under a certain threshold can be filtered.

2.3 Probabilistic Lexical Relationship Discovery

Once we have obtained annotations for schema elements, we can use the probability distributions over the set of possible meanings (i.e. the output of PWSD) to infer probabilistic lexical relationships among the object and attribute terms. As stated

[4] We assume that the probabilities being combined are independent. This assumption is not usually hold, however, factoring out dependencies in WSD context is extremely difficult as they are usually hidden [22].

[5] At present, we combine five WSD algorithms.

in Definition 2, the lexical relationships SYN (Synonym-of), BT (Broader Term)/ NT (Narrower Term) and RT (Related Term) are defined on the basis of the semantic relationships defined in WN among the meanings of two schema elements. To each lexical relationship, it is assigned a probability value that depends on the probability value of the meanings under consideration for the schema elements and it is determined by the formula of the joint probability.

More formally, given two schema elements t and s with the related probabilistic annotations, $PM(t)$ and $PM(s)$, a probabilistic lexical relationship $LexRel$ between t and s with probability P, denoted by

$$< t, LexRel, s, P >$$

is defined iff

1. $\exists < t_{\#i}, P(t_{\#i}) > \in PM(t), \exists < s_{\#j}, P(s_{\#j}) > \in PM(s)$, and $P = P(t_{\#i}) * P(s_{\#j})$
2. and one of the following conditions holds
   (a) $t_{\#i} \equiv s_{\#j}$ and $LexRel = SYN$
   (b) $t_{\#i}$ hypernym of $s_{\#j}$ and $LexRel = BT$
   (c) $t_{\#i}$ meronym of $s_{\#j}$ and $LexRel = RT$

2.4 Experimental Evaluation

In [21], our normalization and annotation methods have been evaluated in order to measure and qualify their performance. They have been integrated within the MOMIS (Mediator EnvirOment for Multiple Information Sources) data integration system [5][6], and have been evaluated on two test cases; the first is a set of three ontologies from the benchmark OAEI 2008[7]; the second is composed of two relational schemata of the well-known Amalgam integration benchmark for bibliographic data[8]. Even if these data sources represent different scenarios w.r.t. Semantic Resource Framework, our previous evaluations can be used to give an idea about the quality of the results obtained by our methods.

To assess the quality of our method, gold standards were created for each normalization step as well as for the lexical annotation and the lexical relationship discovery methods. The gold standards were manually generated by a human expert. Then, we compared the gold standard with the result obtained by using our methods. For each experimental phase, we determined: the true positives, i.e. correct results (TP), as well as the false positives (FP) and the false negatives (FN).

Based on the cardinalities of the TP, FP, and FN sets, the following quality measures are computed:

– *Precision*= $\frac{|TP|}{|TP|+|FP|}$

[6] See`http://www.dbgroup.unimore.it` for references about the MOMIS project.
[7] 101, 205 209 ontologies available at
`http://oaei.ontologymatching.org/2008/benchmarks/`
[8] See `http://dblab.cs.toronto.edu/~miller/amalgam/`

Table 1. Average performance of the lexical annotation and lexical relationship discovery methods with and without normalization

	Precision	Recall	F-Measure
Lexical annotation without normalization	0.63	0.43	0.51
Lexical annotation with normalization	0.62	0.73	0.67
Discovered lexical relationships without normalization	0.49	0.29	0.36
Discovered lexical relationships with normalization	0.81	0.74	0.77

- *Recall*= $\frac{|TP|}{|FN|+|TP|}$
- *F-Measure*= $2 * \frac{Precision*Recall}{Precision+Recall}$

Table 1 shows the average performance of lexical annotation and lexical relationship discovery with and without the normalization method. The experimental results show how the effectiveness of automatic lexical annotation and, as a consequence, the quality of the discovered lexical relationships are improved by the normalization method.

3 Towards Annotated Services in SRF

In this section, we describe an application of our normalization and annotation methods to the Semantic Resource Framework (SRF) described in a previous chapter of this book [6]. SRF is a multi-level (conceptual, logical, and physical level) description of data sources for searching computing applications. It extends the Service Mart model presented in [8] by making such model more expressive and more fitting to the web service description requirements. SRF represents a first step for adding more semantics to web service description.

By using our method, it is possible to enrich the semantics of SRF descriptions by annotating them w.r.t. the lexical resource WN.

Our normalization and annotation methods find application at the *conceptual level* of a Service Mart. The conceptual level includes the object's name and the collection of the object's attributes; all the attributes are typed: they can be atomic (single valued) or part of a repeating group (multi-valued). Moreover, the discovered lexical relationships may suggest useful information at the *logical level* to derive *connection patterns* between Service Marts.

We can easily apply our method by considering the classes and the attributes of a Service Mart [6], as the classes and attributes of a generic object schema as defined in Section 2.

Let us suppose we have a flight booking Service Mart having the following conceptual description:

Booking(CustomerName, BookingNR, FlightNumber, Airline,
DepartureDatetime, DepartureAirport, ArrivalDatetime, ArrivalAirport) (1)

For the service attribute names that do not have an entry in WN, we apply our normalization method.

Table 2. Annotations of some attributes of the "Booking" Service Mart (for the CNs the word representing the head is underlined)

Attribute	Lex. Annotation	Prob.
Airline	$Airline_{\#2}$	0.89
ArrivalAirport	$\underline{Airport_{\#1}}$ FOR $Arrival_{\#2}$	0.9
DepartureAirport	$\underline{Airport_{\#1}}$ FOR $Departure_{\#1}$	0.9
BookingNumber	$Booking_{\#2}$ HAVE $\underline{Number_{\#4}}$	0.8
FlightNumber	$Flight_{\#9}$ HAVE $\underline{Number_{\#4}}$	0.62
	$Flight_{\#2}$ HAVE $\underline{Number_{\#4}}$	0.67

Table 3. WordNet glosses

WN synset	WN gloss
$Airline_{\#2}$	*a commercial enterprise that provides scheduled flights for passenger*
$Airport_{\#1}$	*an airfield equipped with control tower and hangers as well as accommodations for passengers and cargo*
$Arrival_{\#1}$	*accomplishment of an objective*
$Arrival_{\#2}$	*the act of arriving at a certain place*
$Departure_{\#1}$	*act of departing*
$Booking_{\#2}$	*the act of reserving (a place or passage) or engaging the services of (a person or group)*
$Number_{\#4}$	*a numeral or string of numerals that is used for identification*
$Flight_{\#2}$	*an instance of traveling by air*
$Flight_{\#9}$	*a scheduled trip by plane between designated airports*

3.1 Normalization

The normalization method is divided in three steps: preprocessing, abbreviation expansion, and CN annotation. In the first step, the method recognizes as non-dictionary words the following labels:

CustomerName, BookingNR, FlightNumber, DepartureDatetime, DepartureAirport, ArrivalDateTime, ArrivalAirport

As a consequence, these labels are first tokenized (e.g. "CustomerName" as "Customer" and "Name") and then classified as CNs (i.e. "CustomerName, FlightNumber, DepartureDateTime, DepartureAirport, ArrivalDateTime, and ArrivalAirport") and as CNs containing abbreviations (i.e. "Booking NR").

The second step of normalization is focused on expanding abbreviations: the method automatically expands the previously identified CN containing abbreviations "BookingNR" as "BookingNumber".

Table 4. The most relevant annotations of a subset of attributes of the "Flights" Service Mart

Attribute	Lex. Annotation	Prob.
Airport	$Airport_{\#1}$	1.0
FlightStatus.FLTNO	$Flight_{\#2}$ HAVE $Number_{\#4}$	0.89
FlightStatus.Airline	$Airline_{\#2}$	0.89
FlightStatus.ARRAirport	$Airport_{\#1}$ FOR $Arrival_{\#2}$	0.95

The last step of the normalization deals with the CNs. During this step the CNs are interpreted. Let us consider the label "BookingNumber";it is composed by two constituents: "Booking" and "Number" which are automatically annotated by PWSD with, respectively, the synset $Booking_{\#2}$ with probability 0.89, and with $Number_{\#4}$ with probability 0.89, as shown in Tables 2 and 3. Then the *HAVE* semantic relationship is automatically selected and a new WN meaning for the CN is created and inserted in the WN noun hierarchy: we associate the new term "Booking Number" with a gloss given by union of the glosses of "Booking" and "Number" connected by the relationship *HAVE* (i.e. gloss of $Booking_{\#2}$ "HAVE" gloss of $Number_{\#4}$); moreover, we create a hypernym/hyponym relationship between the new synset for "Booking Number" and the synset of "Number", and a Related Term relationship between the new synset and the synset of the modifier "Booking". The probability value associated to the new synset will be the product of the probabilities of the individual annotations $Booking_{\#2}$ and $Number_{\#4}$, i.e. 0.8.

Note that, on this example, only the attribute "Airline" is a WN term, whereas the others are CNs not present.

3.2 Probabilistic Lexical Annotation

After normalization, we perform the probabilistic lexical annotation of all the labels except for the CNs that have been annotated by the normalization method.

For annotating the attribute "Airline", WSD1 selects $Airline_{\#2}$ with a probability of 0.65, WSD2 provides $Airline_{\#2}$ with a probability of 0.7 and WSD3 selects $Airline_{\#1}$ with a probability of 0.6. The Dempster-Shafer's rule of combination applied on these outputs returns the following annotations:

$$PM(Airline) = \{< Airline_{\#1}, 0.11 >, < Airline_{\#2}, 0.89 >\}.$$

By applying a threshold of 0.2 , the annotation $Airline_{\#1}$ is discarded. In the end, "Airline" is thus annotated by $Airline_{\#2}$ with a probability value of 0.89 (see Table 2).

3.3 Probabilistic Lexical Relationship Discovery

After lexical annotation, we can use the probability distributions over the set of possible meanings (i.e. the output of the PWSD) to infer probabilistic lexical relationships among the attributes of the two Service Marts that share the same application domains.

Let us assume, for example, another Service Mart about the scheduled flights departing from an airport, to explain the relationship discovery task:

Table 5. Lexical relationships between "Flights" and "Booking" Service Marts

"Booking"	**Lex. Rel.**	"Flights"	**Prob.**
Airline	SYN	*FlightStatus.AirLine*	0.79
FlightNumber	SYN	*FlightStatus.FLTNO*	0.60
ArrivalAirport	SYN	*FlightStatus.ARRAirport*	0.85
ArrivalAirport	NT	*Airport*	0.9
DepartureAirport	NT	*Airport*	0.9
BookingNR	RT	*FlightStatus.FLTNO*	0.71

Flights(Airport, FlightStatus(FLTNO, Airline, ARRAirport, ScheduledDPTDateTime, EstimatedDPTDateTime)) (2)

The normalization and annotation methods applied on the conceptual description of "Flights" retrieve the annotations shown in Table 4. From the annotations of "Booking" and "Flights", we discover a set of lexical relationships between their attributes (as shown in Table 5).

Let us consider for example, the attribute "ARRAirport" (expanded to "ArrivalAirport") in "Flights"; it is split into its constituents, "Arrival" and "Airport". The constituents are disambiguated as $Arrival_{\#2}$ and $Airport_{\#1}$. Then, the semantic relationship "FOR" between the meanings of the head and the modifier is selected. When we compare the annotation of *Arrival Airport* in "Booking" and the annotation of *FlightStatus.ARRAirport* in "Flights", we discover a SYN relationship as the two elements share the same meaning ($Airport_{\#1}$ FOR $Arrival_{\#2}$.). On the other hand, when we examine the annotations of *Arrival Airport* in the "Booking" description and *Airport* in the "Flights" description, we discover an NT relationship as the new meaning of *Arrival Airport* is an NT of $Airport_{\#1}$.

At the logical level of Service Marts, lexical relationships may suggest useful information to build a *connection pattern* between them. Every connection pattern has a conceptual name and a logical specification, consisting of a sequence of simple comparison predicates between pairs of attributes or sub-attributes of the two services [8].

For example, the SYN relationships discovered between "Flights" and "Booking" (showed in Table 5) might be used to determine a connection between these two Service Marts. This connection could be exploited on previously defined access patterns where the "Airline", "FlightNumber", and "ArrivalAirport" attributes are defined as output parameters in the "Booking" access pattern, while "FlightStatus.AirLine", "FlightStatus.FLTNO", and "FlightStatus.ARRAirport" are input parameters in the "Flights" access pattern.

As a first example, let us suppose we want to define a simple connection between "Flights" and "Booking", checking just the existence of scheduled flights at the arrival airport of a booking: in this case, the lexical relationship *ArrivalAirport SYN FlightStatus.ARRAirport* (as shown on Table 5) helps in the identification of which attribute has to be connected to "ArrivalAirport".

As a consequence, the following connection pattern can be define:

ExistsArrivalAirport(Booking,Flights):[(ArrivalAirport=ARRAirport)]

This means that "Flights" and "Booking" are connected via the connection pattern "ExistsArrivalAirport", which uses a join on arrival airports. The interpretation of joins within connection patterns is existential: if the arrival airport in the "Booking" Service Mart is equal to the ARRAirport of any scheduled flights in the "Flights" description, the predicate is satisfied, and the two instances of "Booking" and "Flights" are composed to form an instance of the result.

Suppose now, we want to define another connection between "Flights" and "Booking" that controls the departure airport in addition to the arrival airport. In this case, selecting the attribute to be connected to "DepartureAirport" is less intuitive. The set of lexical relationships discovered comes to be an important help in this selection. In fact, as shown in Table 5, we found the relationships *DepartureAirport NT Airport* that has a hight probability value (i.e. 0.9). We can, thus, write the connection pattern:

ExistsLink(Booking,Flight):[(ArrivalAirport=ARRAirport) and (DepartureAirport = Airport)]

4 Related Work

Works related to the issues discussed in this chapter are in the area of schema matching, including probabilistic matching and WSD techniques.

The problem of linguistic normalization has received much attention in different areas such as machine translation, information extraction and information retrieval. As observed, the presence of non-dictionary words in schema element labels (including CNs and abbreviations) may affect the quality of *schema elements matching* and requires additional techniques to deal with [11]. Surprisingly, current schema integration systems either do not consider the problem of abbreviation expansion at all or solve it in a non-scalable way by including a *user-defined abbreviation dictionary* or by using only simple *string comparison techniques* [17,2]. Dealing with short forms using a user-defined dictionary only suffers from the lack of scalability: (a) the dictionary cannot handle ad hoc abbreviations; (b) same abbreviations can have different expansions depending on the domain, thus an intervention of a schema/domain expert is still required; and (c) the dictionary evolves over time and it is necessary to maintain the table of abbreviations. Some works have tried to address the limitations of the user-defined dictionary approach by using simple string comparison techniques (e.g. the Similarity Flooding [18]). Syntactical methods are able to detect matches by comparing prefixes and suffixes of literals, however, they are not able to bring to the surface the semantics of abbreviations, thus, in contrast w.r.t. our method, they cannot detect a match between synonyms like "QTY" (short form of quantity) and "amount". Similarly to the abbreviation expansion problem, few papers address the problem of CN interpretation in schema matching area. The CN interpretation is manually executed or relies on a set of manually created rules in most of the work [12,27]. Other schema and ontology matching tools do not interpret nor normalize CNs but they treat the constituents of a CN in isolation [15,27,25]. This oversimplification leads to the discovery of false positive relationships, thus worsens the matching results.

Several language-based methods have been experimented in the context of ontology matching and data integration (H-MATCH [9], CUPID [17]). Some methods rely on string-based techniques only. Other methods make use of external resources, such as dictionaries, to find similarities between terms, but in most of the cases, without performing any disambiguation on the terms. Unlike these methods, our approach is based first of all on the lexical annotation of ontology/schema elements. It is only after this phase that the similarity between elements is computed, thus overcoming the limitation of methods that cannot recognize the meaning of the elements. To the best of our knowledge, [3] is the first work that introduces WSD techniques in an integration process. One of its main limitations is that it does not make use of normalization techniques to process CNs, and this is reflected in a low coverage of the method. In the area of NLP, where WSD is a challenging topic, combination methods have been shown to be an effective way of improving WSD performance, in particular it has been showed that combination systems outperform the behavior of the individual algorithms [7,22].

Modeling uncertainty in probabilistic schema matching has been an active area of research for some years [19]. Our method takes inspiration from [23], where the concept of probabilistic schema mapping is introduced and an algorithm for uncertain query answering is presented. The authors start from initial probabilistic schema mappings, and without dealing with the generation of probabilistic mappings, propose a probabilistic query answering method. The paper describes the requirements of a data integration system to support uncertainty: uncertain schema mappings, uncertain data and uncertain queries.

5 Conclusion

Lexical annotation (i.e. the explicit assignment of meanings to a schema elements w.r.t. a lexical reference) is an effective methodology in the discovery of lexical relationships between schema elements. Normalization helps to improve the performance of lexical annotation by increasing the number of annotable elements.

Starting from our previous works in the context of data integration, in this chapter, we presented how normalization and annotation methods work on generic object schema. Then, we shown how the methods might be applied in the context of search computing in order to enrich the semantics of SRF. We provided an application example showing the effectiveness of our methods: they can profitably be used in SRF to annotate the conceptual level of a service and to identify connection patterns between service descriptions belonging to the same application domain.

References

1. An, Y., Borgida, A., Mylopoulos, J.: Discovering the semantics of relational tables through mappings. In: Spaccapietra, S. (ed.) Journal on Data Semantics VII. LNCS, vol. 4244, pp. 1–32. Springer, Heidelberg (2006)
2. Aumueller, D., Do, H.H., Massmann, S., Rahm, E.: Schema and ontology matching with COMA++. In: SIGMOD 2005, pp. 906–908. ACM, New York (2005)

3. Banek, M., Vrdoljak, B., Tjoa, A.M.: Word sense disambiguation as the primary step of ontology integration. In: Bhowmick, S.S., Küng, J., Wagner, R. (eds.) DEXA 2008. LNCS, vol. 5181, pp. 65–72. Springer, Heidelberg (2008)
4. Benassi, R., Bergamaschi, S., Fergnani, A., Miselli, D.: Extending a Lexicon Ontology for Intelligent Information Integration. In: de Mántaras, R.L., Saitta, L. (eds.) ECAI, pp. 278–282. IOS Press, Amsterdam (2004)
5. Bergamaschi, S., Castano, S., Beneventano, D., Vincini, M.: Semantic Integration of Heterogeneous Information Sources. Data & Knowledge Engineering, Special Issue on Intelligent Information Integration 36(1), 215–249 (2001)
6. Brambilla, M., Campi, A., Ceri, S., Quarteroni, S.: Semantic Resource Framework. In: Ceri, S., Brambilla, M. (eds.) Search Computing II. LNCS, vol. 6585, pp. 73–84. Springer, Heidelberg (2011)
7. Brody, S., Navigli, R., Lapata, M.: Ensemble Methods for Unsupervised WSD. In: ACL. The Association for Computer Linguistics (2006)
8. Campi, A., Ceri, S., Maesani, A., Ronchi, S.: Designing service marts for engineering search computing applications. In: Benatallah, B., Casati, F., Kappel, G., Rossi, G. (eds.) ICWE 2010. LNCS, vol. 6189, pp. 50–65. Springer, Heidelberg (2010)
9. Castano, S., Ferrara, A., Montanelli, S.: Matching ontologies in open networked systems: Techniques and applications. J. Data Semantics V, 25–63 (2006)
10. Cohen, W.W., Ravikumar, P.D., Fienberg, S.E.: A comparison of string distance metrics for name-matching tasks. In: Kambhampati, S., Knoblock, C.A. (eds.) IIWeb, pp. 73–78 (2003)
11. Do, H.H.: Schema Matching and Mapping-based Data Integration: Architecture, Approaches and Evaluation. VDM Verlag (2007)
12. Embley, D.W., Jackman, D., Xu, L.: Multifaceted Exploitation of Metadata for Attribute Match Discovery in Information Integration. In: Workshop on Information Integration on the Web, pp. 110–117 (2001)
13. Miller, G.A., et al.: WordNet: An on-line lexical database. International Journal of Lexicography 3, 235–244 (1990)
14. Euzenat, J., Shvaiko, P.: Ontology matching. Springer, Heidelberg (2007)
15. Le, B.T., Dieng-Kuntz, R., Gandon, F.: On ontology matching problems - for building a corporate semantic web in a multi-communities organization. In: ICEIS (4), pp. 236–243 (2004)
16. Levi, J.N.: The Syntax and Semantics of Complex Nominals. Academic Press, New York (1978)
17. Madhavan, J., Bernstein, P.A., Rahm, E.: Generic Schema Matching with Cupid. In: VLDB, pp. 49–58 (2001)
18. Melnik, S., Garcia-Molina, H., Rahm, E.: Similarity Flooding: A Versatile Graph Matching Algorithm and Its Application to Schema Matching. In: ICDE, pp. 117–128 (2002)
19. Nagy, M., Vargas-Vera, M., Motta, E.: Dssim-ontology mapping with uncertainty. In: Shvaiko, P., Euzenat, J., Noy, N.F., Stuckenschmidt, H., Benjamins, V.R., Uschold, M. (eds.) Ontology Matching. CEUR Workshop Proceedings, vol. 225 (2006), CEUR-WS.org
20. Parsons, S., Hunter, A.: A review of uncertainty handling formalisms. In: Hunter, A., Parsons, S. (eds.) Applications of Uncertainty Formalisms. LNCS (LNAI), vol. 1455, pp. 8–37. Springer, Heidelberg (1998)
21. Po, L., Sorrentino, S.: Automatic generation of probabilistic relationships for improving schema matching. Information Systems 36(2), 192–208 (2011), Special Issue: Semantic Integration of Data, Multimedia, and Services
22. Preiss, J.: Probabilistic word sense disambiguation. Computer Speech & Language 18(3), 319–337 (2004)
23. Sarma, A.D., Dong, X., Halevy, A.Y.: Bootstrapping pay-as-you-go data integration systems. In: Wang, J.T.-L. (ed.) SIGMOD Conference, pp. 861–874. ACM, New York (2008)

24. Shafer, G.: A Mathematical Theory of Evidence. Princeton University Press, Princeton (1976)
25. Shvaiko, P., Giunchiglia, F., Yatskevich, M.: Semantic matching with s-match. In: Semantic Web Information Management: a Model-Based Perspective, vol. XX, pp. 183–202 (2010)
26. Sorrentino, S., Bergamaschi, S., Gawinecki, M., Po, L.: Schema normalization for improving schema matching. In: Laender, A.H.F., Castano, S., Dayal, U., Casati, F., de Oliveira, J.P.M. (eds.) ER 2009. LNCS, vol. 5829, pp. 280–293. Springer, Heidelberg (2009)
27. Su, X., Gulla, J.A.: Semantic enrichment for ontology mapping. In: Meziane, F., Métais, E. (eds.) NLDB 2004. LNCS, vol. 3136, pp. 217–228. Springer, Heidelberg (2004)

Towards an Ontological Representation of Services in Search Computing

Fabian Suchanek[1], Alessandro Bozzon[2], Emanuele Della Valle[2], Alessandro Campi[2], and Stefania Ronchi[2]

[1] INRIA Saclay, Paris, France
[2] Politecnico di Milano, Dipartimento di Elettronica e Informazione, P.za L. Da Vinci, 32. I-20133 Milano - Italy

Abstract. In the Search Computing project, Web services are modeled by the Semantic Resource Framework (SRF). In this article, we argue that the SRF could benefit from ontological concepts borrowed from the Semantic Web. We first present the knowledge representation used in the Semantic Web, notably in the YAGO ontology [14]. We show how this model is used in the ANGIE system [12] to represent Web Services in conjunction with YAGO. We draw parallels to the Service Mart [3] model used in SeCo. We propose a symbiosis of the two models, discussing the challenges and advantages that come with the integrated model.

1 Introduction

The Search Computing project (SeCo) [3] uses the Semantic Resource Framework (SRF) to model Web services. The SRF is a multi-layer model. The higher layers provide an abstract semantic description of the services, building on the notions of *Service Marts* and *Connection Patterns*. The lower layers (*service interfaces* and *access patterns*) are concerned with the physical properties of the services. Ideally, every service belongs conceptually to a Service Mart. A Service Mart is structurally defined by means of attributes. Two Service Marts can be connected by a *Connection Pattern*. At the logical level, each Service Mart is associated with one or more access patterns representing the signatures of the service calls. *Access patterns* contain a subset of the attributes of the Service Mart, which are tagged with either I (input), or O (output). Attributes can also be tagged as R (ranking), to denote attributes that are used for ordering result instances. Ranking is particularly important in SeCo, because it allows mastering the combinatory explosion of multi-domain queries typical in Search Computing.

By design, the creation of a SRF is a bottom-up process, whereby the real world entities modeled by the Service Marts are typically created on the basis of the Web services. In this article, we try to anticipate how SeCo can cope with a large scale deployment scenario with a larger number of administrators and services. We argue that, when the complexity of the knowledge represented in a SeCo deployment increases, the SRF model might benefit from adapting ideas

S. Ceri and M. Brambilla (Eds.): Search Computing II, LNCS 6585, pp. 101–112, 2011.

from the Semantic Web technologies. Therefore, we propose to substitute the topmost level of the SRF (the service marts) with an ontology. We argue that this will reduce the maintenance effort for a SRF with a large number of data sources, and facilitate the interaction of SeCo with ontology-based systems.

In order to motivate why the SRF model used in SeCo may need to be extended with an ontological representation, we start with a example.

1.1 Motivation

The SRF model leaves room for modeling the same data source in different ways. Assume for example the availability of two search services (Figure 1). The first, ws_1, exposes information about movies played in theaters located close to a given location. The second service, ws_2, queries a repository containing information about movies. The access patterns of these two Web Services can be described on the conceptual level of the SRF as two service marts `MOVIE` and `THEATRE` (see Figure 2), where the attribute `Movie` of the service mart `THEATRE` is linked to the attribute `Title` of the service mart `MOVIE` .

Now let's assume the availability of a service ws_3, which, given a place, searches for all the *actors* that were born there and returns a description of the actor, including the list of the movies he played in (Figure 3).

As we explained in the introduction, ideally, each service belongs to one Service Mart. The service ws_3 does not belong to any of the service marts that have already been defined. Therefore, the SRF administrator is confronted with two choices: extend the existing `MOVIE` service mart to accommodate the additional parameters brought by ws_3 (as in Figure 4(a)), or create a new `ACTOR` service mart (as in Figure 4(b)). Since structured attributes can replicate the attributes of other concepts, both options are valid and none of them is a priori better than the other.

If there are multiple SRF administrators, different SRF administrators may choose different (and potentially redundant or incompatible) modeling approaches.

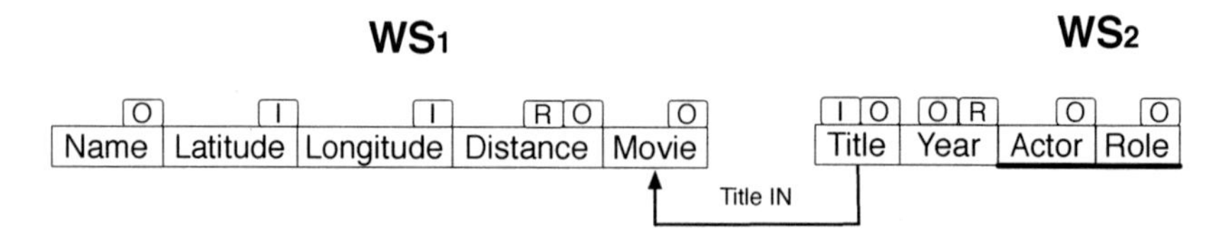

Fig. 1. The Access Patterns for ws_1 and ws_2 on the logical level

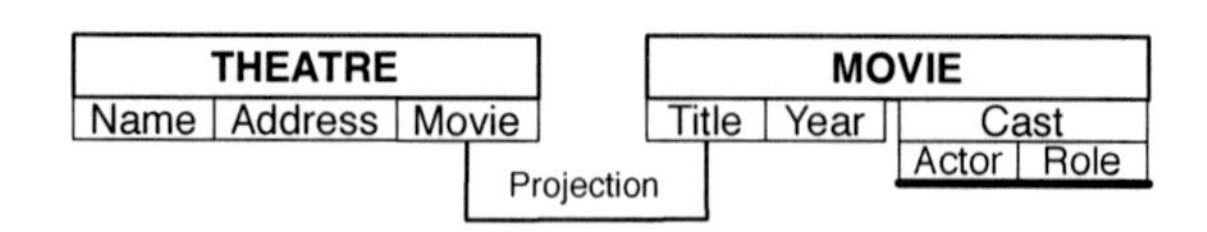

Fig. 2. Two service marts for ws_1 and ws_2 on the conceptual level

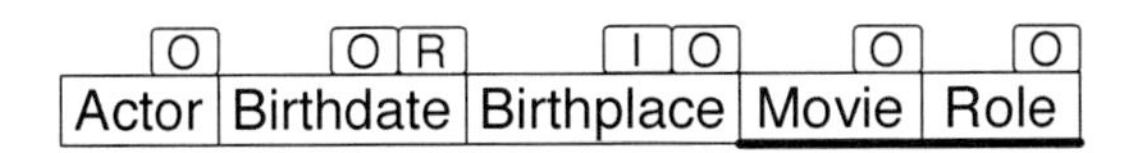

Fig. 3. The Access Pattern for ws_3

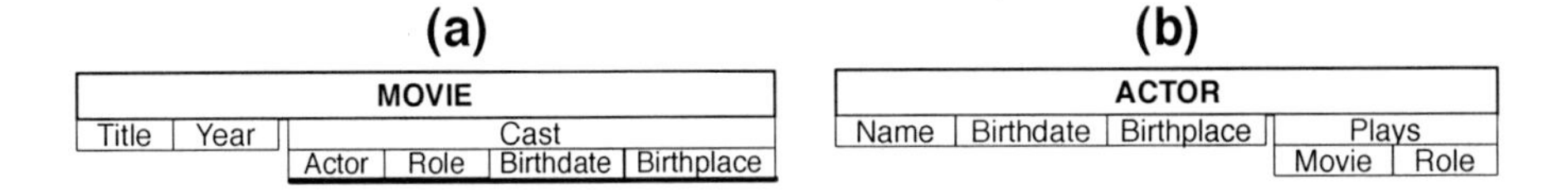

Fig. 4. Examples of alternative evolution of the service marts

In the case of a large scale deployment of a Search Computing system, with potentially hundreds of services and dozens of administrators, these small incompatibilities may quickly add up, leaving the SRF model cluttered with redundancies and inconsistencies. These inconsistencies can lead to major maintenance problems once the size of the SRF surpasses what a single human can assess.

1.2 Contribution

In this chapter, we propose an evolution of the SRF knowledge representation model that will allow it to cope better with some of the maintenance challenges described above. Our proposed model leverages the ontological model that is used in the Semantic Web. It builds on ANGIE [12], a novel approach for Web service description based on the YAGO Ontology [14]. We show that our proposed model not only eases the problem of design alternatives, but also brings additional benefits for the maintenance of the SRF. The new model does not come without challenges of its own. Therefore, this chapter pursues a rather inspirational goal, laying the ground for further investigation.

The chapter is structured as follows: Section 2 discusses the related work; in Section 3, we present YAGO and ANGIE. In Section 4, we propose a merger of the ANGIE knowledge representation model and the SeCo model. In Section 5 we elaborate on the properties of our proposed hybrid model. Finally Section 6 discusses potential future work and concludes.

2 Related Work

The combination of ontologies and services has been proposed before [10,12,13]. The ultimate goal of such an endeavor is to enable a higher degree of automation for the tasks involved in the life-cycle of service based applications. These include (among others) the discovery and selection of services, their composition, their execution and their monitoring. As proposed first in [13], ontologies can be used to model four types of service semantics: *data semantics* (the semantics pertaining to the data contained in the data source), *functional semantics* (the semantics pertaining to the functional capabilities of the service), *non-functional semantics* (the semantics related to the non-functional aspects of the service, such as security or reliability) and *execution semantics* (the semantics related to the invocation, result processing and exception handling of the service).

Several ontology have been proposed to describe these semantics and annotate services. The most well-known ones are OWL-S [2], WSMO [9], SAWSDL [7] and WSMO Lite [16]. Some of them hypothesized the need to semantically describe services at a level of detail that even requires to define a specific ontological language, e.g., the Web Service Modeling Language [6]. All these existing approaches have been perceived unsuited by practitioners in large scale deployments; mostly because of the extra cost of annotating services.

To avoid this pitfall, we decided to look for a light weight approach possibly based on a minimal ontological language. Therefore, we turn our interest to ANGIE [12], a novel approach that is centered on YAGO Ontology [14]. ANGIE leverages the "lightest" of the ontological languages, RDF. This model builds only on typed resources and labeled links. ANGIE uses RDF to describe the input and the output parameters of Web Services and to orchestrate their invocation in order to dynamically extend YAGO with instances loaded from the Web. In the remainder of the chapter we report our initial findings on how ideas from the ANGIE model can ease the definition of new services in SeCo.

3 YAGO and ANGIE

3.1 YAGO

YAGO [14] is a large semantic knowledge base (an *ontology*). It contains 3 millions entities (such as movies, cities, universities and people) and 28 millions facts about them (such as birth dates, appearances in movies, geographical location etc.). YAGO was constructed automatically from Wikipedia and WordNet [8]. In YAGO, knowledge is represented in the RDFS model [4]. This model can be seen as a directed labeled multi-graph, in which nodes represent entities and edges represent relationships between the entities. For example, the node *Apollo Theatre* is linked to the node *Shall we dance* by an edge labeled *shows* (see Figure 5). The labels of the edges are called *relations*. YAGO uses a set of 100 predefined relations. These include a variety of link types, such as *bornIn*, *locatedIn*, *directedMovie* or *isCapitalOf*.

RDF knowledge bases distinguish between instances (such as *Jennifer Lopez*, drawn thin) and concepts, i.e., groups of similar instances (such as *actor* or *movie*, in bold). Instances and concepts are both nodes in the RDF graph. For example, the concept *movie* is a node in the RDF graph. An instance is linked to its concept by the relation *type*. A concept is linked to a more general concept by the relation *subclassOf*. For example, the sub-concept *actor* is linked to the super-concept *person* in this way. Whenever an instance is an instance of a sub-concept, it is automatically an instance of all of its super-concepts.

Relations are by themselves nodes in the RDF model. This makes it possible to talk about properties of relations in the same way as about instances. For example, to say that the domain of *shows* is the concept *theatre*, we can link the node of

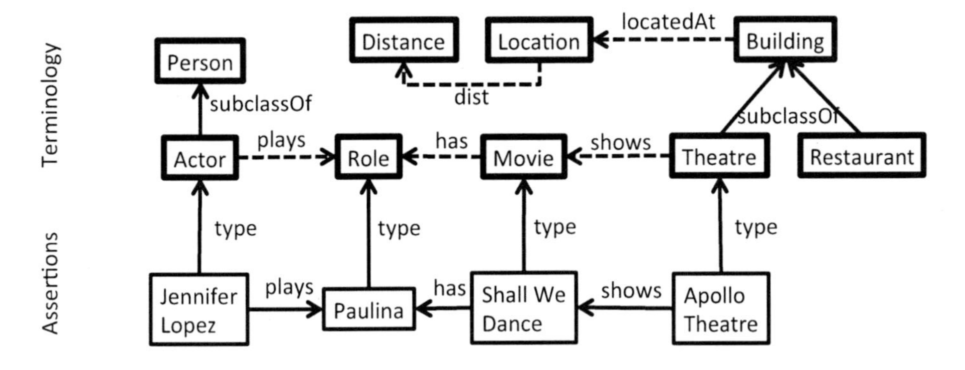

Fig. 5. An excerpt from an RDF ontology

shows to the node *theatre* by the relation *hasDomain* (not shown in the figure). For illustration, we draw the relations as dashed arrows between the domain concept and the range concept (as shown in Figure 5). Every instance of a sub-concept of *theatre* (e.g., every instance of *3d-theatre*) is also an instance of *theatre*. Thereby, the relation *shows* applies automatically to all instances that live somewhere below the concept *theatre*.

Different from classical database models, RDF models are inherently *schemaless*. This means that any instance can be linked to any other instance with any relation, as long as the domain and the range constraint of the relation are not violated.

3.2 ANGIE

ANGIE [12] is a system that uses Web services to extend YAGO. ANGIE requires the manual registration of Web service and a manual mapping of the input and the output of the services to the concepts of the ontology[1]. Once the Web services have been registered, ANGIE allows answering queries on the knowledge base. Whenever the data in the knowledge base is not sufficient to answer a query, ANGIE calls the Web services to retrieve the required additional information. ANGIE can automatically determine the Web services that have to be called to answer the query and it can automatically combine different Web services should that be necessary. This process is transparent to the user, so that the user has the impression of browsing a huge knowledge graph – even though this graph is extended on the fly behind the scenes with the data from the Web services. While ANGIE can deal with arbitrarily-shaped Web services, it cannot deal with ranking of the results.

[1] Such a mapping captures in a very light-weight way the *data semantics*, since all data types exchanged with the Web Service are mapped to ontological concepts. It also captures part of the *functional semantics*, since ANGIE can only model Web Services that provide information about a give topic or entity, and part of the *execution semantics*, since it describes input and output, but cannot handle exceptions. No formal description of the *non-functional semantics* is provided.

ANGIE works on *conjunctive SPARQL queries*. These queries can be thought of as RDF graphs that may have variables in the place of the nodes. For example, a user may ask for all movies of the Apollo Theatre by the query depicted in Figure 6. An answer to such a query is a subgraph of the ontology that is isomorphic to the query. In the example, we can match the query on the ontology depicted in Figure 5 with $?m =$*Shall we dance*. Therefore, $?m =$*Shall we dance* is an answer to the query.

Web Service Orchestration. If a query cannot be matched on the ontology, or if we want to retrieve more answers than the ontology knows, ANGIE can resort to Web services. A Web service is represented just like a query: as an RDF graph that can contain variables in the place of nodes. Each edge is either an *input edge* or an *output edge*. Figure 6 shows a Web service at the top right, which requires as input (solid) a variable $?a$ that must be an theatre and delivers as output (dashed) a fact that the theatre shows some movie $?b$ and that $?b$ is a movie. Before the Web service can be called, all variables in the input edges have to be instantiated. When the Web service returns its results, these are instantiations of the variables in the output edges.

When ANGIE receives a query, it tries to *cover* the query graph with (1) edges from the ontology or (2) output edges of Web services. Consider the example in Figure 6. We start with the query at the bottom. We cover the query with an instantiation of the Web service (on the layer above). The Web service covers the query edge (solid) with an output edge (dashed). It introduces an additional output edge (that $?m$ is of type *movie*, dashed), which we did not ask for. It also introduces an input edge (that the Apollo Theatre has to be a theatre, solid). Now, the procedure is repeated: Every input edge has to covered again, either by the output edge of another Web service or by an edge from the ontology. In the example, the fact that the Apollo Theatre is a theatre is already in the ontology. Therefore, we can cover this input edge (solid) with the edge from the ontology (top layer, dashed). Now, every input edge and the query is covered. If we call the Web services from top to bottom, we will receive answers to the query. In our example, there is only one Web service, but if there are multiple Web services, then the outputs of one service are "piped" into the inputs of another service. ANGIE implements a sophisticated scheduling algorithm that can compute such query covers efficiently and give preference to covers that are likely to return more answers. This works even when the query covers are recursive.

All data retrieved from the Web services is added to the ontology. Thereby, future queries can make use of the knowledge that has already been computed.

Virtual Web Services. In some cases, the way the ontology models the data and the way the Web services return the data may not coincide. Consider for example the Web services depicted in Figure 7. While the first Web service can return the actors for a given movie, the second service can return not only the actors but also their roles. If the user asks only for the movies (as in the figure), Web service 1 can be applied, but Web service 2 cannot, because none of its

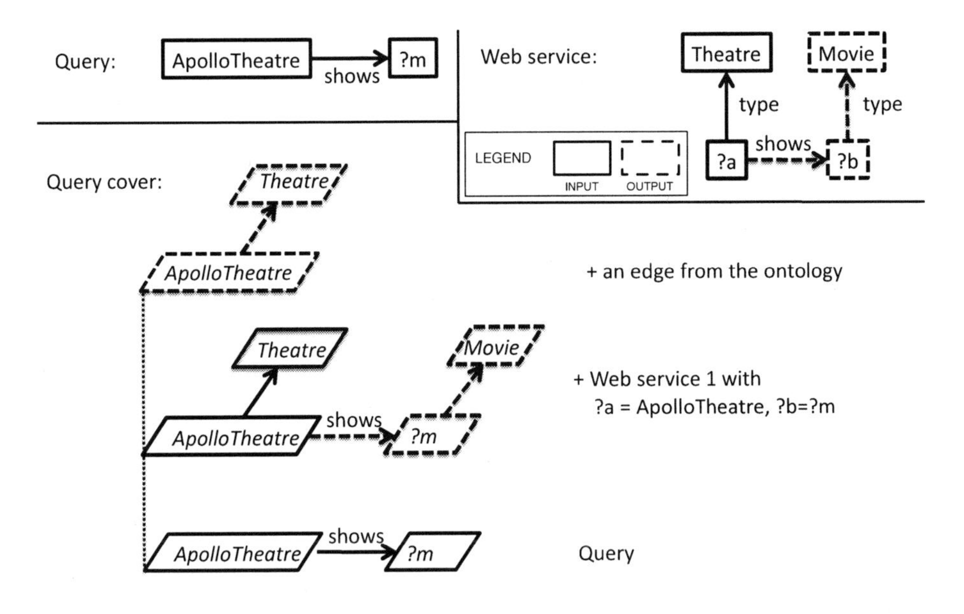

Fig. 6. A user query, a Web service, and a query cover

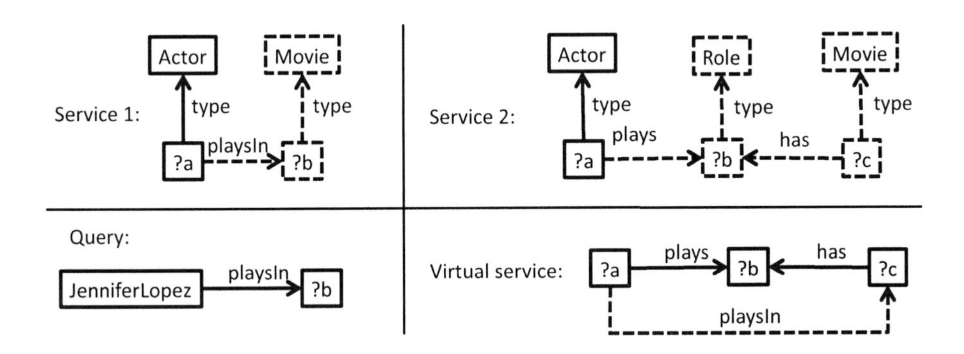

Fig. 7. A virtual Web service

output edges matches the query. Even the edges from the ontology cannot be applied, since the ontology models roles as a separate entity.

To bridge such mismatches, ANGIE supports *virtual services*. A virtual service is a pseudo Web service that does not have a physical Internet service behind it. "Calling a virtual service" means matching the input edges and then adding the output edges to the query cover – without any physical Web service call. In the example in Figure 7, the virtual service allows adding a *playsIn* fact, if there are *plays* and *has* facts for a role.

With the virtual service, our query can be answered even with Web service 2: We first call Web service 2 to retrieve movies together with their roles. Then, we "call" the virtual Web service to transform the output into *playsIn* facts that answer the query. Such a call does not involve any physical Web service call. It is a pure data manipulation on the query edges, which allows us to bridge data modelizations of different granularity.

4 Toward an Ontological Representation of the Access Patterns

Inspired by the ANGIE approach, we propose to substitute the Service Mart layer of the SRF model with the YAGO ontology. The role of Service Marts will be taken over by the concepts and relations of the ontology – just like explained in Section 3.2. Thereby, the conceptual layer of the SRF for the running example will be the ontology depicted in Figure 5.

In addition, we propose to describe the access patterns in the way Web Services are described in ANGIE: Each access pattern becomes a graph. The edges are labeled with relationships from the ontology. The nodes can be either constants from the ontology or variables. Each attribute of an access pattern becomes an edge that is labeled with the corresponding relationship from the ontology. This edge connects to a node with a variable. Structured attributes become star-shaped patterns. As in ANGIE, nodes are labeled as *input nodes* or *output nodes*. The type of a variable node is indicated by an outgoing `type` edge to a concept node.

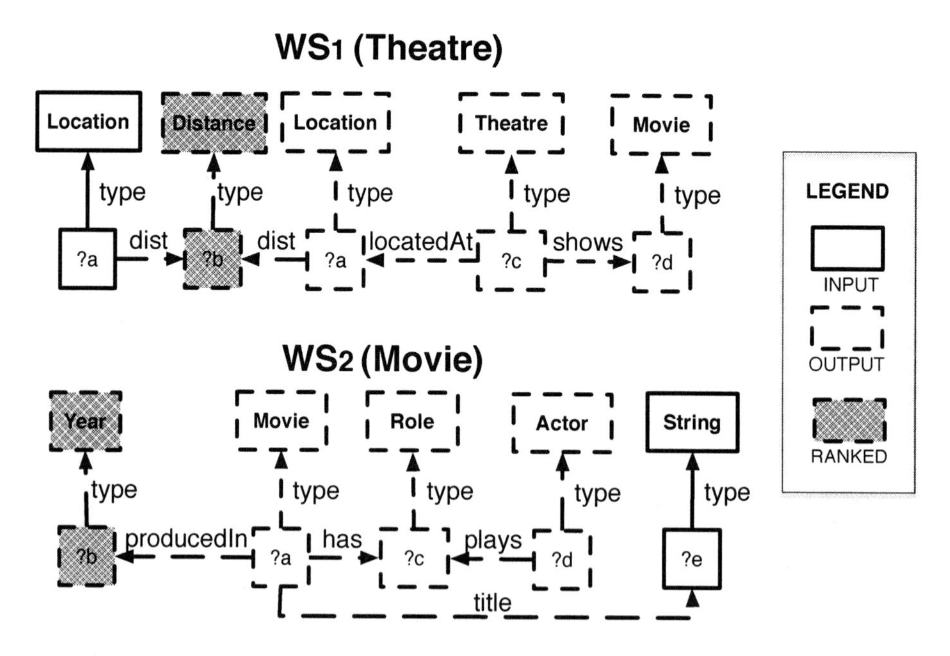

Fig. 8. RDF representation of two Access Patterns shown in Figure 3

To cope with the notion of *ranking* (typical for SeCo), ANGIE's representation of Web services has to be extended so as to allow not only input nodes and output nodes, but also *ranking nodes*. In every access pattern graph, there can be at most one node that is labeled as a ranking node. Such a node has to be an output node. In the illustrations, we represent a ranking node by a filled dashed frame. Figure 8 shows how the access patterns ws_1 and ws_2 from the introduction can be modeled; ws_1, which belonged to the Service Mart *Theater*, requires as input (solid) a variable ?a, a location, and produces as output (dashed) the theaters ?c, located at a distance ?b from the input location, and showing some movie ?d; the results of ws_1 are also ordered according to the distances ?b. Likewise, ws_2, which belonged to the Service Mart *Movie*, requires as input a the title ?e of movie ?a, and it produces as output the actors ?d that acted in the movies with the role ?c; results are ordered according to the production year ?b.

Noteworthily, the novel RDF representation is isomorphic w.r.t. the original one. This assures compatibility with the service description exploited by the other levels of the architecture. Indeed, we remark that all of our proposed changes happen purely at the level of service modeling. No changes to the Web service composition algorithms or the ranking algorithms of SeCo are necessary.

5 Properties of the Proposed Model

We note that the proposed representation helps in coping with the issues mentioned in the introduction: as Figure 9 depicts, there is no ambiguity whether the information brought by ws_3 shall be stored within the service mart `MOVIE` or `ACTOR`. The administrator just has to describe the access pattern as a query on YAGO ontology: the access pattern selects instances of actors (`?a type Actor`) based on their birthplace (`?a bornIn ?b`), ranked by the date of birth (`?a bornWhen ?e`) and lists the movies they played in (`?d type Movie; ?a plays ?c; ?d has ?c`).

Given that every concept and property exists exactly once, there are fewer possibilities for inconsistencies to appear in the ontological model.

In addition to the specific maintenance case that we developed through the paper, the use of an ontological description for the conceptual level brings several advantages: notably, the ontology can evolve independently from the registered data sources. In the SRF model that we are proposing, the ontology models

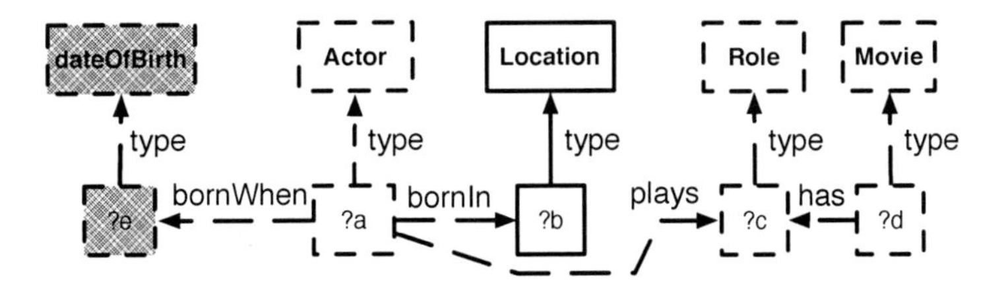

Fig. 9. The Web service ws_3 in the new model

the world and the access patterns model the Web Services. An extension of the ontology will not influence the access patterns. For instance, the subclass `Restaurant` of `Building` can exist in the ontology even if no access pattern refers to it. Vice versa, the addition or removal of an access pattern will not influence the ontology (provided that all necessary relations and concepts are present). Thereby, the roles of the ontology and the access patterns are clearly defined and distinct.

Since the ontology allows for the creation of sub-concepts, more specific concepts can be created without having to redo the work that has been done for the super-concept. The attributes of the sub-concept are automatically consistent with the attributes of the super-concept. By design, a Web service that delivers an instance of the sub-concept can also be used to deliver an instance of the super-concept. Furthermore, through the domains and ranges of relations, the target type of a relation is explicitly defined. There is no need to replicate this information with every link or every Web service. Target types are an inherent part of the model.

Since relations are first-class citizens of the model, the joinability of attributes follows directly from the definition of the relations and the access patterns. For example, if a Web service returns movies, then the joinability with another Web service that returns movies follows from the fact that both output variables are of type *movie*. This also works across different levels in the concept hierarchy: If one Web service returns 3d-movies and the other Web service returns silent movies, these Web services are still joinable, since both concepts are sub-concepts of *movie*. By factoring out this common information from the level of Web services to the level of classes, the ontological model avoids redundancy.

The only maintenance task left to SRM administrators is the extension of the ontology when the registration of a new Web Service requires adding new concepts and relationships to YAGO. For instance, if the relationship `bornIn` is not present in the ontology, but is brought in by a Web service, then this relationship has to be added by the SRF administrator. This is not a trivial task per se, but at least we can rely on established methodological approaches (e.g., METHONTOLOGY [5], On-To-Knowledge [15], and DILIGENT [11]) for ontology development and maintenance.

6 Conclusions and Future Work

In this article, we have discussed some of the maintenance challenges that SeCo will face when more users, services and administrators start using the system in parallel. We have argued that, for this task, the SeCo knowledge representation model would benefit from a more ontological design. We have proposed and studied an ontological adaptation of the SeCo model, inspired by the model used in the ANGIE system. We have shown that the new model mitigates some of the maintenance challenges, thus contributing to SeCo's fitness for going mainstream.

We believe that, apart from easing some of the maintenance challenges, an ontological top layer of SeCo opens the door to a wide range of possible interactions between SeCo and existing ontologies. These include not only YAGO but also the vast resources of the Linking Open Data Project (LOD) [1]. Whilst the LOD has not been the focus of this article, we are confident that the shared knowledge representation will ease knowledge exchange between the LOD and SeCo in the future – e.g., by answering queries with data from the LOD in case a Web service is not available.

We believe that there is much further research potential in the combination of ANGIE and SeCo. For example, the orchestration algorithm of ANGIE can only combine the outputs of one Web Service with the inputs of another Web Service (i.e., it can only perform *pipe joins*, in SeCo terminology). Therefore, some registered Web Service cannot be used directly. SeCo, in contrast, can master the combinatory explosion that appears when results from multiple Web Services are combined, because it uses rank aware parallel join operators.

On the other hand, ANGIE is very good at answering SPARQL queries. This is an avenue on which SeCo could benefit. One possibility to combine the advantages of SeCo and ANGIE would be to register Web Services both in ANGIE and in SeCo. Whenever a user issues a SPARQL query on YAGO, ANGIE could orchestrate (virtual and real) Web Services normally whenever it can pipe the output of one Web Service into the input of another Web Service. When ANGIE faces the problem to perform parallel joins on the results of multiple Web Services, it could delegate the execution of this part of the orchestration to SeCo. This is just one example for the research possibilities that wait to be discovered in further cross-fertilization of these approaches.

Acknowledgement

The work by Fabian Suchanek has been partially funded by the European Research Council under the European Community's Seventh Framework Programme (FP7/2007-2013) / ERC grant Webdam, agreement 226513. `http://webdam.inria.fr/`

References

1. Bizer, C., Heath, T., Berners-Lee, T.: Linked data - the story so far. International Journal on Semantic Web and Information Systems 5(3), 1–22 (2009)
2. Burstein, M., Hobbs, J., Lassila, O., Mcdermott, D., Mcilraith, S., Narayanan, S., Paolucci, M., Parsia, B., Payne, T., Sirin, E., Srinivasan, N., Sycara, K.: Owl-s: Semantic markup for web services. Website (2004)
3. Ceri, S. (ed.): Search Computing. LNCS, vol. 5950. Springer, Heidelberg (2010)
4. World Wide Web Consortium. Rdf primer, w3c recommendation (2004), `http://www.w3.org/TR/rdf-primer/`
5. Corcho, Ó., Fernández-López, M., Gómez-Pérez, A., López-Cima, A.: Building legal ontologies with METHONTOLOGY and webODE. In: Benjamins, V.R., Casanovas, P., Breuker, J., Gangemi, A. (eds.) Law and the Semantic Web. LNCS (LNAI), vol. 3369, pp. 142–157. Springer, Heidelberg (2005)

6. de Bruijn, J., Fensel, D., Kerrigan, M., Keller, U., Lausen, H., Scicluna, J.: Modeling Semantic Web Services: The Web Service Modeling Language. Springer, Berlin (2008)
7. Farrell, J., Lausen, H.: Semantic annotations for wsdl and xml schema (August 2007)
8. Fellbaum, C. (ed.): WordNet: An Electronic Lexical Database. MIT Press, Cambridge (1998)
9. Fensel, D., Holger Lausen, A.P., de Bruijn, J., Stollberg, M., Roman, D., Domingue, J.: Enabling Semantic Web Services: The Web Service Modeling Ontology. Springer, Berlin (2006)
10. McIlraith, S.A., Son, T.C., Zeng, H.: Semantic web services. IEEE Intelligent Systems 16(2), 46–53 (2001)
11. Pinto, H.S., Staab, S., Tempich, C.: Diligent: Towards a fine-grained methodology for distributed, loosely-controlled and evolving engineering of ontologies. In: de Mántaras, R.L., Saitta, L. (eds.) ECAI, pp. 393–397. IOS Press, Amsterdam (2004)
12. Preda, N., Kasneci, G., Suchanek, F., Neumann, T., Yuan, W.: Active knowledge: Dynamically enriching rdf knowledge bases by web services (angie). In: International Conference on Management of Data (SIGMOD 2010). ACM, New York (2010)
13. Sivashanmugam, K., Verma, K., Sheth, A.P., Miller, J.A.: Adding semantics to web services standards. In: Zhang, L.-J. (ed.) ICWS, pp. 395–401. CSREA Press (2003)
14. Suchanek, F.M., Kasneci, G., Weikum, G.: Yago: A Core of Semantic Knowledge. In: 16th International World Wide Web Conference (WWW 2007). ACM Press, New York (2007)
15. Sure, Y., Staab, S., Studer, R.: Methodology for development and employment of ontology based knowledge management applications. SIGMOD Record 31(4), 18–23 (2002)
16. Vitvar, T., Kopecký, J., Viskova, J., Fensel, D.: Wsmo-lite annotations for web services. In: Bechhofer, S., Hauswirth, M., Hoffmann, J., Koubarakis, M. (eds.) ESWC 2008. LNCS, vol. 5021, pp. 674–689. Springer, Heidelberg (2008)

Part 4

Rank Join

The rank-join problem is the search for the top-k combinations of objects coming from different, ranked data sources and matching according to a given join predicate. Objects are assumed to be equipped with a score, or, in general, a score vector. Building joins of ranked objects is the core operation of search computing, therefore studying this problem is essential.

In the first chapter, the theoretical foundations of the rank-join problem are presented. The desired optimality properties that should be featured by rank-join algorithms are stated. After showing that optimality *per se* is generally unachievable, the much-studied notion of instance-optimality is discussed, which represents a relaxed but robust version of optimality. Rank-join algorithms are discussed according to a pattern that consists of two main components: a *pulling strategy* that indicates which data source to select for extraction of a new object, and a *bounding scheme* providing a criterion for stopping the execution. Instance optimality can be attained by resorting to *tight* bounding schemes, which however incur a heavy computational cost. Trade-offs between instance-optimality and efficiency in computing the bound are discussed.

In the second chapter, the problem is generalized to cases of objects also equipped with feature vectors that can be used to compare the objects to one another so as to join them on the basis of a notion of "proximity". The problem becomes then retrieving combinations of objects that have high scores, whose feature vectors are close to one another and possibly to a given feature vector (the query). In this case, too, the bounding scheme (and a tight version thereof) and the pulling strategy play a crucial role to efficiently compute the solution.

In the third chapter, a further generalization of the problem regards rank-join in the presence of uncertainty. In particular, uncertainty in the objects' scores may derive from different levels of reliability of the data sources. It is thus very relevant to model uncertainty in the scores. This can be done in different ways, among which using ranges. The resulting ranking of the object combinations is not univocally determined by the scores because of their uncertainty. This gives rise to new semantics of top-k queries, which must take the probability of a possible ranking into account.

The fourth short chapter discusses the current open problems and trends in rank-join.

The Rank Join Problem

Neoklis Polyzotis

University of California - Santa Cruz

Abstract. In the rank join problem, we are given a set of relations and a scoring function, and the goal is to return the K join results with the highest scores. It is often the case in practice that the inputs may be accessed in ranked order and the scoring function is monotonic. These conditions allow for efficient algorithms that solve the rank join problem without reading all of the input. In this chapter, we review recent efforts in the development and analysis of such rank join algorithms. First, we present some theoretical results that state the inherent complexity of the rank join problem and essentially reveal that any rank join algorithm has to trade off between I/O efficiency and computational efficiency. We then review a specific rank join algorithm that adjusts this trade-off at runtime, depending on the data and the scoring function, in order to strike a balance between I/O overhead and computation.

1 Background and Basic Definitions

A relational ranking query (or a top-K join query) specifies a scoring function over the results of a join and returns the K tuples with the highest scores. As an example, the following query (written in an SQL-like language) retrieves the top 10 hotels and restaurants located in the same city, giving priority to the cheap hotels and the best restaurants with live music.

SELECT *h.name*, *r.name*
FROM *Hotel h*, *Restaurant r*
WHERE $h.city = r.city$
RANK BY $0.4/h.price + 0.4 * r.rating + 0.2 * r.hasMusic$
LIMIT 10

Ranking queries have become increasingly popular in many application domains, from multimedia retrieval [2] to uncertain databases [1], as they allow a user to focus on the most relevant query results.

Rank join evaluation, i.e., computing the top K results of a relational join according to a specific scoring function, form an integral component of ranking query processing. Several recent studies have considered specialized rank join algorithms [1,2,3,4,5,7,8,9], the integration of such operators in the query optimizer [6], and the computation of statistics for query optimization [10]. In what follows, we provide a formal definition of rank joins and some necessary background in order to discuss the problem further.

Consider a natural join of relations $R_1, \ldots, R_n$ where each $\tau_i \in R_i$ is composed of *named attributes* and *base scores*. The base scores are denoted as a vector

S. Ceri and M. Brambilla (Eds.): Search Computing II, LNCS 6585, pp. 115–120, 2011.

$\mathbf{b}(\tau_i) \in [0,1]^{e_i}$ for some $e_i \geq 0$, and signify the importance of the tuple according to criteria specified by the ranking query. Returning to the previous example, we observe that *Restaurant* has two base scores, corresponding to the rating and the music event respectively. Base scores are aggregated using a *scoring function* $\mathcal{S}$ that computes the score of a join result τ as $\mathcal{S}(\mathbf{b}(\tau))$. We may also use $\mathcal{S}(\tau)$ as a shorthand for the score of τ. Following common practice, we assume that $\mathcal{S}$ is monotonic, i.e., $\mathcal{S}(x_1, \ldots, x_e) \leq \mathcal{S}(y_1, \ldots, y_e)$ if $x_i \leq y_i$ for all i.

Let τ be a join result such that $\tau = \tau' \bowtie \rho$ for some intermediate results τ' and ρ. We define $\overline{\mathcal{S}}(\tau')$ to be the value of $\mathcal{S}$ using the base scores of τ', and substituting 1 for any that are missing. The monotonicity of $\mathcal{S}$ implies that $\mathcal{S}(\tau) \leq \overline{\mathcal{S}}(\tau')$ since each base score of ρ is at most 1. Thus we call $\overline{\mathcal{S}}(\tau')$ the *score bound* of τ', since it is an upper bound on the scores of join results derived from τ'.

The rank join problem can now be stated as follows. We are given relations $R_1, \ldots, R_n$ and a monotonic scoring function $\mathcal{S}$, such that each relation is accessed sequentially in decreasing order of $\overline{\mathcal{S}}$, and the goal is to compute the K results of $R_1 \bowtie \ldots \bowtie R_n$ with the highest score ($1 \leq K \leq |R_1 \bowtie \ldots \bowtie R_n|$). We can efficiently implement the particular access model for several scoring functions that are frequently used in practice, by relying on commonly available access methods such as B-trees. In what follows, we use $I = (R_1, \ldots, R_n, \mathcal{S}, K)$ to denote an instance of the rank join problem.

The previous definition requires that at least K join results exist, which guarantees that it is possible to fulfill a request for the top K results. In addition, note that the solution to an instance of the problem may not be unique, due to the existence of ties in the computed score values. However, the *terminal score*, that is, the score of the K-th result, is uniquely determined for a given instance, and is denoted as $\mathcal{S}^{\text{term}}$.

Given an algorithm A that solves the rank join problem, we use $cost(A, I)$ to denote the cost that A incurs on a specific problem instance I. A commonly used cost metric is based on the idea of *depth*. The depth on an input relation R_i is the number of tuples read sequentially from R_i before returning a solution. We use $depth(A, I, i)$ to denote this depth, and define $sumDepths(A, I)$ as the sum of depths on all inputs. Clearly, *sumDepths* is an interesting cost metric as it indicates the amount of I/O performed by an algorithm.

2 Analysis of Rank Join Algorithms

Several recent studies have explored deterministic algorithms to solve the rank join problem [1,2,3,4,5,7,8,9]. In this section, we review the main theoretical results in the complexity and properties of these algorithms. The review is based on the analysis presented in [9].

We begin by stating two desirable optimality properties for a rank join algorithm. Given a class of algorithms $\mathcal{B}$, a class of problem instances $\mathcal{J}$, and a cost metric *cost*, we say that a rank join algorithm $A \in \mathcal{B}$ is *optimal* if $cost(A, I) \leq cost(A', I)$ for all rank join algorithms $A' \in \mathcal{B}$ and problem instances $I \in \mathcal{J}$. An optimal algorithm may not be feasible in specific settings, which leads us to a relaxed form of optimality known as instance optimality. We say that A is *instance-optimal* if there exist constants c_1 and c_2 such that $cost(A, I) \leq c_1 \cdot cost(A', I) + c_2$ for all $A' \in \mathcal{B}$ and $I \in \mathcal{J}$. The constant c_1 is called the *optimality ratio*.

Algorithm template PBRJ($R_1, \ldots, R_n, \mathcal{S}, K$)
Template parameters: pulling strategy P; bounding scheme B
Input: relations $R_1, \ldots, R_n$; scoring function $\mathcal{S}$; result size K
Output: set of K join results with highest score
Data structures: input buffers $HR_1, \ldots, HR_n$; output buffer O
1. $t \leftarrow \infty$
2. **while** $|O| < K$ **OR** $\min_{\omega \in O} \mathcal{S}(\omega) < t$ **do**
3. $i \leftarrow P.chooseInput()$
4. $\rho_i \leftarrow$ next unseen tuple of R_i
5. $R \leftarrow HR_1 \bowtie \ldots HR_{i-1} \bowtie \{\rho_i\} \bowtie HR_{i+1} \bowtie \ldots \bowtie HR_n$
6. Add each member of R to O, retaining only the top K tuples
7. Add ρ_i to HR_i
8. $t \leftarrow B.updateBound(\rho_i)$
9. **end while**
10. **return** O

Fig. 1. PBRJ Template

Given these two properties, we can ask whether there exist (instance-)optimal algorithms within a specific family $\mathcal{B}$. To effectively perform this analysis for several possible classes, we introduce the *Pull-Bound Rank Join* algorithm template that can express any deterministic rank join algorithm. The PBRJ template, shown in Figure 1, is instantiated by a deterministic *pulling strategy* P and a deterministic *bounding scheme* B. On each loop iteration of PBRJ, the pulling strategy P chooses a relation R_i to read, and the new tuple ρ_i is stored in an input buffer HR_i (typically a hash table). New join results are generated by joining ρ_i with the tuples in the other input buffers HR_j for $j \neq i$. The generated results are pushed to an output buffer O that holds the top K results seen so far. After each tuple is processed, it is given to the bounding scheme B via the method *updateBound*, which returns a new upper bound on the score of unseen join results. The results are returned when the K-th buffered score is at least as large as the bound t provided by the bounding scheme, since this indicates that the buffered results cannot be improved by reading more tuples.

It is straightforward to see that PBRJ is correct if we require that P returns the index of an un-exhausted relation, and that B returns a correct upper bound on the scores of join results that use at least one unread tuple. Furthermore, PBRJ can model any deterministic rank join algorithm by an appropriate choice of P and B, and hence it makes a convenient vehicle for the analysis of rank join algorithms.

The analysis in [9] considered two choices for the bounding scheme B, namely, the *corner bound*, which is used in the HRJN [5] family of rank join algorithms, and the *feasible region bound*, which was introduced in [9]. There were also two choices for P, namely, *round-robin* and *corner-bound adaptive*. The latter is specific to the corner bound and prioritizes access to the inputs based on the information maintained in this specific bounding scheme. The main results of the analysis can be summarized as follows:

- Within the family of PBRJ instantiations with the corner bound, both round-robin and adaptive pulling yield an instance-optimal algorithm. Moreover, the adaptive strategy becomes optimal under the absence of a specific type of score-value ties related to $\mathcal{S}^{\text{term}}$.
- Within the same family, corner-bound adaptive is always no worse than round-robin in terms of the per-input depth metric.
- No PBRJ instantiation that uses the corner-bound is instance optimal within the extended family of all PBRJ instantiations.
- The instantiation of PBRJ with the feasible region bound and the round-robin strategy yields an instance optimal rank join algorithm within the family of all deterministic rank join algorithms.

The last two results indicate that the corner bound may yield an unpredictably high *sumDepths* metric, whereas the feasible region bound enables a property of robustness (again, with respect to *sumDepths*). The basic difference between the two bounding schemes is that only the feasible region bound is *tight*, i.e., it computes a score value that can be actually achieved by the unseen data, which in turn allows PBRJ to stop as early as possible without committing mistakes. However, as shown in [9], robustness in terms of *sumDepths* comes at a price in terms of computational complexity. Essentially, computing a tight bound is provably hard–the corresponding decision problem is NP-Complete. In other words, the rank join problem exhibits an inherent trade-off between computational and I/O efficiency.

3 The Trade-Off between I/O Robustness and Computational Efficiency

The previous theoretical results raise an interesting question: Can we design a rank join algorithm that explores the trade-off between I/O robustness and computational efficiency? In this section, we review some initial developments in this direction, based on the results presented in [4].

We first repeat the formalization of I/O robustness as instance optimality with respect to the *sumDepths* metric (see previous section). We say that a rank join algorithm A is robust, if there exist constants c_1 and c_2 such that $sumDepths(A, I) \leq c_1 \cdot sumDepths(A', I) + c_2$ for any other algorithm A' and rank join instance I. As a starting point in our exploration of robust and efficient rank join algorithms, we can examine the actual performance of PBRJ using the feasible region bound. The experimental study presented in [4] shows that PBRJ performs very badly in terms of total execution time, even though the algorithm does less I/O due to instance optimality. There are two reasons for this overall inefficiency: the costly computation of the feasible region bound, and the "blind" access to inputs from the round-robin pulling strategy. Therefore, one possible direction is to optimize PBRJ along these two dimensions. Indeed, there exists an alternative (yet equivalent) definition of the feasible region bound that performs far fewer computations than the original variant. Moreover, the feasible region bound can be coupled with an adaptive pulling strategy, which is the counterpart of the corner-bound adaptive strategy, that allows PBRJ to prioritize access to its inputs. Overall, these modifications yield an instantiation of PBRJ that remains instance optimal, does fewer pulls than the round-robin strategy, and is more computationally efficient.

The previous improvements can reduce the overhead of the feasible region bound, but they cannot lift the complexity barrier of computing a tight bound. Hence, the next step is to consider alternative bounding schemes that can explore the trade-off between tightness and computational efficiency. The study proposes the *adaptive feasible region bound* that achieves precisely this property. The new bounding scheme follows the same logic as the original feasible region bound, but it performs its computations on quantized base scores. The level of quantization determines directly the space complexity of the feasible region bound, and in effect its time complexity. When the level of quantization is infinitely small, the bound is essentially the same as the feasible region bound. This means that it is provably tight but also potentially costly to compute. A coarser quantization implies lower overhead, but it also means that the bound is no longer tight. An interesting property is that the adaptive bound coincides with the corner bound at the coarsest quantization. By utilizing this hybrid bounding scheme, PBRJ can essentially regulate the overhead of bound computation, and can explore adaptively the space between instance-optimality and computational efficiency.

Experimental results demonstrate that the instantiation of PBRJ with the hybrid bounding scheme and the adaptive strategy really offers the best of both worlds. The algorithm's I/O performance is the same as an instance optimal rank join algorithm for the class of inputs where the tight bound is cheap to compute, and degrades gracefully in other cases. In overall efficiency, the new algorithm outperforms other instantiations of PBRJ that correspond to existing rank join algorithms, thus validating the idea of an adaptive trade-off between I/O robustness and computational complexity.

References

1. Agrawal, P., Widom, J.: Confidence-aware join algorithms. In: International Conference on Data Engineering, pp. 628–639 (2009)
2. Fagin, R.: Combining fuzzy information from multiple systems. J. Comput. Syst. Sci. 58(1), 83–99 (1999)
3. Fagin, R., Lotem, A., Naor, M.: Optimal aggregation algorithms for middleware. J. Comput. Syst. Sci. 66(4), 614–656 (2003)
4. Finger, J., Polyzotis, N.: Robust and efficient algorithms for rank join evaluation. In: Proceedings of SIGMOD (2009)
5. Ilyas, I.F., Aref, W.G., Elmagarmid, A.K.: Supporting top-k join queries in relational databases. International Journal on Very Large Databases (VLDBJ) 13(3), 207–221 (2004)

6. Li, C., Chang, K.C.C., Ilyas, I.F., Song, S.: RankSQL: query algebra and optimization for relational top-k queries. In: Proceedings of ACM SIGMOD, pp. 131–142 (2005)
7. Mamoulis, N., Yiu, M.L., Cheng, K.H., Cheung, D.W.: Efficient top-k aggregation of ranked inputs. ACM Transactions on Database Systems 32(3), 19 (2007)
8. Natsev, A., Chang, Y.C., Smith, J.R., Li, C.S., Vitter, J.S.: Supporting incremental join queries on ranked inputs. In: Proceedings of VLDB, pp. 281–290 (2001)
9. Schnaitter, K., Polyzotis, N.: Evaluating rank joins with optimal cost. In: Proceedings of the 27th Symposium on Principles of Database Systems, pp. 43–52 (2008)
10. Schnaitter, K., Spiegel, J., Polyzotis, N.: Depth estimation for ranking query optimization. In: Proceedings of VLDB, pp. 902–913 (2007)

Proximity Rank Join in Search Computing

Davide Martinenghi and Marco Tagliasacchi

Politecnico di Milano

Abstract. Rank join can be generalized to sets of relations whose objects are equipped with a score and a real-valued feature vector. Such vectors can be used to compare the objects to one another so as to join them based on a notion of "proximity". The problem becomes then that of retrieving combinations of objects that have high scores, whose feature vectors are close to one another and possibly to a given feature vector (the query). Traditional rank join algorithms may read more input than needed when solving proximity rank join. Such weakness can be overcome by designing new algorithms for which, as in classical rank join, bounding scheme (and a tight version thereof) and pulling strategy play a crucial role to efficiently compute the solution.

1 Introduction

Proximity rank join [9] is the problem of finding the best combinations of objects with the highest aggregate score, in which objects are coming from different services, and each object is equipped with both a score and a real-valued feature vector. The score and the feature vector can be compared across different objects in order to establish their respective quality and similarity. Indeed, the feature space may represent, in a broad sense, the "geometry" of the objects and can thus be used in the computation of the overall score of a combination. Contrast with the traditional rank join problem [7], where the overall score of a combination depends only on the scores of the single objects.

The proximity rank join problem can therefore capture many interesting multi-domain scenarios relevant for Search Computing, namely all those in which a vector of reals suitably represents aspects of an object that can be conveniently compared. The fields in which proximity rank join problems naturally emerge include information retrieval, bioinformatics, multimedia databases, and many more. For instance, in the context of Web search, such problems include finding sequences of events which are close both in space and in time, or news about the same event extracted from different news services. The same pattern is also applicable to classical information retrieval (finding two or more similar documents, each extracted from a different data source, most similar to a given query, expressed as a collection of keywords) or multimedia retrieval (finding two or more similar images from different archives most similar to a sample image) or domain-specific search (discovering orthologous genes from a different organisms given a target annotation profile). Therefore, proximity rank join

S. Ceri and M. Brambilla (Eds.): Search Computing II, LNCS 6585, pp. 121–127, 2011.

solves a class of important problems, present in many scientific fields, whose relevance is growing with the increasing availability of services providing accesses to independent data sources.

In all the mentioned cases, proximity plays a crucial role, as it captures the mutual relationships between the objects in the feature space. Moreover, there is an intrinsic notion of distance between objects whose definition varies depending on the scenario at hand. For instance, when comparing objects in which the feature vector expresses their geo-localization, Euclidean distance might be the most relevant criterion (or, alternative, walking distance on a map, if available). Instead, if textual documents or images are being compared, Euclidean distance might be replaced by a cosine similarity measure. Different definitions of distance will, in general, require different specialized algorithms in order to properly address proximity rank join.

Typically, objects on the Web can be retrieved according to their distance, in ascending order, from a given vector (the query). Alternatively, it is also common to retrieve objects sorted by score in descending order. As was pointed out in the previous chapter, exploiting the ordering allows pruning the search for the best combinations early, without having to scan all the objects. Although existing rank join algorithms may address proximity rank join, they are doomed to behave sub-optimally, as they do not leverage the geometry of the problem.

The main contribution of this chapter is to describe *proximity rank join* in the context of Search Computing. The problem will be presented as the search for the best combinations of objects coming from different services, where each object is equipped with both a score and a real feature vector. The aggregation function assigning a score to a combination depends on the individual scores, on the proximity of the individual vectors to a given vector (called query) and on their mutual proximity, according to some notion of distance in the feature space.

2 Problem Definition

We consider a set of relations (services) $R_1, \ldots, R_n$ where each tuple $\tau_i \in R_i$ comprises, besides named attributes, a real-valued feature vector $\mathbf{x}(\tau_i) \in \mathbb{R}^d$, and a score $\sigma(\tau_i) \in \mathbb{R}$.

The two most common access kinds that arise in practice are *distance-based access*, where the relations are accessed sequentially in increasing order of distance from a given feature vector, and *score-based access*, where the order is decreasing in the score of the objects.

Let $\tau_i^{(r_i)} = R_i[r_i]$ indicate the r_i-th tuple extracted from R_i according to the available access kind, and $P_i \subseteq R_i$ the ordered relation containing the tuples already extracted from R_i. Also, let $\tau = \tau_1 \times \cdots \times \tau_n \in R_1 \times \ldots \times R_n$ denote an element of the cross-product of the n relations, hereafter called *combination*.

Proximity rank join searches consist of a constant vector $\mathbf{q} \in \mathbb{R}^d$ called *query* together with a metric distance $\delta(\mathbf{x}(\tau), \mathbf{q})$ between vectors $\mathbf{x}(\tau)$ and $\mathbf{q}$, and a monotonic aggregation function f defining the *aggregate score* $\mathcal{S}(\tau)$ of a combination τ as

$$\mathcal{S}(\tau) = f\left(\mathcal{S}(\tau_1), , \ldots, \mathcal{S}(\tau_n)\right) \quad (1)$$

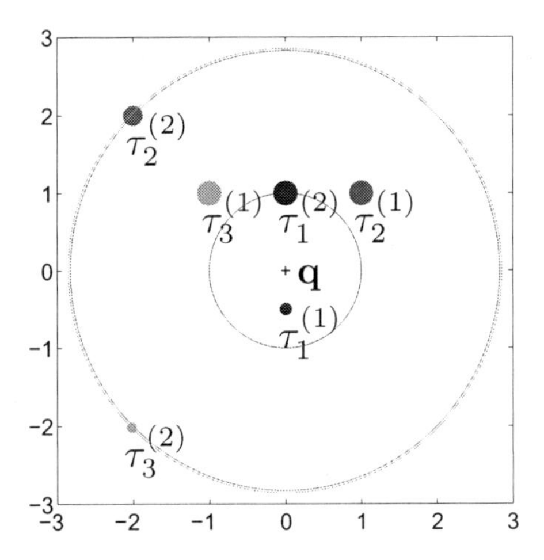

Fig. 1. Location of tuples with feature vectors in $\mathbb{R}^2$ represented as discs whose radius is proportional to the score

where $\mathcal{S}(\tau_i) = g_i\left(\sigma(\tau_i), \delta(\mathbf{x}(\tau_i), \mathbf{q}), \delta(\mathbf{x}(\tau_i), \boldsymbol{\mu}(\tau))\right)$, and $\boldsymbol{\mu}(\tau) \in \mathbb{R}^d$ denotes the *centroid* of a combination. The functions g_i must behave in such a way that the *proximity weighted score* $\mathcal{S}(\tau_i)$ increases with the score and decreases with the distance from the query vector $\mathbf{q}$ and from the combination centroid $\boldsymbol{\mu}(\tau)$. In top combinations, then, the constituent tuples have high scores, are close to the query vector, and are close to each other.

Example 1. A possible example of aggregation function, in which Euclidean distance is used between objects in the feature space, is the following:

$$\mathcal{S}(\tau)=\sum_{i=1}^{n} w_s \ln(\sigma(\tau_i)) - w_q\|\mathbf{x}(\tau_i)-\mathbf{q}\|^2 - w_\mu\|\mathbf{x}(\tau_i)-\boldsymbol{\mu}(\tau)\|^2 \qquad (2)$$

Weights (w_s, w_q and w_μ) in the different components of the function allow expressing different user preferences. Figure 2 shows an example with three relations and the location of their tuples in $\mathbb{R}^2$ represented as discs whose radius is proportional to the score (blue for R_1, red for R_2, green for R_3). The three circles in the figure have a radius equal to the distance of the last accessed tuple from each relation. There are $2 \cdot 2 \cdot 2 = 8$ combinations in total. The top-1 combination using (2) with $w_s = w_q = w_\mu = 1$ turns out to be $\tau_1^{(2)} \times \tau_2^{(1)} \times \tau_3^{(1)}$.

Proximity rank join is the problem of determining an ordered relation O containing the top K combinations from $R_1 \times \ldots \times R_n$ ordered by $\mathcal{S}$ (defined as in (1)), where the elements of $R_1, \ldots, R_n$ can be retrieved either by distance-based access or by score-based access.

The proximity rank join problem can be tackled by adapting the *Pull/Bound Rank Join* (*PBRJ*) template originally introduced for rank join in [11]. *PBRJ* at each step decides which relation to access next (the deciding component is called *pulling strategy*). After each step, if at least K combinations are formed, *PBRJ* computes (according to a so-called *bounding scheme*) an upper bound on the aggregate score of the unseen combinations. If such bound is exceeded by the score of the current K-th best combination found, the algorithm terminates and reports the K best combinations found as the actual top-K.

The adaptation of *PBRJ* to proximity regards both the computation of the upper bound and the pulling strategy, as discussed in the next section.

The assessment of limitations and potential of proximity rank join algorithms will refer, as customary in top-k query answering, to the notion of instance-optimality [3] to characterize I/O efficiency. Roughly, an algorithm is *instance-optimal* if its I/O cost for any input is within a constant factor of the cost of any other algorithm on the same input.

3 Proximity Rank Join Algorithms

We now discuss how bounding schemes and pulling strategies conforming to the *PBRJ* scheme can be used to address proximity rank join.

3.1 Bounding Scheme

Under distance-based access, an upper bound t_c can be computed by keeping track of the distance from the query $\mathbf{q}$ of the first and last accessed tuple from each relation, i.e., $\delta(\mathbf{x}(R_i[1]), \mathbf{q})$ and $\delta(\mathbf{x}(R_i[p_i]), \mathbf{q})$, respectively:

$$t_c = \max\{t_1, \ldots, t_n\}, \text{ with } t_i = f\left(\bar{\mathcal{S}}_1, \ldots, \underline{\mathcal{S}}_i, \ldots, \bar{\mathcal{S}}_n\right) \tag{3}$$

where $\bar{\mathcal{S}}_j = g_j\left(\sigma_j^{\max}, \delta(\mathbf{x}(R_j[1]), \mathbf{q}), 0\right)$ is an upper bound on the proximity weighted score that can be attained by a tuple $\tau_j \in R_j$ ($\sigma_j^{\max}$ is the maximum score possible for R_j), and, similarly, $\underline{\mathcal{S}}_i = g_i\left(\sigma_i^{\max}, \delta(\mathbf{x}(R_i[p_i]), \mathbf{q}), 0\right)$ by an unseen tuple $\tau_i \in R_i - P_i$. The case of score-based access is handled similarly.

The bound in (3) is a *corner bound* [11]. We say that a bounding scheme is *tight* [11] if the upper bound can be the actual score of some combination using at least an unseen tuple for the problem at hand (or for any problem coinciding with it on the seen tuples).

Bound (3) is not tight (not even with only two relations), since a possible combination whose aggregate score is equal to the bound might not exist. The lack of tightness of the corner bound t_c entails non-instance-optimality.

The problem of finding a tight bound can be cast as a maximization problem subject to constraints due to the access mode, i.e., finding the feasible locations of the unseen tuples that maximize the aggregate score. Solving such maximization problem might be difficult, depending on the aggregation function and the distance in use. Cases with Euclidean distance, such as the aggregation function in (2), allow efficient solutions of the problem. In particular, thanks to the geometry of the problem, finding a tight bound for (2) requires solving a convex quadratic program with linear constraints, which can be done using off-the-shelf solvers. Cases with cosine similarity do not seem to admit efficient solutions, unless approximations are allowed.

Further efficiency gains can be achieved by considering that the entire set of (partially formed) combinations does not need to be fully explored each time the bound needs to be found, as some of the partial combinations can be tagged as dominated, in the sense that their score will never reach the upper bound. Dominance can be checked by solving a feasibility linear program. Although checking dominance is costly, a good heuristics consists of periodically checking it after a fixed number of accesses so as to save time during the calculation of the tight bound.

3.2 Pulling Strategy

A pulling strategy identifies the relation R_i from which the next tuple is retrieved. The simplest pulling strategy accesses the inputs in a *round-robin* fashion (e.g., in the order $R_1, \ldots, R_n$). Tightness of the bounding scheme and a round-robin strategy are sufficient to guarantee instance optimality.

Alternatively, one can adopt more refined strategies. A *potential adaptive* given in [4] for 2 relations can be generalized as follows. Let pot_i denote the *potential* of relation R_i, defined as the upper bound on the aggregate score of combinations that can be formed with unseen tuples from R_i. The potential adaptive strategy requires then to access relation R_i such that pot_i is maximal (among $pot_1, \ldots, pot_n$), breaking ties in favor of the relation with the least depth p_i, then the relation with the least index i.

With a tight bound, the efficiency of a potential adaptive strategy is always at least as good as that of a round robin strategy, and thus instance optimality is also achieved.

4 Related Work

Proximity rank join is a significant extension of rank join, a class of problems that has lately been receiving a great deal of attention [6,12,11,4,7]. A comprehensive survey on the subject is found in [7].

The idea of using an upper bound comes from the threshold-based stopping condition of the well-known *threshold algorithm* [3] (TA). TA addresses rank aggregation, which is the problem of combining several ranked lists of objects to produce a single consensus ranking.

The *PBRJ* template encompasses the well-known *HRJN* and *HRJN** operators [6]. As was discussed, some of the results on instance optimality in [11] do not carry over to proximity rank join, due to the geometry of the problem (absent in rank join).

In [4], the authors propose an efficient way for computing an approximation of the tight bound, in order to find a good trade off between I/O cost and CPU cost.

We have considered a scenario in which the objects' feature vectors are deterministic. Others have addressed related problems with uncertainty [1].

Weighted proximity join is introduced in [14], however only for 1-dimensional spaces. Similarity join is promoted to a first-class operator in [13].

The study of join predicates that depend on the spatial proximity among the objects has been thoroughly investigated in the past literature [5,2,10,8]. Nevertheless, in such works, the authors assume that input relations are indexed by structures of the R-tree family. This contrasts the presented setting, where indexes are unavailable.

5 Conclusion

Proximity rank-join is the problem of finding the best combinations of heterogeneous objects that are close to a given target object (the query) and to each other. We presented a general formulation of the problem and discussed instance optimality properties and other desiderata.

Future lines of investigation involve instances of the problem with richer access modes, e.g., random access, or even mixtures of both score- and distance-based access. Further specific distances, useful for practical cases, need to be studied or better understood, as is the case with cosine similarity, for which no efficient and exact solution is known.

References

1. Beskales, G., Soliman, M.A., Ilyas, I.F.: Efficient search for the top-k probable nearest neighbors in uncertain databases. PVLDB 1(1), 326–339 (2008)
2. Corral, A., Manolopoulos, Y., Theodoridis, Y., Vassilakopoulos, M.: Algorithms for processing k-closest-pair queries in spatial databases. Data Knowl. Eng. 49(1), 67–104 (2004)
3. Fagin, R., Lotem, A., Naor, M.: Optimal aggregation algorithms for middleware. J. Comput. Syst. Sci. 66(4), 614–656 (2003)
4. Finger, J., Polyzotis, N.: Robust and efficient algorithms for rank join evaluation. In: SIGMOD Conference, pp. 415–428 (2009)
5. Hjaltason, G.R., Samet, H.: Incremental distance join algorithms for spatial databases. In: SIGMOD Conference, pp. 237–248 (1998)
6. Ilyas, I.F., Aref, W.G., Elmagarmid, A.K.: Supporting top-k join queries in relational databases. VLDB J. 13(3), 207–221 (2004)
7. Ilyas, I.F., Beskales, G., Soliman, M.A.: A survey of top- query processing techniques in relational database systems. ACM Comput. Surv. 40(4) (2008)
8. Mamoulis, N., Theodoridis, Y., Papadias, D.: Spatial joins: Algorithms, cost models and optimization techniques. In: Spatial Databases, pp. 155–184 (2005)
9. Martinenghi, D., Tagliasacchi, M.: Proximity rank join. PVLDB 3(1), 352–363 (2010)
10. Papadopoulos, A.N., Nanopoulos, A., Manolopoulos, Y.: Processing distance join queries with constraints. Comput. J. 49(3), 281–296 (2006)
11. Schnaitter, K., Polyzotis, N.: Evaluating rank joins with optimal cost. In: PODS, pp. 43–52 (2008)

12. Schnaitter, K., Spiegel, J., Polyzotis, N.: Depth estimation for ranking query optimization. In: VLDB, pp. 902–913 (2007)
13. Silva, Y.N., Aref, W.G., Ali, M.H.: The similarity join database operator. In: ICDE (2010)
14. Thonangi, R., He, H., Doan, A., Wang, H., Yang, J.: Weighted proximity best-joins for information retrieval. In: ICDE, pp. 234–245 (2009)

Uncertainty in Rank Join

Ihab F. Ilyas

University of Waterloo

Abstract. At the core of the query processing engine of a search computing system are operators that retrieve, filter, join and aggregate results from these Web services. The main goal is to deliver relevant and multi-domain answers to user queries. In these scenarios, users usually expect a ranked list of relevant answers in contrast to the full answer set. Hence, ranking query results in the presence of uncertainty is a fundamental query processing challenge in search computing environments.

Rank-join is a basic relational operator that reports the top-k join results as soon as possible, avoiding the expensive materialize-then-sort approach. Due to the early-out and pipelined nature of rank-join, it acts as one of the major building blocks in compiling execution plans for multi-domain queries (also knows as liquid queries). In this chapter, we discuss the implication of data uncertainty on the semantics and implementation of rank-join operators, and we survey some of the recent techniques to address these challenges.

1 Introduction

In the search computing framework, uncertainty appears at multiple levels. For example, on the data extraction level, lack of reliability of data sources, conflicting information among different sources, partial matches among corresponding schemas, privacy and annonamization constraints, and presentation formats are all common sources of data uncertainty. Similarly, we expect to experience uncertainty at the level of source availability under the Web services approach adopted by search computing. Other types of uncertainty are at query formulation phase; since search computing allows for non-exact match of the surface query and the matched Web services guided, for example, by sources availability, due to run-time constraints, and more importantly because of the exploration and interactive nature of query interfaces. Considering these sources of uncertainty, deciding on the semantics of querying against the underlying uncertain data becomes crucial in deciding on the soundness and the goodness of query results, as well as in designing efficient query processing algorithms to implement these semantics.

Due to Web interactive and dynamic nature, Web scale search computing triggers the need to rank large volumes of constantly-changing search results with respect to some preference measures. Ranking acts as an effective and intuitive data exploration tool in this scenario. Traditionally, ranking queries compute the top-k query results based on a given scoring function. A join query augmented with ranking specifications (i.e., a scoring function and a parameter k), usually referred to as "rank join query" [1,2,3], reports the top-k join results based on the computed scores. The central idea of rank join is to allow for early query termination by making use of sorted inputs and scoring function monotonicity to upper bound the scores of non-materialized join results.

S. Ceri and M. Brambilla (Eds.): Search Computing II, LNCS 6585, pp. 128–134, 2011.

Example 1. Consider a Web user planning to spend a vacation in New Zealand. The user would like to find a hotel with a good reputation and reasonable prices. For credibility, The user has decided to obtain hotel pricing and rating information from two independent sources: `www.vianet.travel`, a source that provides pricing information and online booking, and `www.tvtrip.com`, a source that provides hotel ratings from travelers' reviews. The search scenario is thus (1) **extracting** hotel records from both sources; (2) **matching** records based on hotel name; and (3) **ranking** matched results based on some function of *price* and *rating*. We refer to such process as *extract-match-rank*.

Manual processing of the *extract-match-rank* task requires navigating through many pages, while memorizing and ranking interesting matches, which is clearly infeasible. The problem is far more complicated if matching across more sources is involved. An automated *extract-match-rank* system can probably formulate the task in the following SQL-like rank join query, which prefers hotels with low prices and high ratings, and reports only the top-k hotels.

```
SELECT *
FROM vianet, tvtrip
WHERE vianet.HotelName ~ tvtrip.HotelName
ORDER BY 500-vianet.Price+ 100* tvtrip.Rating
LIMIT k
```

Different modules of the search computing framework focus on the many challenges raised in the aforementioned example:

- *Data Extraction.* Structured records need to be extracted from such sources to apply rank join techniques. Search computing applies various techniques for the discovery, matching and invocation of the available Web services published for the involved services to extract timely structured data records.
- *Interleaving Extraction with Processing.* Users do not tolerate long waiting times, where expensive extraction must complete before query processing starts. Moreover, exhaustive extraction does not leverage early-out nature of ranking. We thus need to interleave asynchronous data retrieval and extraction with query processing, and avoid unnecessary extraction operations based on ranking requirements. The run-time platform in search computing tries to address these problems by, for example, exploiting the incremental invocation capabilities of the underlying Web services to extract only the needed information to report the top-k results.

One remaining important challenge is to handling uncertainty in the extracted data records, especially in the scores involved in the provided raking function. The scoring models adopted by current rank join techniques assume that all scores are exact, which yields a unique top-k answer (score ties are resolved using a tie-breaking criterion). In situations where the underlying data do not conform to these assumptions, the semantics and processing techniques of current rank join methods become inapplicable.

Possible approaches to rank-join multiple inputs with uncertain scores include the following:

(1) Fall back to exact scores. For example, we use expected score values. While suitable in some settings, this approach can be quite unreliable due to the inability to reflect

variance of score ranges. Exact score representation of multiple ranges may coincide, or become very close to each other, even though ranges are considerably different. For example, assume 3 records, t_1, t_2, and t_3 with uniform score ranges $[0, 100]$, $[40, 60]$, and $[30, 70]$, respectively. All expected scores are equal to 50, and hence all orderings are equally likely. However, based on how score distributions overlap, the likelihood of different orderings can be computed as nested integrals [4], which results in different probabilities: $\Pr(\langle t_1, t_2, t_3\rangle) = .25$, $\Pr(\langle t_1, t_3, t_2\rangle) = .2$, $\Pr(\langle t_2, t_1, t_3\rangle) = .05$, $\Pr(\langle t_2, t_3, t_1\rangle) = .2$, $\Pr(\langle t_3, t_1, t_2\rangle) = .05$, and $\Pr(\langle t_3, t_2, t_1\rangle) = .25$. That is, some orderings are more likely than the others, even though score ranges are uniform with equal expectations.

(2) Compute all, then rank. The problem of ranking with uncertain scores coming from a single input has been addressed in [4]. However, when scores are computed online (e.g., by aggregating scores of joined Web sources), applying the techniques in [4] requires computing all uncertain scores before ranking. Moreover, [4] assumes the independence of the random variables representing tuples' scores, which does not apply to join results' scores that are intrinsically correlated.

In this chapter, we focus on the effect of data uncertainty on the semantics of ranking queries, in particular, the rank join queries.

2 Uncertain Rank Join

We mainly summarize the latest contributions in designing uncertain rank join algorithms in the context of uncertain mashups [5]. We briefly summarize the data model used and an overall description of an uncertain rank join algorithm, URANKJOIN proposed in [5] .

2.1 Uncertain Scoring Model

Without loss of generality, the score of tuple t_i as a random variable with possible values in the interval $[lo_i, up_i] \subseteq [0, 1]$. The higher score values are preferred. Single-valued scores are represented as score intervals with equal bounds. The score random variable of tuple t_i has a PDF P_i encoding the likelihood of possible scores of t_i. The random variables representing the scores of base tuples are assumed to be independent. Uncertain score is the interval-based score representation described above.

The authors in [6] assume a similar model, where *generating functions* are used to formulate and compute ranking queries efficiently on continuous score distributions. The work in [5] mainly addresses consequences of assuming such model in the case of joins.

Definition 1. [Score Dominance] *We say that tuple t_i dominates another tuple t_j, denoted $(t_i \succ t_j)$, iff $lo_i \geq up_j$.* □

When $t_i \succ t_j$, t_i is ranked on top of t_j. However, when $t_i \not\succ t_j$ and $t_j \not\succ t_i$, the relative order of t_i and t_j needs to be defined using additional means and semantics. Furthermore, for two different tuples t_i and t_j with equal single valued scores (i.e., $lo_i = up_i = lo_j = up_j$), a tie-breaker $\tau(t_i, t_j)$ that gives a deterministic relative order is assumed. That is, $\tau(t_i, t_j)$ decides whether $t_i \succ t_j$ or $t_j \succ t_i$. One example for such tie-breaker is ordering based on unique tuples IDs.

It is straightforward to see that score dominance is non-reflexive (i.e., $t_i \not\succ t_i$), transitive (i.e., $((t_i \succ t_j), (t_j \succ t_k)) \Rightarrow (t_i \succ t_k)$), and asymmetric (i.e., $(t_i \succ t_j) \Rightarrow (t_j \not\succ t_i)$). It follows that a partial order holds on tuples with uncertain scores [4].

Uncertain scores induce a space of tuple orderings. Specifically, given a relation $R = \{t_1, \ldots, t_n\}$, let ω be an ordering of R tuples, where $\omega[t_i]$ is the rank of t_i in ω. ω is a valid ordering of R iff $(\omega[t_i] < \omega[t_j]) \Rightarrow (t_i \succ t_j)$ or $(t_i \not\succ t_j \wedge t_j \not\succ t_i)$. The valid orderings are equivalent to possible linearizations (topological sorts) of a partial order. The number of possible orderings is exponentially large in general. The orderings space is generated by a probabilistic process that draws, for each tuple t_i, a score $s_i \in [lo_i, up_i]$ based on the density P_i. Ranking the drawn scores gives an ordering whose probability is the joint probability of drawn scores. That is, the probability of an ordering $\omega = \langle t_1, t_2, \ldots t_n \rangle$ is computed as follows:

$$\Pr(\omega) = \int_{lo_1}^{up_1} \int_{lo_2}^{x_1} \ldots \int_{lo_n}^{x_{n-1}} \mathcal{P}(x_1, x_2, \ldots, x_n)\, dx_n \ldots dx_1 \tag{1}$$

where $\mathcal{P}(.)$ is the PDF of the joint distribution of $P_1, \ldots, P_n$. When the score random variables are independent, we have $\mathcal{P}(x_1, \ldots, x_n) = \Pi_{i=1}^{n} P_i(x_i)$.

Under this uncertain scoring model, multiple probabilistic ranking semantics have been proposed in the literature. Example semantics include the following:

- *Expected Scores*: uncertain score intervals are translated into a single value computed as the expected score value. Ranking is then carried out deterministically (assuming a given tie breaker). For example in Figure 1, based on expected scores, $\omega^* = \langle t_5, t_1, t_2, t_3, t_4, t_6 \rangle$, assuming that the tie between t_1 and t_2 is resolved in favor of t_1.
- *Expected Ranks* [7]: The expected rank of a tuple is computed across all possible orderings (worlds). Tuples are then ordered w.r.t their expected ranks resolving ties deterministically. In Figure 1, based on expected ranks, $\omega^* = \langle t_5, t_2, t_1, t_3, t_4, t_6 \rangle$.
- *Most Probable Ordering* [4]: the ω^* is defined as $argmax_{\omega \in \Omega} \Pr(\omega)$, where $\Pr(\omega)$ is computed using Equ 1. For example in Figure 1, ω^* is the ordering $\omega_1 = \langle t_5, t_1, t_2, t_3, t_4, t_6 \rangle$.

Example 2. Figure 1 shows a relation R with uniform uncertain scores. The relation R has 7 possible orderings $\{\omega_1, \ldots, \omega_7\}$. The probabilities of ω_i's are computed by evaluating Equ 1 using Monte-Carlo integration, while assuming independence of score densities. E[t] and ER[t], shown in Figure 1, are the expected score value and the expected rank of the tuple [7], respectively.

2.2 URankJoin

We assume a monotone user-defined scoring function $\mathcal{F}$ to be the source of the scores of join results (i.e., $\mathcal{F}(x_1, \ldots, x_n) \geq \mathcal{F}(\acute{x}_1, \ldots, \acute{x}_n)$ whenever $x_i \geq \acute{x}_i$ for every i). Typical scoring functions, such as summation, multiplication, min, max, and average, are monotone functions.

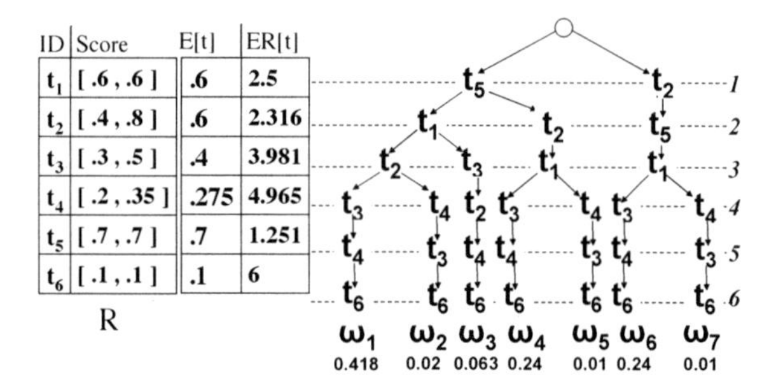

ID	Score	E[t]	ER[t]
t_1	[.6 , .6]	.6	2.5
t_2	[.4 , .8]	.6	2.316
t_3	[.3 , .5]	.4	3.981
t_4	[.2 , .35]	.275	4.965
t_5	[.7 , .7]	.7	1.251
t_6	[.1 , .1]	.1	6

Fig. 1. Orderings space for tuples with uniform scores

URANKJOIN [5] returns a total order, ω^*, of the set tuples that have less than k other dominating tuples (we call this set $\mathcal{J}_k$). The total order is computed according to one of the possible ordering semantics (e.g., one the semantics mentioned in Section 2.1). Note that computing $\mathcal{J}_k$ does not require knowledge of the density functions P_i's of base or join tuples, since $\mathcal{J}_k$ is based on score dominance only. However, computing ω^* requires knowledge of P_i's.

Example 3. In Figure 2, URANKJOIN($\{R, S\}$ with a scoring function $\mathcal{F} = (R.a_1 + S.a_1)/2, 3)$, where the join condition is equality of attribute 'jk', returns a total order of join tuples in $\mathcal{J}_3 = \{(r_1, s_2), (r_3, s_1), (r_2, s_2)\}$, since all join tuples in $\mathcal{J}_3$ are dominated by less than 3 join tuples, and all join tuples not in $\mathcal{J}_3$ (only (r_4, s_3) in this example) are dominated by at least 3 tuples. Based on the monotonicity of $\mathcal{F}$, the lo and up scores of join tuples are given by applying $\mathcal{F}$ to the lo and up scores of the corresponding base tuples. For example, the score of (r_1, s_2) is given by $[\mathcal{F}(.7, .3), \mathcal{F}(.8, .4)] = [\frac{.7+.3}{2}, \frac{.8+.4}{2}] = [.5, .6]$.

A common interface to most rank join algorithms, is to assume input relations sorted on per-relation scores, while output (join) relation is generated incrementally in join scores order. URANKJOIN uses a generic rank join algorithm (such as HRJN [1] as a building block to compute $\mathcal{J}_k$ incrementally.

URANKJOIN assumes two sorted inputs (e.g., indexes) L^i_{lo} and L^i_{up}, for each input relation R_i, giving relation tuples ordered on lo and up scores, respectively. By

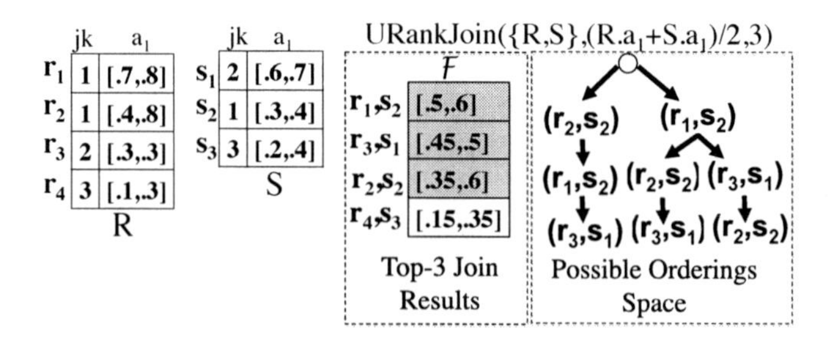

Fig. 2. Example

processing the lo and up inputs simultaneously, URANKJOIN incrementally computes $\mathcal{J}_k$. This is done by using two instances of HRJN, denoted HRJN$_{lo}$ and HRJN$_{up}$ where HRJN$_{lo}$ rank-joins tuples on their overall lo scores to find exactly k join results, while HRJN$_{up}$ rank-joins tuples on their overall up scores to find all join results with up scores above the kth largest score reported by HRJN$_{lo}$. The execution of HRJN$_{lo}$ and HRJN$_{up}$ is interleaved. Tuples in $\mathcal{J}_k$ are reported in up scores order to allow for incremental ranking.

A URANKJOIN operator is a logical operator that accepts two inputs each has two sorted access paths, corresponding to the lo and up score orders of the two input relations. The operator produces two output tuple streams corresponding to sorted join results based on lo and up scores.

Example 4. Figure 3 gives an example logical URANKJOIN query plan. The shown plan rank-joins three relations R, S, and T with uncertain scores x, y, and z, respectively. The bottom URANKJOIN operator uses indexes on the lo and up scores in Relations R and S as its input access paths, while the top URANKJOIN operator uses indexes on Relation T and the output of the bottom URANKJOIN operator as its input access paths. The ULIMIT operator consumes both lo and up inputs from the top URANKJOIN operator.

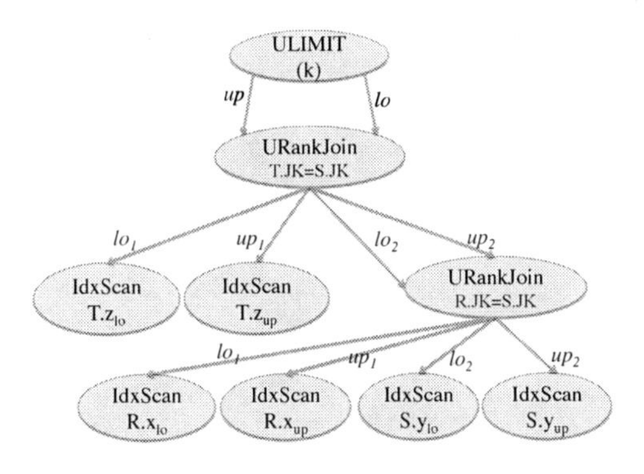

Fig. 3. A logical URANKJOIN query plan

3 Conclusion

Uncertainty is a distinguishing characteristic in common search computing scenarios. In this chapter, we briefly discussed the challenges of ranking query results assembled by joining multiple sources in the presence of score uncertainty. We surveyed some of the proposed probabilistic ranking semantics and briefly described the state-of-the art uncertain rank join algorithm URANKJOIN.

References

1. Ilyas, I.F., Aref, W.G., Elmagarmid, A.K.: Supporting top-k join queries in relational databases. VLDB Journal 13(3), 207–221 (2004)
2. Natsev, A., Chang, Y.-C., Smith, J.R., Li, C.-S., Vitter, J.S.: Supporting incremental join queries on ranked inputs. In: Proceedings of the 27th International Conference on Very Large Data Bases, pp. 281–290 (2001)
3. Schnaitter, K., Polyzotis, N.: Evaluating rank joins with optimal cost. In: PODS, pp. 43–52 (2008)
4. Soliman, M.A., Ilyas, I.F.: Ranking with uncertain scores. In: ICDE, pp. 317–328 (2009)
5. Soliman, M.A., Ilyas, I.F., Saleeb, M.: Building ranked mashups of unstructured sources with uncertain information. PVLDB 3(1), 826–837 (2010)
6. Li, J., Deshpande, A.: Ranking continuous probabilistic datasets. PVLDB 3(1), 638–649 (2010)
7. Cormode, G., Li, F., Yi, K.: Semantics of ranking queries for probabilistic data and expected ranks. In: ICDE, pp. 305–316 (2009)

Trends in Rank Join

Ihab Ilyas[1], Davide Martinenghi[2], Neoklis Polyzotis[3], and Marco Tagliasacchi[2]

[1] University of Waterloo
[2] Politecnico di Milano
[3] University of California - Santa Cruz

Abstract. This chapter reports the main findings of a panel that was moderated by Davide Martinenghi in which Ihab Ilyas, Neoklis Polyzotis, and Marco Tagliasacchi shared their thoughts with the attendees of the Second SeCo Workshop regarding current and future issues related to what was presented during the session on rank join. The topics touched upon during the discussion regarded relevance for Search Computing, pertinence of optimization, multi-way joins, approximate answers, and uncertainty.

1 Relevance of Rank Join for Search Computing

Rank join [3] is recognized as a well-established optimization problem that is relevant for top-k query scenarios. As such, its importance in Search Computing is apparent [4]. Its relevance is even higher when the objects returned by data sources on the Web are equipped with scores, although rank join might even find application when scores are opaque.

2 To Optimize or Not to Optimize?

A nowadays common opinion is that there are cases and contexts in which optimization is unnecessary and not worth the effort, so one should not blindly start spending energies and resources on uncalled-for optimization – to this end, Knuth's famous quote about early optimization immediately comes to mind[1]. Yet, rank join is certainly one of those problems for which optimization still makes a lot of sense. It is often worthwhile to spend a little extra CPU time to save time-consuming accesses to remote data sources whose cost is usually orders of magnitude higher than that used for optimization. Several claims providing evidence in this direction were contributed in the previous chapters as well as in the literature on rank join [2,1,5].

[1] "Premature optimization is the root of all evil", in Knuth, Donald. Structured Programming with go to Statements, ACM Journal Computing Surveys, Vol 6, No. 4, Dec. 1974. p.268.

S. Ceri and M. Brambilla (Eds.): Search Computing II, LNCS 6585, pp. 135–137, 2011.

3 Multi-domain Queries

Handling complex queries that aggregate results coming from several relations, possibly more than two, is one of the main objectives of Search Computing. In this respect, the construction of an n-ary rank join operator that coordinates the accesses to each single data source and combines the results seems the most promising approach to address this issue.

4 Exact vs. Approximate Answers

In some application scenarios, finding the exact top-k results for a query may be prohibitively expensive or even impossible without exploring the entire search space. In such cases, a softened top-k semantics seems more appropriate. Such alternative semantics, sometimes referred to as "good-k" [6], usually provides less guarantees (the good-k results are only guaranteed to be in some top-k' set, with k' greater than k) but requires a smaller amount of computation to obtain some results. The question arises then whether good-k is enough for practical purposes. All things considered, a good-k semantics seems to be a satisfactory approximation of top-k when computing the exact top-k answers is too expensive. Moreover, determining the exact top-k results is often not particularly meaningful, as both the scores and the aggregation function are imprecise. In addition, it is usually possible to show how far approximate answers are from the actual top-k answers and to present this indication to the user who issued the query.

5 Uncertainty in Rank Join

The role of uncertainty [7] in Search Computing as regards rank join is a significant issue that will have an impact on how research on this topic should be continued. A promising line of research is to consider uncertainty in the aggregation function, which reflects the uncertainty the user him/herself has in the model (s)he wants to use to pose a top-k query. Uncertainty might even be brought to an extreme by dismissing the aggregation function completely. In such a case, research on skylines may be the answer. Yet, it can be debated whether the search space reduction power of skylines may be of help in crucial, market-oriented applications. Indeed, it can be observed that, in such cases, all objects (competitors) naturally place themselves on the skyline in order to avoid being excluded by the market itself.

References

1. Finger, J., Polyzotis, N.: Robust and efficient algorithms for rank join evaluation. In: SIGMOD Conference, pp. 415–428 (2009)
2. Ilyas, I.F., Aref, W.G., Elmagarmid, A.K.: Supporting top-k join queries in relational databases. VLDB, 754–765 (2004)

3. Ilyas, I.F., Beskales, G., Soliman, M.A.: A survey of top- query processing techniques in relational database systems. ACM Comput. Surv. 40(4) (2008)
4. Ilyas, I.F., Martinenghi, D., Tagliasacchi, M.: Rank-join algorithms for search computing. In: SeCO Workshop, pp. 211–224 (2009)
5. Martinenghi, D., Tagliasacchi, M.: Proximity rank join. PVLDB 3(1), 352–363 (2010)
6. Martinenghi, D., Tagliasacchi, M.: Top-k pipe join. In: ICDE Workshops, pp. 16–19 (2010)
7. Soliman, M.A., Ilyas, I.F., Ben-David, S.: Supporting ranking queries on uncertain and incomplete data. VLDB J. 19(4), 477–501 (2010)

Part 5
Query Processing

While rank-join theory fuels search computing with methods and algorithms, such fuel requires engines to be capable of applying methods and algorithms and transforming them into efficient computations over Web sources. Efficient query processing is foundational in any data-driven computation, and search computing makes no exception. This part discusses the role of query processing in search computing as a collection of abstractions, models, and techniques for the analysis and execution of multi-domain queries on the Web, together with the functionalities, the internal structure, and the planned improvements to the existing prototype.

The first chapter overviews the search computing query processor. At the highest level of abstraction, one language fits for all, and this is SeCoQL, an SQL variant chosen as the most compact and readable conjunctive formulation for both experts and developers, easily generated by the UI modules and easily parsed by the underlying modules. Queries are then expressed at the logical level in the form of acyclic invocation workflows after a compile-time analysis that decides a cost-driven scheduling of service invocations. At the physical level, queries are then translated into an executable specification that distinguishes between the data flow and the control flow, provides support for parallelism, accounts for stateless and stateful computation tasks, and supports backwards and forward control.

The second chapter presents a study on how run-time adaptivity can be achieved in the context of search computing. Proposals for adaptive join processing can be classified as *plan preserving*, where adaptation takes place over an essentially stable representation of a query, or *plan changing*, where adaptation changes the query plan, and thus adaptation involves stopping the execution, re-optimizing the plan, and resuming execution. Plan changing proposals must take account of the (partial) work done at the point when re-optimization takes place. After a general overview of the available methods, several options for both plan preserving and plan changing adaptation in search computing are considered.

Efficient Computation of Search Computing Queries

Daniele Braga, Michael Grossniklaus,
Francesco Corcoglioniti, and Salvatore Vadacca

Dipartimento di Elettronica e Informazione, Politecnico di Milano
P.za L. Da Vinci, I-20133 Milano, Italy
{braga,grossniklaus,corcoglioniti,vadacca}@elet.polimi.it

Abstract. This chapter gives a high-level overview of how query processing is carried out in SeCo. At the highest level of abstraction, queries are expressed in a conjunctive declarative query language over service interfaces, named SeCoQL, chosen to be a compact and readable formulation to serve both experts users and system developers. Queries are then expressed at a *logical* level in the form of acyclic invocation workflows, after a compile-time analysis that decides a cost-driven scheduling of service invocations. At a lower, *physical* level queries are then translated into executable specifications that distinguish between the data flow and the control flow, support parallelism, account for stateless and stateful computation tasks, and support backward and forward control. The query engine is implemented as an interpreter of these physical plans. A workbench and testing environment is also available in the form of a tool, to monitor the processing of complex queries by inspecting all phases of their analysis and execution, at all levels of abstraction.

1 Introduction

The query processor deals with the problem of scheduling service calls, taking into account their invocation constraints, pursuing some optimization objectives at compile-time, and then dealing at run-time with their actual possibly unexpected behavior during the execution.

The chapter presents all the abstractions and assumptions used by the query planner and by the execution engine, in the way they are currently supported by the implementation. These concepts are here addressed in top-down order, and the chapter concludes with a short description of the implementation of the query engine and of the characteristics of the query workbench that is available as a demo of the system. The core concepts, abstractions, and system components related to query processing are the following.

- **Multi-domain queries:** independently of the higher-level languages and representations in which queries are formulated at the user interface level, any query is, from the engine's perspective, the specification of a collection of services to be invoked and a set of conjunctive conditions over their

S. Ceri and M. Brambilla (Eds.): Search Computing II, LNCS 6585, pp. 141–155, 2011.

results. SeCoQL, the declarative textual language chosen to represent abstract queries at this stage, serves well as matching point between different components, is easily generated by the UI modules and easily parsed by the underlying modules, and is compact and readable enough to be convenient also for expert users and developers.

- **Logical plans:** a compile-time analysis of the SeCoQL query performs a cost-driven optimization of the scheduling of service invocations. The input of this stage is a SeCoQL query in which the Service Interfaces to be invoked have already been chosen by higher layers, and the join types (pipe vs parallel) are therefore already fixed. The planner exploits the remaining available degrees of freedom to decide the topology, the number and sequence of service invocations, and the join strategies. The output is a logical plan, i.e., a specification of a workflow with quantitative estimates of the size of partial results and of the number of invocations to be performed on each service in order to produce these results.
- **Physical plans:** logical plans are then translated into query plans that are directly executable by the query engine. These plans are expressed in Panta Rhei, a unit-based language with support for parallelism, stateless and stateful execution, and backwards and forward control. Panta Rhei was designed to bridge the gap from the compile-time analysis performed by the query planner at the logical level to the run-time enactment of the query. It was designed with the objective of providing a clear specification of the engine behavior and also enabling runtime adaptivity in the form of reactions to events that do not match the expectations of the user or the assumptions made by the system at compile-time. Distribution, parallelization, and replication issues have also been considered.
- **Query Execution:** the query engine is implemented as an interpreter of Panta Rhei plans. The execution of a query is based on the simple assumption that any query consists of either (a) a simple invocation of a service interface, or (b) the combination of the results of two subqueries. In the former case, the engine supports the invocation of service interfaces that wrap data sources of many different kinds. As for the latter case, the results of the sub-queries can only be joined in series (pipe join) or in parallel (parallel join), and the interleaving of invocations performed on the sub-queries is handled by interpreting the signals sent by operators dedicated to this task, called strategy units. Also, the execution of a query can be monitored in all its stages by means of a workbench tool; a preview of this tool is shown in a demo video on the project website [1].

Figure 1 shows the conceptual architecture of the project and places the three query representation formats in the data flow between modules, from the User Interface (UI), in the upper part of the architecture, down to the invocation of actual services. The rest of the chapter is organized in four sections, dealing with the four items mentioned above.

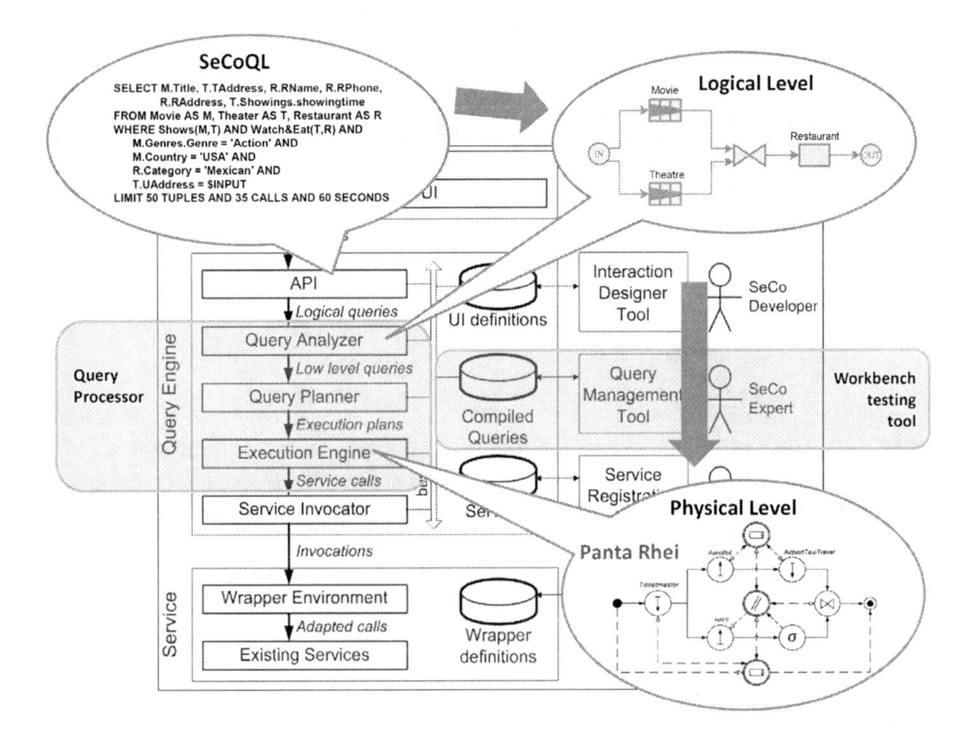

Fig. 1. Query representation formats in the SeCo framework

2 Background on Multi-domain Queries

We recall here the notion of multi-domain query by means of a classical running example: a query that searches for a good and recent American action movie in a theater not too far from the user's home and with a good Mexican restaurant nearby. The query involves three Service Marts (*Movie*, *Theater*, and *Restaurant*), and specifically three Service Invocations that implement three specific Access Patterns of these Marts, having the following schemata, where superscripts I, O, and R respectively identify parameters in input, in output, and in output with ranking.

- **Movie**($Genres.genre^I$, $MCountry^I$, $Title^O$, $Director^O$, $Score^R$, $Year^O$, $Language^O$, $Actor.name^O$)
- **Theater**($UAddress^I$, $TName^O$, $TAddress^O$, $TPhone^O$, $Distance^R$, $Showings.title^O$, $Showings.showingtime^O$)
- **Restaurant**($UAddress^I$, $Category.name^I$, $RName^O$, $RAddress^O$, $RPhone^O$, $MapUrl^O$, $Distance^O$, $Rating^R$)

Note that these Access Patterns are all relative to data sources that return ranked results. Movies are returned in order of the scores assigned by users, theaters are returned in distance order from the user address that is specified in input, and restaurants are returned in rating order. Reataurants are indeed also retrieved based on a distance range from the address specified in input, but in the case of this access pattern the distance condition only acts as a filter and produces a readable value of distance (the *Distance*^{O} attribute), while the actual order in which the results are returned is that of the restaurant rating.

Join conditions on attributes can be denoted by means of connection patterns [2] between service marts, such as *Shows* and *Watch&Eat*, whose definition in terms of equality between attributes is given below.

- **Shows**(*Movie* M,*Theater* T): [(M.Title=T.Showings.title)]
- **Watch&Eat**(*Theater* T,*Restaurant* R): [(T.TAddress=R.UAddress)]

Note that the former connection pattern checks equality between an atomic value (M.Title) and a set of values (T.Showings.title) taken from a multi-valued attribute, thus expressing the containment of the former in the latter.

SeCoQL has a declarative SQL-like syntax, in which the query result is defined as the concatenation of the tuples qualifying from the services listed in the `FROM` clause by means of the evaluation of the predicates listed in the `WHERE` clause and projected according to the list of attributes of the `SELECT` clause. More precisely, every service mentioned and associated to an alias in the `FROM` clause represents one call to a specific Service Interface, and there may be more than one call to the same Interface in the same query, in which case two different aliases are required. SeCoQL supports the specification of join conditions by means of connection patterns, but in the normal form of the language they are always expanded into their explicit form of conjunction of equalities over attributes. The query of the running example may have the following formulation in SeCoQL.

```
SELECT M.Title, T.TAddress, R.RName, R.RPhone, R.RAddress,
       T.Showings.showingtime
FROM Movie AS M, Theater AS T, Restaurant AS R
WHERE Shows(M,T) AND Watch&Eat(T,R) AND
      M.Genres.Genre = 'Action' AND
      M.Country = 'USA' AND
      R.Category = 'Mexican' AND
      T.UAddress = $INPUT
LIMIT 50 TUPLES AND 35 CALLS AND 60 SECONDS
```

Open parameters represent unbound variables whose values are assigned by the execution environment at runtime. All other input variables in the query must either have values specified in the conditions (as it is the case for Genre = "Action" in the example) or get their value from an output attribute (as it is the case for TAddress passed from Theater to Restaurant). In the example, there is only one open parameter, `$INPUT`, that represents the geographic point chosen by the user in order to center the search for theaters. For readability concerns, we restrict open parameters to be specified with identifiers whose first character is a $.

The `LIMIT` clause, that is fully optional, fixes the query limitations in terms of three limit conditions: number of combinations in the result (keyword `TUPLES`), number of calls to services during query execution (keyword `CALLS`), and number of seconds after which the execution is halted by time-out (keyword `SECONDS`). Query execution terminates as soon as one of the limit conditions is reached.

Based on the access patterns of the service interfaces used in the query, it is already possible at this stage to fix the join types that will be used in the query plan. The planner instantiates pipe joins whenever attributes that are in output from a service are used as input for other services, and instead parallel joins whenever the attributes involved in a join condition are both output attributes. In the example, the join between Movie and Theater can only occur in parallel, and Restaurant must always be invoked after Theater, in order to feed the pipe join with TAddress values, but there are no precedence constraints on the mutual position of Movie and Restaurant in the plan.

The computation of a fully determined topology for the query plans is the main responsibility of the Query Planner, that transforms the SeCoQL query into a physical plan that specifies a workflow of invocations. Also, based on some relevant profile figures taken from the service mart repository, the "optimization" of the query takes place, in a two-step approach (*Logical* and *Physical planss*), as described next, so as to estimate the amount of service calls to be overall performed in order to produce a target number of results.

3 Logical Query Plans

The optimization problem considered in the generation of logical plans, and addressed in this section, is the following: given a SeCoQL query, find the query plan topology that minimizes the *expected* execution cost according to a given *cost metric* in order to obtain k answers. The process of generating a plan starts from the conjunctive query expressed in SeCoQL and ends with a fully instantiated invocation schedule. The choice between alternatives is guided by heuristics.

3.1 Cost Metrics

A cost metric is a function that associates a cost to each query plan. We mainly consider two cost metrics: (a) the *execution time metric*, which measures the (expected) time elapsed from query submission time to the production of the k-th answer, and (b) the *sum cost metric*, which computes the cost of a plan for producing k answers as the sum of the costs of each operator used in the plan. In the former, the time required for producing k tuples takes into account the number of invocations of each unit and the expected elapsed time for the execution of that unit in order to obtain a given number of results, and the cost must account for the slowest path flowing tuples from the input to the output of the plan. For the latter, examples of costs for a service invocation are the cost of computing joins or the cost charged by the service. A special case of the sum cost metric is the request-response cost metric, which consists of considering only the cost of service invocations required to execute the plan, omitting to consider operation execution costs. This metric is particularly relevant when the transfer of data over the network is the dominating cost factor.

Other cost metrics of interest, though so far not considered in detail in the project are the *bottleneck cost metric*, which gives the execution time of the slowest service in the plan and is relevant in contexts of pipelined execution of continuous queries, and the *time-to-screen* cost metric, which measures the time required to present the user with the first result. The former metric is suitable to contexts with homogeneous services that respond to invocations with "continuous" streams of results. In these cases, the time spent to initialize and load the pipelines is negligible w.r.t. the time spent in stable regime of execution, during which the overall throughput is limited by the throughput of the "slowest" service. This metric is hardly applicable in our context, where search services rarely produce all their tuples and the execution is normally limited to reaching k answers rather than being run as continuous queries. The latter metric is potentially more relevant to the project, as it is suitable for settings in which the user expects a prompt interaction, and would enable an optimization based on estimates of the time that users are ready to spend waiting for the first results, before quitting the task.

3.2 Heuristic Planning: Topology and Join Strategies

The optimization method adopted to determine the less costly plan explores the combinatorial solution space of all possible translations of the declarative query into fully instantiated invocation schedules. The exploration is organized by means of an incremental construction of the query plans, that takes place in two phases, imposing a discipline in the order in which alternative plans are generated and considered. A detailed description of the approach and of the background for optimization is given in our previous work [3,4]. The graphical notation used in the figures is also taken from these previous works, to which the interested reader is directed.

The first phase in the incremental construction of a logical plan is the selection of a topology for the logical query plan that is compatible with the given choice of service interfaces. This phase fixes the order of invocation of the services, as well as the data flow and the details of join operations. Even if at this stage all access patterns have been fixed, and therefore the nature of joins (pipe vs parallel) is fully determined, there may still be alternative topologies compatible with the precedence constraints that they enforce on the invocation order. The second phase in the construction is the choice of the number of fetches to be performed over all chunked services. This phase allows to fully determine the execution schedule and the join strategies, and therefore to compute its cost according to a given metric. For each phase, there are alternative heuristics for effectively building efficient plans. Once more, the reader is directed to [3] for details. The default heuristics currently adopted are "parallel is better" for the topology and "square is better" for the fetching ratios, that are respectively meant to minimize the time-to-screen of the first results and to maximize the diversification of results.

In the running example, there are two alternatives. Textually, these topologies can be denoted as "(M//T)|R" and "M//(T|R)", where "//" and "|" stand for parallel and pipe join, respectively. The two logical plans that implement the alternative topologies are shown in Fig. 2.

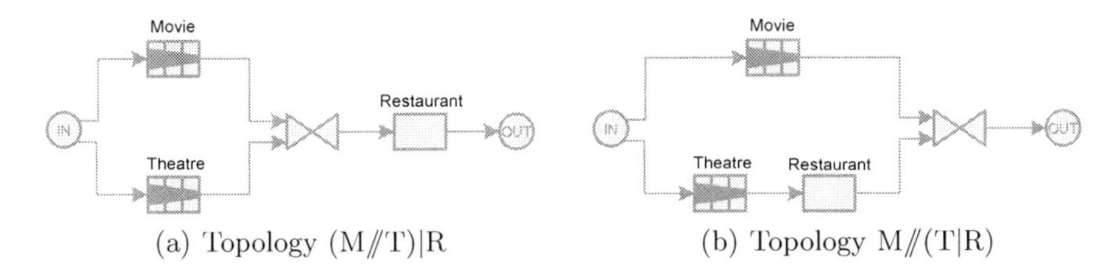

(a) Topology (M//T)|R (b) Topology M//(T|R)

Fig. 2. Alternative query plan topologies

4 Physical Query Plans

Physical plans are expressed in Panta Rhei and are interpreted by the query engine. The design of Panta Rhei follows a well-defined set of design principles. Plans need to be *composable*, i.e., complex plans can be formed from simpler sub-plans. They must lend themselves to *parallelization*, to leverage concurrency as much as possible, and must be *distributable* over a number of computing nodes. We start by introducing the underlying data and control models, and then describe the topology of the plans, that are graphs with nodes that represent physical operators (or units) and edges that represent the data and control flow in the plan. We then define all types of edges and units in detail. Finally, we describe how physical plans can be composed by giving a minimal set of recursive rewriting rules to define the concept of a *well-formed* plan.

4.1 Data and Control Model

The *data model* of the execution engine is based on the Service Mart framework [2] which associates each service with a flat relational schema extended with a controlled use of multi-valued attributes. The schema of the tuples of the results is simply a subset of the schema obtained by concatenating the schemes of all the services that are involved in a query. Some of the attributes are initialized with the constants specified by the user. Result tuples are progressively composed by using service results as the query evaluation progresses.

The *control model* of the execution engine addresses the fact that, in Search Computing, plans need to be highly configurable at compile-time and, to a certain extent, capable of adaptation at run-time. The requirement for configurability stems from the fact that search services may have very different and time

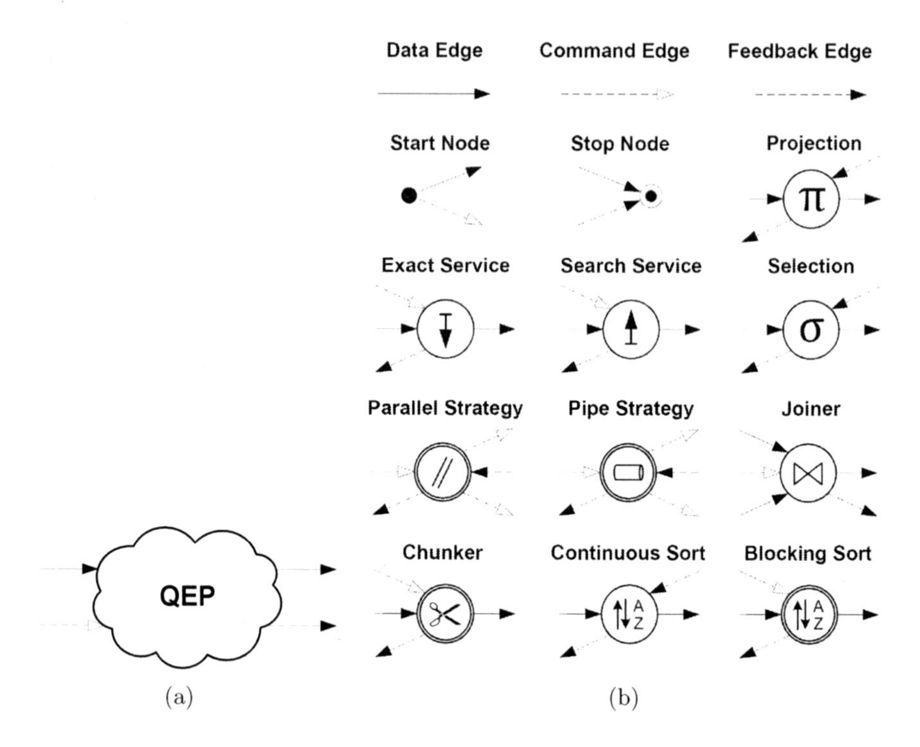

Fig. 3. (a) Query execution plans (QEPs) and (b) their components

varying computational complexity, execution time, or monetary cost. Adaptation is motivated by the fact that the properties of the search services and the data distribution encountered at run-time may be significantly different from the assumptions made by the query planner and the optimizer at compile-time, that are necessarily based on statistics derived from previous observations. Moreover, plans which want to guarantee optimality (top-k) must adapt their behavior to the actual ranking values that are progressively read from service results.

Data Flow. The *data flow* of a physical plan consists of data edges that form a directed acyclic graph. Every data edge carries tuples whose schema is obtained as the concatenation of all the schemas of the services invoked by antecedent nodes of that edge. Data messages contain chunk identifiers and chunks of tuples.

Control Flow. The *control flow* of a physical plan comprises *command edges* and *feedback edges* to support both forward and backward scheduling. Command edges transport command messages that can be further distinguished into *fetch messages* and *join messages*.

– A fetch message specifies which tuple of which chunk carried by a data edge should be used as input to the query execution plan for performing the nth fetch operation, where a fetch operation is next defined. We use the shorthand "`3@A2`" to denote that the third fetch is done with tuple 2 from chunk `A`.

- A join message defines which pair of chunks from two data edges in input to a join operator should be joined. Join messages are used to implement different exploration strategies in parallel and pipe joins.

Feedback edges transport feedback messages containing information about the execution of a physical plan. Feedback messages are transmitted after the plan has processed the commands given in input and serve primarily as acknowledgements. Feedback messages can be further classified into *statistics* messages and *exception* messages. Statistics messages contain data such as the cardinality of the computed result, the window of encountered ranking scores, the distribution of monitored attributes (e.g., join attributes), and the EOF marker if no more results can be obtained from a plan. Exception messages are transmitted by a plan in the case it observes an abnormal behavior.

The control flow is, therefore, bidirectional: the forward control flow transports instructions to a plan indicating how the tuples in input must be considered by the plan, the backward control flow reports as feedback statistical data characterizing the plan execution. As a consequence, the execution engine can support a *purely forward* scheduling of plans, as well as a *mixed forward and backward* scheduling of plans. In the former case, the query optimizer determines the entire execution strategy and configures the engine accordingly. In the latter case, the query execution strategy is dynamically determined based on selected attributes that are monitored by the execution engine.

4.2 Query Execution Plans

A query execution plan (QEP) is a well-formed and executable physical query plan, modeled as a graph, that accepts in input chunks of tuples and control messages, denoted by means of incoming data flow edges and control flow edges respectively, and produces in output chunks of result tuples and feedback messages, denoted by means of outgoing data flow edges and control flow edges respectively. Graphically, a QEP is represented as shown in Fig. 3(a). The incoming data edge transmits chunks of tuples in input to the QEP, while the outgoing data edge transmits chunks of tuples to a downstream QEP or to the stop node. The incoming control edge carry messages that regulare how the tuples in input should be processed within the QEP. The outgoing control edge transmits feedback data about the execution of the QEP.

We next discuss the nodes that can appear within QEPs, listed in 3(b), then we give simple compositional rules which explain how well-formed query plans can be generated.

Each node in the plan represents a processing units of one of the following kinds; the behavior of nodes is determined by its input and its state.

- **Start/Stop nodes:** The *start node* injects the constant values specified by the query into the query execution plan along the data flow edge. Additionally, it transmits the start command along the control flow edge. There is only one start node per query. The *stop node* collects the results of a query execution plan and makes them available to clients of the execution engine. Additionally, it acts as a sink for all feedbacks. There is only one stop node per query.
- **Service invocation nodes:** The *exact service invocation node* completes the tuples in input by invoking an exact service. Exact services produce a finite set of tuples that represent the exact (and thus complete) response to the service call query given the input parameters. The output tuples are not ranked. The *search service invocation node* completes the tuples in input by invoking a search service. Search Services exhibit a behavior similar to Web search engines: results are unbound, ranked and chunked, and normally there is no interest in obtaining a complete result, but only in obtaining the first chunks. Both types of service nodes return one chunk of tuples in output for every invocation.
- **Strategy nodes:** A *parallel strategy node* controls two QEPs that are scheduled in a parallel join, while a *pipe strategy node* controls two QEPs that are scheduled in a pipe configuration. Both these unit are assigned a "budget", i.e., a number of invocations that they are allowed to "spend". Spending the budget means breaking it down and passing it to the controlled QEPs to be consumed by considering the commands received in input and distributing the invocation commands to the connected sub-plans accordingly. Optionally, these units also receive feedback from the controlled QEPs, that they may use to tune the spending of the remaining invocation budget.
- **Joiners:** The *joiner node* is always controlled by a parallel strategy node. It receives chunks of tuples on its two different incoming data edges, and evaluates the join predicate in order to join them into new chunks of tuples, according to the join messages received from the strategy node. The joiner per se is a stateless unit whose role is limited to the evaluation of a simple predicate, but coupled with the strategy unit it provides the flexibility to implement many different logical join operations at the physical level.
- **Data flow modifiers:** A *selection node* filters chunks of tuples according to a selection predicate. Since the selection unit does not re-chunk the tuples, the chunk size can decrease in the order of the selectivity of the given predicate. A *projection node* shrinks the schema of the result to the specified list of attributes only. The *chunker node* intercepts chunks of tuples flowing along a data edge and recombines them sequentially into chunks of a given size until it finds the EOF marker. A *continuous sort node* sorts data on a per chunk basis as they flow along a data edge, whereas a *blocking sort node* caches all chunks flowing along a data edge until it receives the EOF marker and then sorts the tuples across all chunks. Both sort nodes are configured with a sort function. While a continuous sort node produces a same-sized output chunk for every input chunk, a blocking sort node is additionally configured with an output chunk size.

Parallel and pipe joins are implemented in a QEP by different stategy nodes, whose parameter setting is determined according to the service interface specifications (their chunk sizes, average response times, and invocation costs). Figure 4 shows the plan configuration for pipe and parallel joins between two generic QEPs.

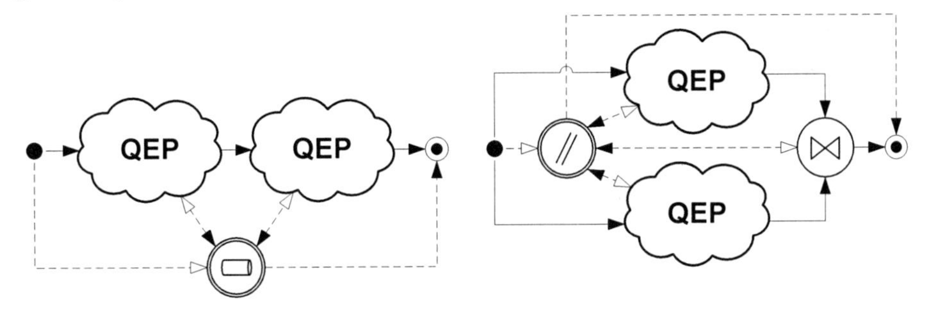

Fig. 4. QEPs for pipe and parallel joins

4.3 Plan Composition

The following set of production rules defines how QEPs can be recursively composed to form more complex plans. The axiom plan consists of a single QEP having a start node as predecessor and a stop node as successor. Rules indicate that the pipe or parallel composition of two QEPs gives a QEP, and that a QEP can be composed with any unary modifier operator (i.e., units with a single input and output data and control flow) yielding a QEP. Plans obtained by arbitrary applications of these rules to the axiom are called well-formed and have associated well-defined semantics.

$$QEP := (\uparrow) \tag{1}$$

$$QEP := (\downarrow) \tag{2}$$

$$QEP := QEP \bowtie_{pipe} QEP \tag{3}$$

$$QEP := QEP \bowtie_{parallel} QEP \tag{4}$$

$$QEP := QEP \cup \{(\times), (\updownarrow), (\updownarrow), (\sigma)\} \tag{5}$$

An example of a QEP composed using all five rules is shown in Fig. 5. Based on Rule 3, the outer-most QEP can be deconstructed into two sub-plans that are combined using a pipe join. The first sub-plan can then be further decomposed based on Rule 4 into two sub-plans that are combined using a parallel join. Applying Rule 1, each of these sub-plans can be substituted by a search service (S_1 and S_2). The second sub-plan of the pipe join can be decomposed into a query execution plan followed by a modifier (selection) based on Rule 5. Finally, this last query execution plan can be substituted by an exact service (S_3) according to Rule 2.

If we instantiate S_1, S_2, S_3 as Movie, Theater and Restaurant respectively, ignoring the application of Rule 5, we also have the Panta Rhei plan for the running example.

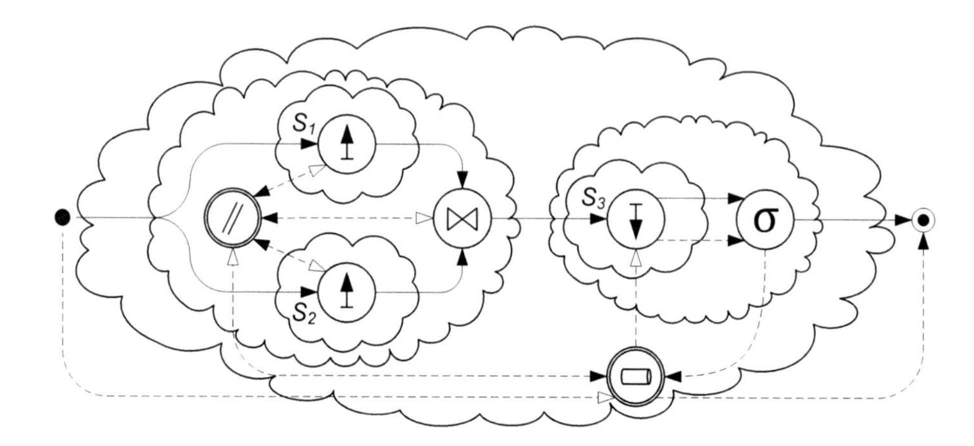

Fig. 5. Nested QEPs

5 The Experimental Workbench

A prototype of an execution engine implementing the functionalities provided by Panta Rhei is currently used for internal tests of the Search Computing framework. The engine is built on top of the Service Mart invocation and wrapping environment shown in Fig. 1, which exposes heterogeneous services, such as Web services, custom services, and relational databases, whose description is out of the scope of this paper.

The system provides both synchronous and asynchronous search mechanisms. End users of the engine can either wait for search results to be produced or be notified of their production. In the first case, the execution engine is directly connected to a user interface, whereas, in the latter case, the execution engine itself is bundled as a service. Fine-grained control is also provided, to allow for interaction with the search process, in order to dynamically orchestrate it, for example to react to the results being produced.

The engine is implemented as a multi-threaded environment that uses a thread pool to support both inter and intra-query parallelism. To support inter-query parallelism, queries can share computation and storage resources as well as reuse cached partial results of already executed invocations, if still valid. Intra-query parallelism is achieved by spawning multiple instantiations of replicable units to distribute the workload of computationally complex and time consuming tasks, such as invocations of services with high latency.

A scheduling algorithm activates the various units, assigning them to threads to handle incoming control messages, according to the priority of the search process. The control flow is implemented by means of message queues, and, in the

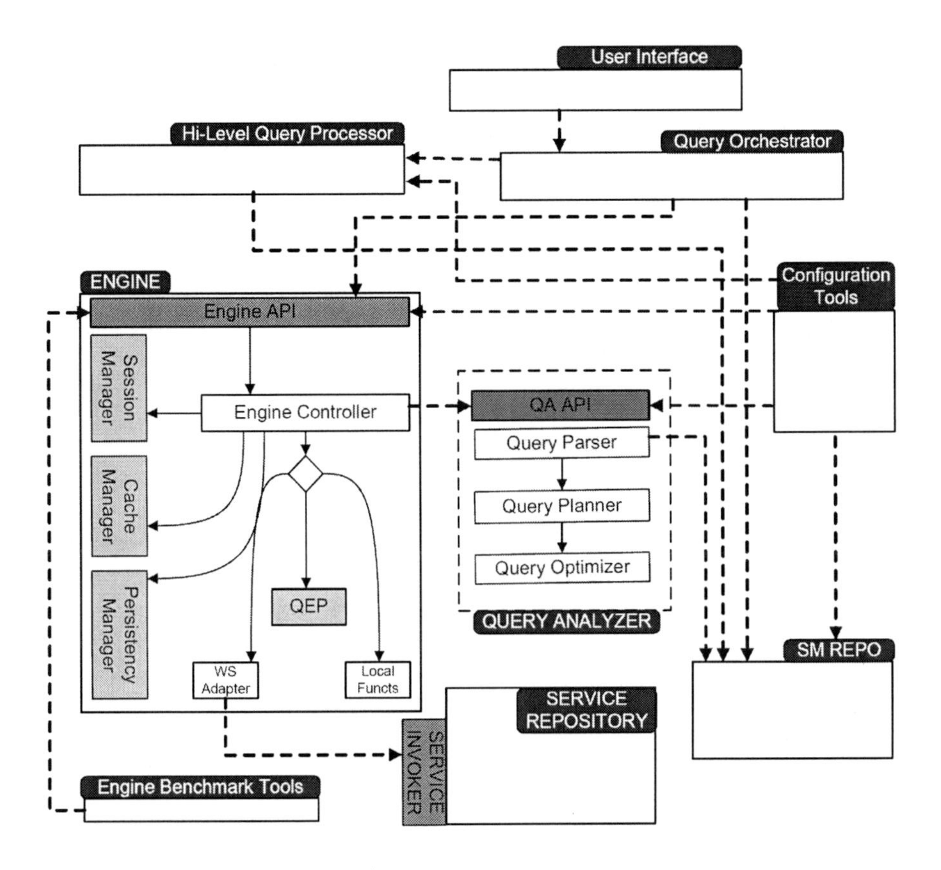

Fig. 6. The architecture of the engine and planner implementation

case of replicable units, messages can be delivered to multiple unit instances, even out-of-order. The data flow is implemented by means of buffers shared between units, according to a producer/consumer paradigm. The implementation of join strategies is fully decoupled from the implementation of the strategy units. This allows new strategies to be plugged in with no impact on the implementation of the units themselves.

Also, we adopt a two-level caching mechanism. On the one hand, we take advantage of the cache implemented at the service level, so as to avoid repeated invocations with the same input. On the other hand, we also cache partial query results at the engine level, which impacts on the performance of multiple executions of the same process and also of the same sub-plan shared between different search processes or reused by the same query in a later execution.

Figure 6 shows the modules of the query engine and the query planner in the context of the overall software architecture of the system. The invocation of the execution of a query via the Engine API causes the instantiation of a Controller for the incoming query. The controller is assigned to the current search session and is given access to both (a) the Cache Manager, that handles the session-based cache and connects to the inter-session caching system, and possibly (b) to the Persistency Manager, that is in charge of the materialization of query results into a persistent data storage, whenever queries are registered as permanent data sources that need to be periodically updated. The interpretation of a query by the controller follows a simple recursive scheme that is based on the internal structure of the query into QEPs and sub-QEPs. The main functionality of the Controller is to instantiate the threads corresponding to the units in the QEP currently under evaluation, and in turn activate an execution context for any sub-QEP possibly contained into the current QEP. Invocation units are interpreted performing a supervised access to the corresponding Web Service (through an adaptor that connects to the Service Repository) or resorting to Local Functions in some special cases in which the invocation units represent access to locally available computation resources (as it may be the case, e.g., for the transcoding of two geo-referenced locations and the computation of their distance).

Queries can be submitted to the API in the form of already instantiated executable query plans, and in this case they are immediately interpreted, or in the form of SeCoQL statements. In the latter case, they are passed to the Query Analyzer to undergo the transformations described in Sections 3 and be converted in Panta Rhei plans as defined in Section 4.

A demo that consists in a walk through the abstractions presented here is available on the project website [1]. A snapshot of the workbench is shown in Fig. 7. In the shown panel the tool allows the user to control the query results as they are progressively produced and the execution timeline that tracks the activation of the different units.

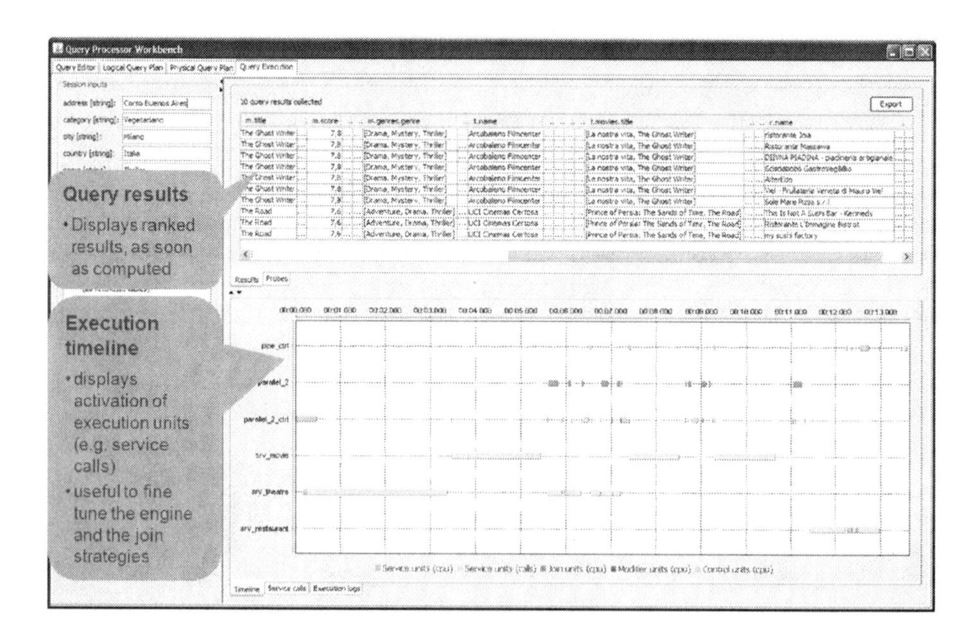

Fig. 7. A snapshot of the Query Processor Workbench

6 Conclusion

In this chapter, we have given a high-level overview of how query processing is carried out in SeCo, addressing the formulation and execution of multi-domain queries at different levels of abstraction, from the expression in SeCoQL, a declarative query language over service interfaces, down to the interpretation of physical query plans by means of a query engine that support parallelism and distribution. We also described the implementation of the query engine as the interpreter of these physical plans, as it is demonstrated in the demo available on the project website [1].

References

1. The Search Computing Project: Demonstrator of the Execution Engine (June 2010), `http://www.search-computing.org/demo/qp`
2. Campi, A., Ceri, S., Gottlob, G., Maesani, A., Ronchi, S.: Service marts. In: Ceri, S., Brambilla, M. (eds.) Search Computing. LNCS, vol. 5950, pp. 163–187. Springer, Heidelberg (2010)
3. Braga, D., Ceri, S., Daniel, F., Martinenghi, D.: Optimization of Multi-domain Queries on the Web. PVLDB 1(1), 562–573 (2008)
4. Braga, D., Ceri, S., Grossniklaus, M.: Join Methods and Query Optimization. In: Ceri, S., Brambilla, M. (eds.) Search Computing. LNCS, vol. 5950, pp. 188–210. Springer, Heidelberg (2010)

Run-Time Adaptivity for Search Computing

Daniele Braga[1], Michael Grossniklaus[1], and Norman W. Paton[2]

[1] Dipartimento di Elettronica e Informazione, Politecnico di Milano
{braga,grossniklaus}@elet.polimi.it
[2] School of Computer Science, University of Manchester
npaton@manchester.ac.uk

Abstract. In Search Computing, queries act over internet resources, and combine access to standard web services with exact results and to ranked search services. Such resources often provide limited statistical information that can be used to inform static query optimization, and correlations between the values and ranks associated with different resources may only become clear at query runtime. As a result, search computing seems likely to benefit from adaptive query processing, where information obtained during query evaluation is used to change the way in which a query is executing. This chapter provides a perspective on how run-time adaptivity can be achieved in the context of Search Computing.

1 Introduction

In contrast to traditional data sources, such as databases and digital libraries, service-based data sources like web services and search engines are more challenging to characterize in terms of average response time, expected result size, data distribution, and similar features that are typically used to perform query planning and optimization according to consolidated techniques. A traditional optimization approach alone is therefore expected to be less effective in a setting where most data is accessed via service invocations and by means of search engines.

Nevertheless, as these data sources are central to Search Computing, it is crucial to have a query processing paradigm that is (a) sophisticated enough to compute an "off-line optimal" query plan at compile-time, and (b) flexible enough to adapt that plan at run-time, in response to deviations from the assumptions made at compile-time on the expected behaviors.

More specifically, a query execution plan embodies the decisions made at compile-time by the query optimizer that generated it. These decisions broadly consist of

1. the order of evaluation of operators, whenever alternative schedules can be considered equivalent;
2. the choice of alternative algorithms and auxiliary data structures to implement the evaluation of operators (as can be the case for different join strategies and different levels of caching);
3. the level of partitioned parallelism, so as to control if the same operation can be performed in parallel by different computational units on different data fragments; and

S. Ceri and M. Brambilla (Eds.): Search Computing II, LNCS 6585, pp. 156–166, 2011.

4. the allocation of plan fragments to available computational resources, so as to balance the computational load and avoid loss of performance due to bottlenecks.

An adaptive query processor may revise any of the above decisions at query run-time, in the light of feedback received, e.g., on actual rather than predicted selectivities, on the presence of delays and idle computational units, or on the correlation between rankings from different searches.

Proposals for adaptive query processing can be classified as *plan preserving*, where adaptation takes place by tuning the strategy used to execute an essentially stable representation of a query, or *plan changing*, where adaptation changes the query plan, and thus involves stopping the execution, re-optimizing the plan, and resuming execution. Plan changing proposals, in particular, must take account of the (partial) work done at the point when re-optimization takes place.

This chapter will discuss the plan preserving and plan changing opportunities for adaptation in Search Computing. Section 2 outlines the key features of plan changing and plan preserving approaches to adaptation, and discusses how these are affected by ranked data. Section 3 outlines opportunities for plan changing adaptivity in the context of the Panta Rhei execution model, described in the previous chapter. Section 4 concludes the chapter with an outlook on future work.

2 Run-Time Adaptive Query Processing

This section details techniques that have been developed to exploit both *plan preserving* and *plan changing* adaptation, with specific reference to adaptation for rank-aware queries. In this section, an approach is considered to be *plan preserving* if the query optimizer is not invoked at query runtime to identify a new plan, and *plan changing* where the optimizer is invoked at query runtime to generate a plan that may then be used to complete query execution.

2.1 Plan Preserving Adaptation

Plan preserving query adaptation modifies some property of query execution at run-time in response to information about the environment in which a query is executing (e.g. machine loads) or some information about the query (e.g. operator selectivity),

Table 1. Plan preserving adaptations

Proposal	Property Adapted
Eddies [1]	Route followed by tuples between operators
Flux [2]	Distribution policy in partitioned parallelism
DITN [3]	Presence of redundant fragments
Scrambling [4]	Query schedule

while leaving the basic structure of the plan unchanged. Table 1 lists several plan preserving approaches along with the property that is adapted. To take these in turn:

Eddies [1]: Unlike classical query plans, in which the order of operator execution is reflected in the structure of the plan, in this proposal an *eddy* operator is introduced that routes data to the operators that actually evaluate the query. The *eddy* then monitors these operators, for example to ascertain their selectivities, and changes the order in which tuples are routed to operators on the basis of this feedback. This has the effect, for example, of changing the join order at runtime.

Flux [2]: In partitioned parallelism, a single operator may be assigned to multiple computational nodes, and a *distribution policy* indicates what fraction of the data should be directed to each node. In this proposal, the distribution policy may be revised at query runtime, with a view to managing load imbalance, and a protocol is provided for moving operator state to reflect changes in distribution.

DITN [3]: In the Data in the Network (DITN) proposal, it is assumed that queries are being evaluated on non-dedicated (e.g. scavaged) resources, and thus that the resources have unpredictable loads and that there may be surplus resource available. In parallel query scheduling, where there is a delay in completing a query fragment, a redundant copy is run on different nodes, and the first of the copies to produce results is preferred over the other.

Scrambling [4]: In query scrambling, adaptations seek to accommodate delays in sources generating tuples. One of the adaptations changes the order in which query fragments are scheduled, by identifying runnable fragments that are then executed and their results materialized.

Although there is considerable diversity in plan preserving techniques, there are a number of recurring themes; the designers of plan preserving adaptive strategies must:

- Identify the problem to be addressed – this is typically much more specific than *improve query response time*, as the adaptation to be made is generally targeted at a particular problem. For example, the detection of load imbalance may result in a change in workload allocation.
- Identify the monitoring information that is required both to diagnose the problem and to parameterize the adaptation. For the most part, monitoring information is that required to parameterize a query cost model [5], but the monitoring information can be also used to select responses as well as to analyse progress (e.g. for learning the properties of different scheduling algorithms [6]).
- Define the adaptation that is to take place, identifying: (i) any constraints on when the adaptation can safely be applied (e.g. moments of symmetry in eddies [1]); (ii) any additional state that is required to support the adaptation (e.g. State Modules that support efficient query processing by eddies [7]); and

(iii) any changes or additions to query compilation or optimization that are required to support adaptation (e.g. a specialized data distribution policy is used by DITN [3]).
- Integrate the adaptation into the query processor; this may be as an adaptive operator (e.g. [1,2]) or a controller that is notified by the query (e.g. [4]).

Overall, the plan preserving approach has been applied in a wide range of contexts to carry out a diverse collection of plan changes. In principle, several different plan preserving approaches can be in play at the same time, although in practice controlling the interplay between different adaptations may be challenging.

2.2 Plan Changing Adaptation

In defining *plan changing* approaches as those in which the optimizer is called at runtime, we note that some adaptive strategies generate multiple plans before query execution, and then may swap between these at runtime (e.g. [8,9]), and thus might be felt to fall within a gap in our classification. Henceforth in this section, however, we consider proposals in which the adaptation involves stopping the plan, re-optimizing the plan and resuming execution.

Because plan changing proposals may make substantial changes to the way in which a query is being evaluated, they must take account of the work done on the evaluation of a query at the point when re-optimization takes place. As such, the designers of plan changing adaptive strategies must:

- Identify the monitoring information that is required both to establish that it may be useful to consider adapting and to inform the construction of a new plan by the optimizer. For example, adaptation may only take place when changes to statistics have been detected beyond some threshold (e.g. [10]), or where the statistics have moved outside the range for which the current plan was optimal when compiled (e.g. [11]).
- Identify when the plan can safely be stopped to support adaptation, and characterize the work done to date so that the new plan does not unnecessarily repeat work. For example, *coarse grained* techniques may reuse complete results of operator evaluation (e.g. [11]), whereas *fine grained* techniques are able to reuse partial results from operators (e.g. [12,13]). Such fine grained techniques must be able to characterize precisely how the partial result of an operator relates to the data consumed, and thus what work remains to be done [13].
- Modify the optimizer so that it takes account of updated statistical data and the state computed by the current plan in exploring new strategies. In doing this, the previous plan reconsidered with the updated statistics may serve as an upper bound to the cost function to guide the new search for optimality.
- Integrate the adaptation into the query processor; this may be as an adaptive operator (e.g. [11]) or a controller (e.g. [13]).

In designing a strategy following the above steps, it may be considered to be a good thing if as many as possible of the following non functional requirements can be satisfied:

- Queries can be stopped for re-optimization at many points in their evaluation; coarse grained techniques are typically only able to reuse work at materialization points when operators have completed execution, whereas fine grained techniques may support a more agile reuse of cached temporary results.
- Auxiliary or repeated work after re-optimization is minimized; some strategies consider discarding work in order to enable adaptation before operators have completed (e.g. [11]), and some use *stitch-up plans* to complete operator evaluation (e.g. [14]).
- A wide variety of operators can be used; some strategies support fine-grained reoptimization by restricting the collection of operators that can participate in adaptive plans (e.g. [12]). A particularly relevant case is that of rank-aware operators, that typically build on thresholds derived from the data already processed and make assumptions on the data that is still to be processed. Rank-aware operators are addressed in the next section.

Another reason for deciding to change the plan could be explicit or implicit feedback from the user. In all settings in which query sessions are interactive and executions can be manually stopped and resumed directly by the users, or users can evaluate the quality of the results obtained to date, there is potential for reoptimization. This is a relevant scenario for Search Computing, as described in detail in Chapter [15] of the present collection, dedicated to the study of the "Search as a Process" paradigm for Search Computing.

2.3 Adaptive Query Processing for Ranked Data

Where query processing includes ranking, standard query processing operators are supplemented with rank-aware operators [16]. For example, a rank-aware join operator consumes ranked inputs from its operands where each input tuple has a *rank score*, applies a *ranking function* to the rank scores of matching input tuples to compute rank scores for matching tuples, and returns matching tuples in order of their scores.

Adaptivity is potentially important for queries that include ranking; the reasons for adapting in non-ranked queries carry forward to rank-aware queries, where additional challenges result from difficulties in predicting the relationships between ranked inputs. For example, are highly ranked tuples more or less likely to match than randomly selected tuples?

Adaptive rank-aware query optimization has been investigated by Ilyas *et al.* [17], where their plan preserving approach is described in most detail. The focus is on adaptations involving a rank-aware symmetric hash join [16]. In this algorithm, when a tuple is read from an operand, it is: (i) stored in the hash table for its operand indexed by join attributes; (ii) probed against the hash table for the other operand, and the rank score is calculated for each matching tuple, which is inserted into a *rank queue*. A result tuple r can be returned from the rank queue when it is known from the properties of the ranking function and the scores of the tuples read from the ranked inputs that no result tuple with a higher score than that of r can be generated from as-yet-unread input tuples.

In the adaptive query processing strategy, runtime monitoring either detects delays or unexpected correlations between scores. When these are detected, the optimizer is rerun to construct a new plan, and the state associated with each operator in the new plan is either copied from the previous plan or reconstructed from scratch. In essence, state can be copied from a node n in the original plan P to a node n' in the new plan P' when n and n' have the same leaves. This ability to reuse some operator state means that work need not be redone during evaluation, but individual adaptations may still be expensive because operator state that did not exist before adaptation must be computed to generate the state that would have existed if P' had been run from the beginning.

3 Adaptation in Search Computing

Adaptive query processing is relevant to Search Computing

(a) because the actual behavior of data sources can diverge from the expected one in many ways (e.g., unexpected delays, or unpredictable cardinality of the results of exact services), but also
(b) because some computations intrinsically need to react and adapt to run-time evidence, as is the case for the generation of results with top-k guarantees in the presence of unpredictable score information for search services, and also
(c) because of the intervention of the end user in the query execution process, which requires the ability to reuse the results available to date and to reshape them to enable the continuation of the search session.

Panta Rhei is the unit-based language used for the specification of physical query plans in Search Computing. We briefly summarize its features so as to make this chapter self-contained. A physical query plan consists of a directed graph of nodes, corresponding to processing units, and edges, forming the data and control flows. Results are progressively created by joining and pruning the data returned by invoking data sources. The most important processing nodes in Panta Rhei are service invocation units, for extracting ranked data from data sources, and strategy units, that are used to control and synchronize the behavior of invocation units that are in parallel and pipe joins.

Join units are rank-preserving according to either *top-k* or *good-k* join strategies. Joins that are regulated by a top-k strategy produce results according to an ordering imposed by a score aggregation function which is a weighted sum of partial scores. Joins that are regulated by a good-k strategy produce results in an order that is as consistent with the ordering given by the aggregation function as can be formed with the best tuples from each source. The actual order in which tuples are output in the latter case is given by combining the best results from each single source as soon as possible, and the more the partial rankings are correlated with the aggregate one the better the good-k results will approximate top-k results. Also, top-k joins are blocking, because a tuple can be output only when it is guaranteed to have a ranking that is at least as high as any that can possibly be formed with data that has yet to be extracted. Good-k joins, instead, are non-blocking by construction, and therefore their output tuples are produced as soon as they are available.

In a physical plan, operators are assigned a *budget* that expresses the number of service invocations that they are allowed to "spend". More details of Panta Rhei can be found in the previous chapter [18] of this collection.

We can classify the actions performed in order to achieve run-time adaptivity not only as plan-preserving or plan-changing, but also according to the events that trigger the actions. In particular, we distinguish between *user-generated* and *system-generated* actions.

User-generated actions reflect reactions to unsatisfactory results, further classified as:

- *more all* commands, given when the execution halts and the tuples presented in the results are too few with respect to the user's expectations;
- *more one* commands, given when the results associated with one specific source are unsatisfactory, either because they are too few, or because the user believes that there may be additional relevant results which have not yet been extracted, characterized by a lower local ranking but higher capability of forming interesting combinations when joined with results from other services;
- changes to the *score weights in the goal function*, because the current result does not reflect the user's preference, as it over-emphasizes or de-emphasizes one of the services that instead is considered as most relevant for the user;
- *lowering* of the *level of guarantees* that are expected on the result, because a top-k enabled execution leads to marginal results or excessively delays the output of the first combinations, and a good-k approach seems preferable;
- *raising* of the *level of guarantees* that are expected on the result, when a good join execution produces results that cannot be sufficiently trusted, and top-k join strategies should instead be used.

System-generated actions, instead, can be triggered by events and anomalies that are monitored and detected automatically.

3.1 Plan-Preserving Adaptations in Search Computing

In the following cases, plan adaptation occurs in the context of the same plan, whose execution continues after adaptation by reusing the current results.

A. More One and More all Directives. These commands are typically performed by the user when results are perceived as insufficient, and they call for adding more results, either selectively from one service, or globally. It is important to note that Panta Rhei physical plans are intentionally built with the objective of giving few tuples in the result, because users are normally interested only in the top tuples of ranked results. Therefore, allowing within Panta Rhei the structures for a seamless continuation of the execution of a query plan is an important choice, with consequences on the architecture and on the implementation.

B. Change of Weights in the Score Aggregation Function. The change of weights affects top-k strategies, as it changes the threshold values used to assess that a given combination belongs to the top-k results. A change of weights may be followed by a total change of the result which is displayed to the user, but such a change occurs without recomputing the query: a different result set can be selected from the buffered query results by means of a recomputation of threshold values. Such result set may be too small, and in such case the user can continue query execution by means of more one and more all directives.

C. Budget Redistribution. In Panta Rhei, controller units spend their budget of invocations relative to the units they are responsible for, thus implementing the "forward" control logic as decided by compile-time planning and optimization. Then, at run-time, suspend and resume signals are propagated "backward" in order to re-synchronize execution when anomalies are detected, such as delays in one of the controlled units. Therefore, forward controls determine producer-consumer relationships according to the query plan, and backward controls optionally condition those producer-consumer relationships that deviate too much from the optimal plan determined at query optimization time.

The simplest form of system-controlled plan adaptation is a run-time budget redistribution, that consists in moving budget to those operations that have exhausted their budget when execution halts before producing a result with the estimated number of tuples. Redistribution within the same query does not change the global load associated with query execution, and therefore its relationships to other, concurrent query executions. While backward control occurs in the context of the control signals between the units, and is therefore part of the normal adaptive behavior of plan, budget redistribution is properly classified as an adaptation of the plan that occurs on the initiative of a monitoring system module.

D. Service Replacement without Replanning. If a service is unavailable at execution time, the system may detect this fact and autonomously decide to resort to a different service; if the new service has the same signature (or, in SeCo terms, is associated with the same access pattern) then the service can be replaced without changing the plan. Of course the invocation of the new service can have very different delay, cost and output tuples compared to the original service, but such differences can be dealt with by backward controls and budget redistributions.

If the service becomes unavailable during query execution, results presented to the user prior to adaptation are correct, but incomparable with results presented to user after the adaptation, because the first invocation of the replacing service produces ranked results starting from a rank value that has no relationship with the last invocation of the replaced service.

E. Changing Guarantees During Execution. The last plan-preserving adaptation action that we consider here occurs when the user decides to raise or lower the level of guarantees of the joins of a running plan, e.g. because the user decides to accept all good results in a situation in which the system produces too few top-k results. This is a borderline case, because the plan topology does not change, but the controller of join operations changes substantially.

When the level of guarantee is raised (from good to top), a total change of the result which is displayed to the user is needed, but such a change occurs without recomputing the query: a subset of tuples in top-k order can be selected from the buffered query results by means of the computation of threshold values. Such a result set may be too small, and in such case the user can continue query execution by means of more one and more all directives.

When the level of guarantee is lowered, the system initially presents to the user all the result tuples which are buffered in the order in which they were extracted and generated. Such a result set, although larger than the previous result set, may still be too small, and also in such case the user can continue query execution by means of more one and more all directives.

3.2 Plan-Changing Adaptation in Search Computing

In this subsection we sketch some preliminary cases of plan-changing adaptivity. Supporting this form of adaptation is much more complex, as the new plan can be substantially different from the old plan; the new plan (or a portion of it) needs to be installed as a replacement of the old plan. Moreover, results presented to the users prior to adaptation may be incomparable with results presented to user after adaptation, because the execution of the new plan may restart tuple production for some services participating in the plan.

A. Service Replacement with Replanning. If a service is unavailable at execution time, the system may have to resort to a different service with a different signature. This normally causes a change in the plan, as the service signatures dictate the type (pipe vs parallel) of join that connects the service execution unit to the other units; such changes, however, may be known in advance and rapidly installed with minimum perturbation of the rest of the plan. Of course the new plan can have very different performance, and thus may benefit from reoptimization. If a service becomes unavailable during query execution, results presented to the users prior to adaptation are correct, but incomparable with results presented to user after the adaptation.

B. Selection of Equivalent Alternative Plans. Case A is a particular case of Case B, which occurs when a complete plan is substituted by another equivalent plan. Such a situation may occur when the actual availability and performance of services suggests a radical strategy change. Technically, alternative plans can be determined at execution time by marking certain services as unavailable and then seeing if an alternative plan exists; then, such plan should be installed and executed.

4 Conclusion

In this paper, we considered query adaptation, first by reviewing its general properties, and classifying adaptation into the broad classes of plan-preserving and plan-changing. Such a classification can be applied to explore opportunities for adaptive query processing in Panta Rhei, the SeCo execution engine. We have shown that relevant cases of adaptations may be due to explicit user interactions: users adapt to the progressive presentation of results by issuing simple commands, which may have a great impact upon the result computation strategies. Most cases of plan-preserving adaptation are rather straightforward and will be supported by the first releases of the query engine, while plan-changing adaptation is only sketched in this paper and will be the subject of future work.

References

1. Avnur, R., Hellerstein, J.M.: Eddies: Continuously Adaptive Query Processing. In: SIGMOD Conference, pp. 261–272 (2000)
2. Shah, M.A., Hellerstein, J.M., Chandrasekaran, S., Franklin, M.J.: Flux: An Adaptive Partitioning Operator for Continuous Query Systems. In: ICDE, pp. 25–36 (2003)
3. Raman, V., Han, W., Narang, I.: Parallel querying with non-dedicated computers. In: Proc. VLDB, pp. 61–72 (2005)
4. Urhan, T., Franklin, M.J., Amsaleg, L.: Cost Based Query Scrambling for Initial Delays. In: SIGMOD Conference, pp. 130–141 (1998)
5. Gounaris, A., Paton, N., Fernandes, A., Sakellariou, R.: Self-monitoring query execution for adaptive query processing. Data Knowl. Eng. 51(3), 325–348 (2004)
6. Sutherland, T.M., Zhu, Y., Ding, L., Rundensteiner, E.A.: An adaptive multi-objective scheduling selection framework for continuous query processing. In: IDEAS, pp. 445–454 (2005)
7. Raman, V., Deshpande, A., Hellerstein, J.M.: Using State Modules for Adaptive Query Processing. In: Proc. ICDE, pp. 353–364 (2003)
8. Babu, S., Bizarro, P., DeWitt, D.: Proactive Re-Optimization. In: Proc. ACM SIGMOD, pp. 107–118 (2005)
9. Bizarro, P., Babu, S., DeWitt, D.J., Widom, J.: Content-based routing: Different plans for different data. In: VLDB, pp. 757–768 (2005)
10. Kabra, N., DeWitt, D.J.: Efficient Mid-Query Re-Optimization of Sub-Optimal Query Execution Plans. In: SIGMOD Conference, pp. 106–117 (1998)
11. Markl, V., Raman, V., Simmen, D.E., Lohman, G.M., Pirahesh, H.: Robust Query Processing through Progressive Optimization. In: SIGMOD Conference, pp. 659–670 (2004)
12. Li, Q., Shao, M., Markl, V., Beyer, K., Colby, L., Lohman, G.: Adaptively Reordering Joins during Query Execution. In: Proc. ICDE, pp. 26–35 (2007)
13. Eurviriyanukul, K., Paton, N.W., Fernandes, A.A.A., Lynden, S.J.: Adaptive Join Processing in Pipelined Plans. In: Proc. EDBT, pp. 183–194 (2010)
14. Ives, Z., Halevy, A., Weld, D.: Adapting to Source Properties in Data Integration Queries. In: Proc. SIGMOD, pp. 395–406 (2004)
15. Bozzon, A., Brambilla, M., Ceri, S., Fraternali, P.: Exploring the Web with Search Computing. In: Ceri, S., Brambilla, M. (eds.) Search Computing II. LNCS, vol. 6585, pp. 10–25. Springer, Heidelberg (2011)

16. Ilyas, I.F., Aref, W.G., Elmagarmid, A.K.: Supporting top-k join queries in relational databases. VLDB J. 13(3), 207–221 (2004)
17. Ilyas, I.F., Aref, W.G., Elmagarmid, A.K., Elmongui, H.G., Shah, R., Vitter, J.S.: Adaptive Rank-Aware Query Optimization in Relational Databases. ACM Trans. Database Syst. 31(4), 1257–1304 (2006)
18. Braga, D., Corcoglioniti, F., Grossniklaus, M., Vadacca, S.: Efficient Computation of Search Computing Queries. In: Ceri, S., Brambilla, M. (eds.) Search Computing II. LNCS, vol. 6585, pp. 141–155. Springer, Heidelberg (2011)

Part 6

Tools and Mashups

Developing and configuring modern information systems requires new models, methods, and tools, with the objective of implementing systems which are easier to build, that better incorporate user's requirements, and are more reusable. In this context, an important role can be played by mashup editors, which lower the complexity of expressing search computing queries through visual compositions of modules.

The first chapter reports on the tools developed to support the complex search computing application lifecycle. We present a toolsuite structured as an online development platform, in which developers can selectively access tools based on their role. Service registration tools consist of a set of facilities for allowing normalization of service interfaces and their registration as service marts. Query configuration tools allow designers to compose applications consisting of sets of connected service marts. Query plan refinement tools consist of a visual modeling environment that allows search computing experts to edit query plans specified according to the Panta Rhei notation. The tools produce a complete application configuration that is automatically deployed and made available to final users. The availability of the tool suite as an online mashup-like platform aims at increasing search computing application design productivity, reducing the time to deployment, and avoiding the burden of downloading and installing software.

The second chapter presents an overview of the most recent approaches to mashup-based development, that consider distributed mashup, dealing with the configuration both of the client (i.e., the user interface behavior) and the server (i.e., the composition of services), based on the same paradigm. The chapter is based upon several projects ongoing in Trento University.

Finally, a discussion chapter addresses the fitting of end-user oriented design interfaces to different classes of users, dwelling into the distinction of declarative and imperative mashups, on the kind of skill expected by mashup users, on the need of tailoring the design tool to a specific sector, and on the possible interactions and cross-fertilizations between the mashup approaches and the SeCo project.

Tools Supporting Search Computing Application Development

Marco Brambilla and Luca Tettamanti

Politecnico di Milano, Dipartimento di Elettronica ed Informazione,
V. Ponzio 34/5, 20133 Milano, Italy
{fistname.lastname}@polimi.it

Abstract. Search computing provides a solution to the problem of multi-domain, exploratory search. To manage the complex set of subsystems and configurable options, the proper set of development and configuration tools is needed. In this chapter we describe the development process phases envisioned for designing search computing applications and also a unified tool suite that aggregates a set of design and configuration tools. The tools cover the phases of service registration, service annotation, and application configuration. The latter in turn is organized in query specification, query plan refinement, and definition of user interface options.

Keywords: Search computing, tools, software engineering, mashup, development process.

1 Introduction

Search computing provides a powerful solution to the problem of multi-domain, exploratory search. However, to cope with such a complex problem, several techniques and algorithms need to be devised to solve the diverse sub-problems, including query specification, query planning and optimization, query execution, service invocation, user interface rendering, and so on. This leads to a complex software architecture [3], incorporating a large set of subsystems. As a consequence, building and configuring search computing applications may become a non-trivial task for a designer. To ease his work, the proper set of development, configuration and monitoring tools are needed. The tools should support all the development phases necessary to support all the aspects of a multi-domain search application, according to sensible development process guidelines.

The basic development process envisioned for designing search computing applications has been already presented in [1] and a model-driven description of the concepts involved in the development process (together with the transformations they undergo) has been given in [6].

The contributions of this chapter are: (1) the identification of the possible application design scenarios; (2) a refined description of the development process, with a precise definition of the sub-phases involved in each design task, with special

S. Ceri and M. Brambilla (Eds.): Search Computing II, LNCS 6585, pp. 169–181, 2011.

attention to service registration; and (3) the presentation of the tool set that has been developed for supporting the designer in all the development phases.

The development process is structured into the following phases: service registration, service annotation, and application configuration; the latter in turn is composed by query specification, query plan refinement, and user interface definition.

Besides the design and configuration tools, also monitoring tools are crucial for checking the behavior and the performances of the designed search computing applications. These tools are out of the scope of this chapter. Instead, some examples of monitoring tools for the search computing execution engine can be found in [4].

This chapter is organized as follows: Section 2 discusses the development scenarios for search computing applications and summarizes the main roles and processes for SeCo application development; Section 3 details the service registration phases and tools and Section 4 details the application configuration phases and tools; Section 5 provides an overview of the toolsuite architecture; and Section 6 concludes the chapter.

2 Background and Related Approaches

Our work on configuration tools stands in the middle between two classes of tool solutions: mashups and MDD (Model Driven Development) tools. We do not aim at covering the entire background of these two topics here (detailed background discussions are provided in [7] about mashups and in [1] about web development tools and modeling approaches); instead, we report on the evolutions of the field and on our positioning with respect to the recent trends.

Model Driven Development (MDD) aims to raise the level of abstraction for software development by providing more powerful concepts for capturing and reusing development knowledge. MDD approaches are gaining a foothold in industrial practice, thanks to two main trends: the increased support and adoption of universal modeling languages such as UML, and the birth and growth of domain specific languages (DSL) that provide design constructs and notations for focused application domains. The former is being facilitated by the Object Management Group (OMG), which leads the standards development effort (MDA/MOF/UML, ...) for MDD. Leading software tool vendors such as IBM and Microsoft, as well as many smaller vendors and open source projects, are developing technologies to support MDD. In particular, we are recently witnessing a convergence of different domains toward MDD practices. For instance, the BPM (Business process management) field is becoming more and more aware of the advantages of MDD: IBM has been recently pushing MDD approaches to BPM within its WebSphere platform [9].

Our approach fully exploits MDD philosophy, by providing a set of tools for the configuration of Search Computing systems completely based on conceptual models [1]. In particular, a detailed model-driven description of our approach, in terms of models and their transformations has been given in [6].

Mashups natively adopt interaction paradigms much closer to the end users. An early survey of the potential and limitations of mashups is provided in [13].

Some major proposals in the field of end user mashups have been discontinued in the last two years. Among them, we can mention Google MashupEditor, Microsoft Popfly and IBM QEDWiki. In this segment only Yahoo Pipes [12] survives. This may be read as a symptom of perceived weakness of this kind of approaches. However, we think the issue is more related to the target user profile than to the approaches themselves.

Other tools for enterprise mashup development and execution are currently more widely adopted. Enterprise mashups are software applications that consume and combine data and applications across the enterprise, gathering information from a variety of sources, often performing logical or mathematical operations as well as presenting data. The development paradigm and languages they use sit in between mashups and service compositions, so as to be rich enough to automate integration of enterprise processes and interfaces. The added complexity due to the increased richness in the language is compensated by means of highly usable programming UIs, programming and prototyping advices, and the facilitated reuse of programming logic.

IBM Mashup Center [8] is the unified IBM solution for enterprise mashup which integrates IBM InfoSphere MashupHub and IBM Lotus Mashups together in a single product that can be installed with a single click. Jackbe Presto [10] is a complete platform for mashup development, covering all the phases from source wrapping, to infrastructure setup and management, to application design and deployment (with also an AppStore facility).

Some efforts are currently spent towards a unified vision and language for mashups: the Open Mashup Alliance is proposing EMML (Enterprise Mashup Markup Language) [11], an XML markup language for creating enterprise mashups. EMML is an open language specification whose primary benefits are mashup design portability and interoperability of mashup solutions. These benefits are expected to accelerate the adoption of enterprise mashups by creating transferable skills for software developers and reducing vendor lock-in.

Our work is in line with the trend towards enterprise mashup, because the target user of our tools is currently a developer in charge of setting up search computing solutions for specific industrial scenarios.

3 Development Process and Scenarios

3.1 Development Process

The design process for search computing applications can be split into two macro-phases: service registration and application configuration. **Service registration** comprises all the phases that aim at making third party search services available in a search computing platform (i.e., registration and semantic annotation of Service marts, Access patterns, Service interfaces, and Connection patterns), while **Application configuration** covers the phases required to configure targeted queries and applications (Query specification, Query generation, Query plan refinement, and Interface configuration). More details on the development process can be found in [1].

3.2 Needs and Scenarios

The way the development phases are actually executed depends on the search requirements that need to be fulfilled. These needs can be classified in two main scenarios: **open, interactive applications** for information exploration; and customized, **vertical search applications**. Table 1 summarizes the design phases and their roles within the two scenarios, which basically differ only in application configuration time, while they exploit the same tools and approaches both at service registration time and application execution time:

- *Interactive applications*: this scenario addresses the need of flexibility and information exploration of a user who wishes to exploit the entire set of resources available in SeCo and wants to build dynamically his/her own exploration path through several, incremental search steps. In this case, the configuration of the SeCo framework simply consists in defining correctly the resource graph, i.e., the set of searchable concepts and corresponding search services. Once this is done, the user can browse the graph and query any search service in any combination he/she wants. This can be done with exploration interfaces like the flexible Liquid Query one presented in [2]. In this way, the user builds his/her own composition of search services and immediately executes it, like in a mashup design approach;
- *Custom applications*: this is a more traditional scenario in which a developer designs a customized search application (e.g., to be embedded in vertical portals or enterprise sites), by configuring the search computing framework to work on some predefined domains, through a selected set of search services. Then, end users will search and browse the configured application. Information exploration can be still performed, but only according to the initial choices of the application designer.

Table 1. Search computing design phases and roles within the scenarios

Design Phases	Scenarios	
	Interactive application	Custom application
Service registration - Service mart definition - Access pattern definition - Connection pattern definition - Service interface registration - Semantic annotation	Registration process and tools	Registration process and tools
Application configuration - Query specification - Query generation - Query plan refinement - Interface configuration	No Configuration	Configuration process and tools
Application execution - Query submission - Information exploration	Liquid Query (whole information space)	Liquid Query (on vertical domain)

4 Service Registration

Service registration is a complex task that comprises several substeps, corresponding to the different abstraction levels that describe the resources: Service Mart (SM), Access Pattern (AP), and Service Interface (SI), as defined in [5]: SMs characterize real world entities (structurally defined by means of attributes, and their relationships) at the conceptual level; APs describe the access to the conceptual entities in terms of data retrieval patterns at the logical level; and SIs represent the mappings of these patterns to concrete Web Service Interfaces at the physical level. Based on the different situations, a top-down or bottom up registration approach can be adopted. Fig. 1 summarizes the different registration cases:

1. *Registration of a service associated to a new real world entity:* if the concept describing that entity is not yet defined in the repository, all the levels (SM, AP, SI) need to be registered. In this case a top-down registration is the best suited solution;
2. *Registration of a service with a new access pattern, but associated to an existing concept:* in this case, an intermediate approach is adopted: a new access pattern is created and mapped to an existing service mart; then, the new service is registered and mapped to the AP;
3. *Registration of a new service perfectly homogeneous to an already registered one:* in this case, the new service can be registered and then mapped to an existing Access Pattern of choice, with a bottom-up approach.

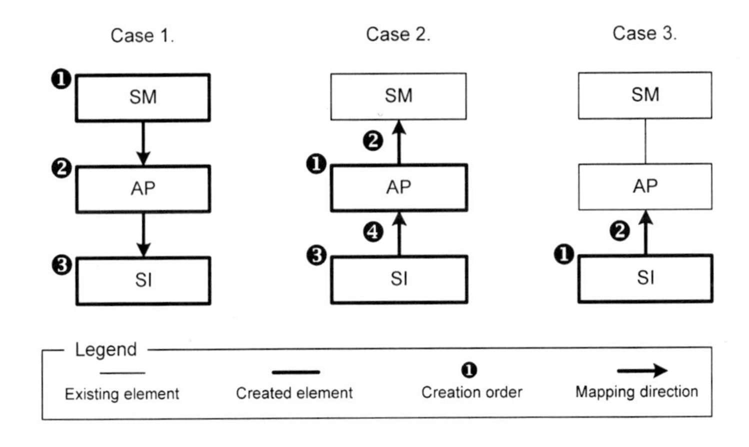

Fig. 1. Service Mart (a) and Access Pattern (b) rendering in the service registration tool

The following subsections describe each phase in details. The discussion will follow the basic top-down scenario (case 1), and the other options will be mentioned by difference with respect to the main case.

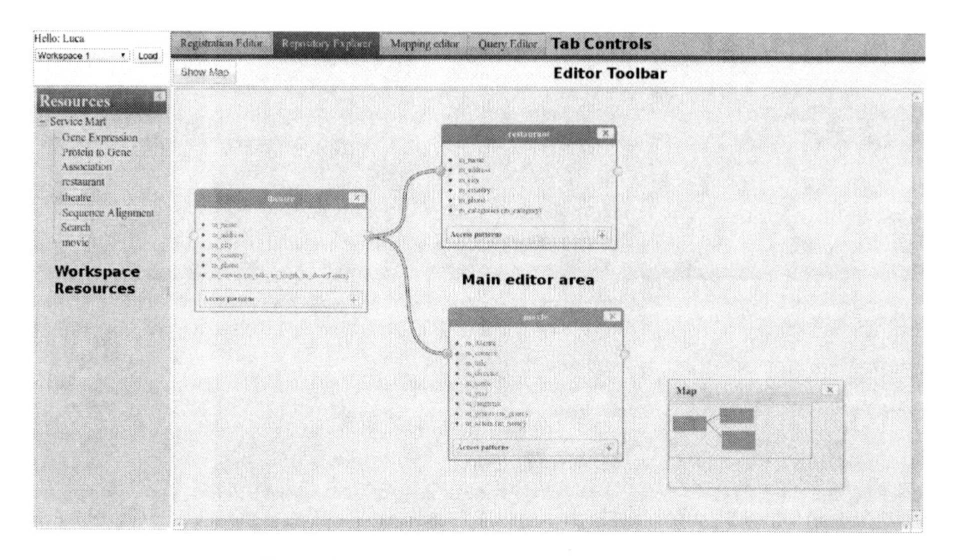

Fig. 2. Overview of the toolsuite user interface

According to the search computing paradigm, the service registration task is needed in all the scenarios and is assigned to a developer and domain expert role. To ease this job, a set of editors have been devised within the search computing tool suite to cover all the registration phases. An overview of the web interface of the toolsuite is shown in Figure 2.

The interface of the web based application is organized into different functional areas: the topmost portion is used for workspace control, tab control and other generic (i.e. editor independent) commands. The left part (which is foldable) contains the resource tree which hosts whatever resources are relevant for the editor currently active. The remaining part of the screen is split between a toolbar (editor specific) and the main working area which is again controlled by the editor component currently active.

4.1 Service Mart Definition

The first step in the registration process is to identify which Service Mart can be used to categorize the service that is going to be registered. In case no suitable Service Mart is found the user shall take a top-down approach to the registration. Service Marts - the highest level of abstraction in our model - represent "real world" objects and are used to hide the underlying modalities of data access and implementation details. The Service Mart is composed by a name, a human readable description and the schema, i.e., a collection of attributes that define the represented real word entity (see Fig. 3. for a sample rendering of a SM in the tool).

All these properties must be specified when a service mart is created. The attributes in the schema can be: *simple attributes* described by a name, an optional description and a data type (which is used as a default type in the underlying layers); or *repeating groups*, composed by a list of simple attributes (only a single level of nesting is allowed).

Furthermore the user may modify the definition of a Service Mart to accommodate new Access Patterns or services; the addition of a new attribute to the schema is always allowed, removal instead is more delicate since the attribute might be referenced in many Access Patterns connected to the current Service Mart; for this reason the tool allows the deletion of an attribute if and only if it is unreferenced.

If an existing Service Mart can be adopted as conceptual description of the new service, the user can start from the lower level, by selecting a suitable Access Pattern (or by creating a new one if none is found) and defining a Service Interface.

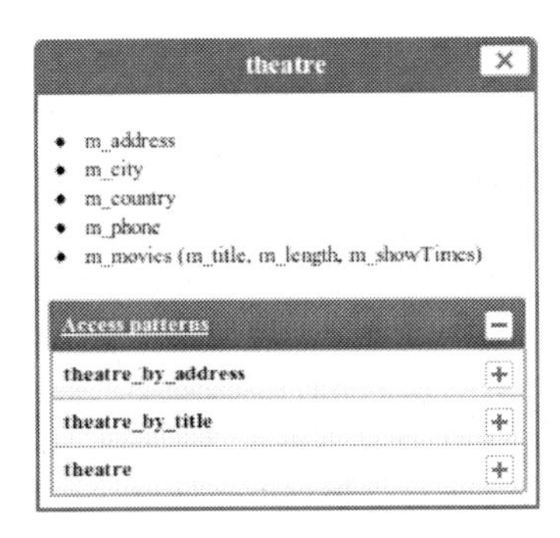

Fig. 3. Service Mart and Access Pattern list rendering in the service registration tool

4.2 Access Pattern Definition

An Access Pattern represents the modality to access a given Service Mart; of course multiple patterns may be defined for each Service Mart. Fig. 3. shows the rendering of the list of Access Pattern for a Service Mart registered in the tool.

In the pure top-down approach, a new Access Pattern is created from the definition of a Service Mart: starting from the schema of the Service Mart, the user can choose to remove attributes that are not relevant for the pattern, add new ones that are not present in the definition of the Mart (the so called *external attributes*[1]), and finally proceed to tag each attribute as "input", "output", or "ranked" (see Fig. 4.a). One input attribute can also be marked as the "selector", i.e. an attribute whose value is used to select the most appropriate Service Interface at run-time[2]. If needed, the user can also override the data types indicated in the corresponding Service Mart. If a bottom-up approach is adopted, the mapping between attributes of the SM and the AP must be specified explicitly, as shown in Fig. 4.b.

---

[1] *External attributes* are "new" attributes appearing at the logical level within the Access Pattern definition, while not being present at the conceptual level; they support object access and ranking.

[2] *Selector attributes* support the selection of specific service implementations. A specific SI can be selected on the basis of the value of the attribute, according to a guard condition. For instance, different search services can be selected for the same concept (e.g., a theatre) according to the user's country.

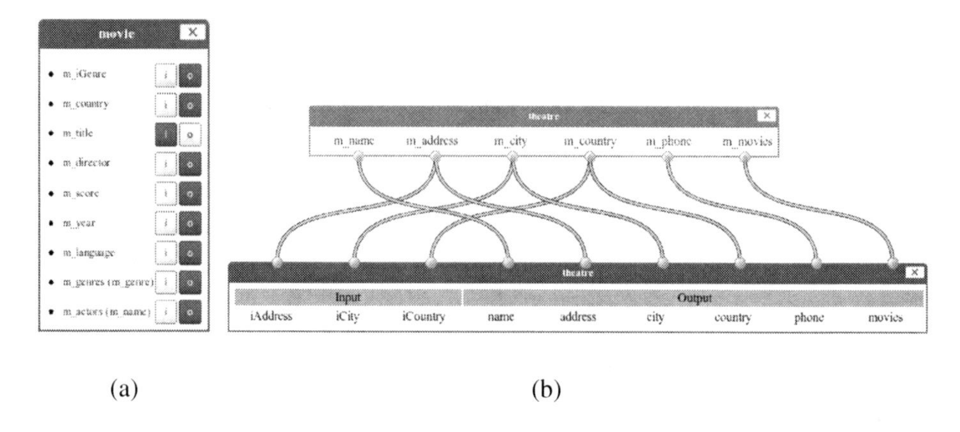

(a) (b)

Fig. 4. Configuration of the Access Pattern inputs and outputs (a) and mapping view between the Service Mart and Access Pattern levels (b)

4.3 Connection Pattern Definition

At this point the user can create a connection between the new Access Pattern and other Access Patterns in the repository, as shown in Fig. 5; these links, called Connection Patterns, will be used later to compose the queries on the system. The connections are specified by the type of the join (either "pipe" or "parallel"), by the couples of attributes involved in the join, and by a list of possible predicates that can be applied to perform the join. For example a Connection Pattern may indicate that the "address" of AP1 should be "near" or "equal" to the "address" of AP2.

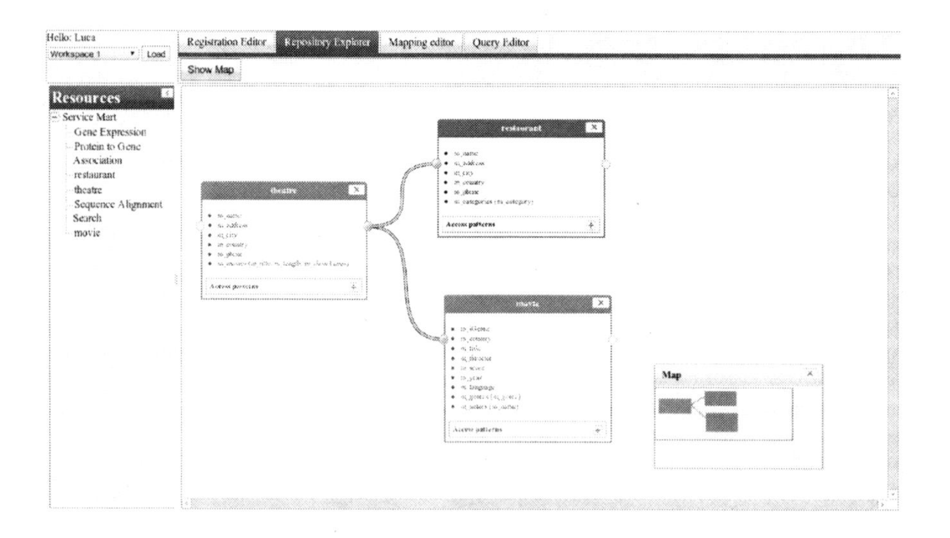

Fig. 5. Tool UI, including a set of SMs connected through connection patterns

Abstract connection patterns can be defined and drawn also at the Service Mart level, but they have no specific semantics: they only represent an aggregation of actual connection patterns between corresponding APs.

Connection patterns can be defined beforehand with respect to the query definition, because they represent the semantically significant exploration and navigation options between the concepts.

4.4 Service Interface Registration

The Service Interface definition, describing the actual search services, is composed by four basic blocks:

- **the input/output schema**, with the mappings towards the corresponding Access Pattern. Each attribute is described by its name, data type, and the mapping toward the corresponding attribute in the Access Pattern. Output attributes might be repeating groups. The schema must be compatible with the Pattern.
- **the quality of service** (QoS) properties, with a first list providing the expected values of the parameters (used by the engine to efficiently plan the query execution and schedule the service invocation) and a second one listing the properties that are monitored at run-time. The QoS dimensions currently considered include: time to live (maximum time the cached results can be considered up to date); average chunk size; ERSPI (expected result size per invocation); service decay factor (the number of expected useful results); service initialization time (time needed to produce the first chunk of the answer); service fetch time (time to complete a fetch of a chunk).
- **the score function** used to compute a global score for the tuples provided by the service;
- **the implementation details** for the service invocation, with a classification of the service based on the implementing technology (e.g., Sparql, Yql, SeCo, …).

4.5 Registration Updates

The tools allow finer grained modifications to the model; there are however several constraints on the modifications that may be performed on the various resources. In general the addition of new attributes at any level of the model is always allowed. More in details, the creation of a new attribute in the SM is allowed unconditionally (although the new attribute is not automatically used in the lower levels); a new attribute in an AP will be either external or connected to an attribute in the corresponding SM if such an attribute is available; in the SI the only constraint is that the new attribute may be marked as input if and only if it can be mapped to an input attribute of the AP; finally, unmapped output attributes are allowed (the attribute is present in the physical implementation of the service, but is not relevant for the given pattern). Deletion – either of attributes or entire objects – is disallowed if the item is referenced elsewhere in the model.

Update rules are more complex but, like deletion, they are always allowed if the element under modification is not referenced elsewhere and the outcome is compatible with the upper layers; even though strictly enforcing, these rules result in tedious and error prone round trips between the various components to fix up all the dependencies, so the tool is also able to propagate some modification automatically. Among the notable update cases we can mention:

- the transformation of a single-value attribute into a repeating group: provided that the attribute is never used as an input, this operation can be done top down (starting from the SM) and the underlying resources are automatically transformed;
- the transformation of a normal attribute into a selector: this operation requires two steps: the creation of the selector itself and the definition of the predicates for selecting the most appropriate service interface. Besides this, the operation is not further constrained;
- the splitting of a composite attribute into its components or vice versa the integration of several attributes into a composed one: these operations are always allowed if the involved attributes (both components or composite) are not repeating groups. Indeed, if they are, possible cardinality inconsistencies may arise: therefore, consistency must be checked before composition/decomposition;
- hierarchies of attributes with semantic description/typing: inheritance can be defined upon types (e.g., Geo-location is-a address, city is-a address), but then consistency must be checked on all the usages of the attributes and on their relationships.

5 Application Configuration

This phase consists in configuring a customized query and associated exploration paths for a vertical domain.

5.1 Query Specification

The specification of the query consists in defining the information needs that one wants to be covered by the vertical application. This concretely means to select from the service repository a set of service marts and connection patterns that describe the real world entities and their relationships that should be included in the query. Furthermore, the designer can select a set of extra service marts usable by the end user to expand the current query (e.g., to expand a query on movies by means of a service that joins selected movies to their reviews and the theatres where they are programmed to nearby restaurants). After that, a specific service implementation can be selected for those service marts associated with multiple sources, together with the way in which they can be invoked in the current query (e.g., selecting the inputs requested to the user and the ones that are hidden).

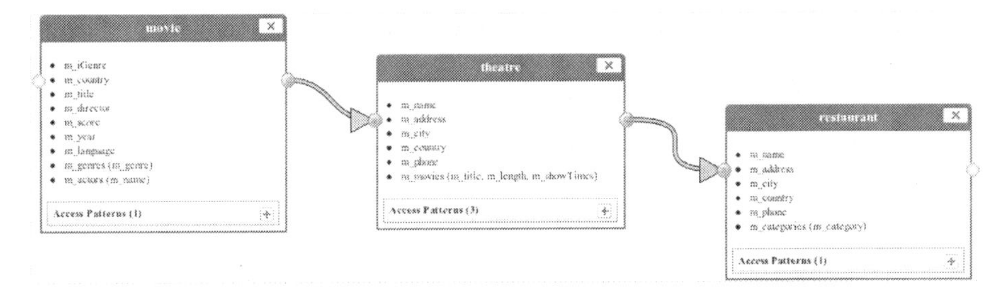

Fig. 6. Tool user interface for query configuration

The query configuration tool enables the user to work on two orthogonal dimensions: the definition of the components of the query (in terms of Service Marts) and the refinement of these components (at first in terms of access patterns and then of service interfaces); the tool also provides a set of filters for quickly narrowing down the candidates access patterns and service interfaces.

5.2 Query Generation

This phase consists in building a logical query, specified in the SeCoQL syntax, starting from the conceptual query, described as a graph. This step takes in input the query definition and performs the translation automatically. The transformation is pretty straightforward, since the SeCoQL query syntax is just a rewriting of the graphical representation available in the tool. In the actual system, the output of the query configuration is not provided to the planner as a SeCoQL specification; instead, a JSON serialization of the query structure and objects is generated. This avoid a back and forward transformation and parsing of the textual SeCoQL language.

5.3 Query Plan Refinement

This phase consists in manually refining the optimized query plan produced by the SeCo platform starting from the query specification. In general, query plans are self-tuned, and therefore this phase is usually not needed. However, an expert can decide to manually override the optimal plans to comply with complex queries, or manually select alternative data sources, or improve scalability through parallelism, or provide customized choices not covered by the optimization.

The query plan refinement tool consists of a visual modeling environment that allows SeCo experts to edit query plans specified according to the Panta Rhei notation. This task can be performed by an expert search computing developer who needs to be well aware of the Panta Rhei language, query plan strategies, and optimization issues.

5.4 User Interface Configuration

The GUI configuration activity requires customizing the structure of a generic interface, by choosing: 1) optional selection predicates to restrict the objects retrieved by the query (e.g., a maximum price target for events or flights); 2) default ranking criteria for the results; 3) visual preferences on the display of the result set (e.g., sorting, grouping, and clustering attributes or the size of the result list). Application configuration tools allow *defining* the interface of the query submission form and of the result set, together with the default settings for the application and the allowed Liquid Query operations.

6 Toolsuite Architecture

To support the complex SeCo lifecycle, we have designed a toolsuite that comprises a variety of instruments, structured as an online development platform in which developers can login and, according to their role, access the right set of tools for building SeCo applications. Developers can decide to work at a single model level (e.g., only at the service mart level), or can define their own workspaces, that provide a coordinated view of all the models of a complex project (e.g., the service mart, query plan, and user interface models for a given application).

The availability of the tools as online applications aims at increasing SeCo application design productivity, reducing the time to deployment, and avoiding the burden of downloading and installing software. The tools produce a complete application configuration that is automatically deployed and made available to final users.

Fig. 7 shows an overview of the architectural design of the application. It is developed using JavaScript and uses the ReST API to interact with the backend and is structured according to the MVC pattern, supported by the JavaScriptMVC library; the UI is rendered using YUI2 (common widgets) and WireIt (boxes and arrows, the design canvas) along with a few custom-built widgets. The application provides a generic framework for managing the models (both application specific – like the workspaces – and the SeCo objects – like SM, AP, SI, etc.) which is used by a number of plugins that provide the design logic; the two components are loosely coupled and the interaction is done only using notifications (i.e. the framework responds to messages sent by the active plugin and in turn dispatches other notifications to

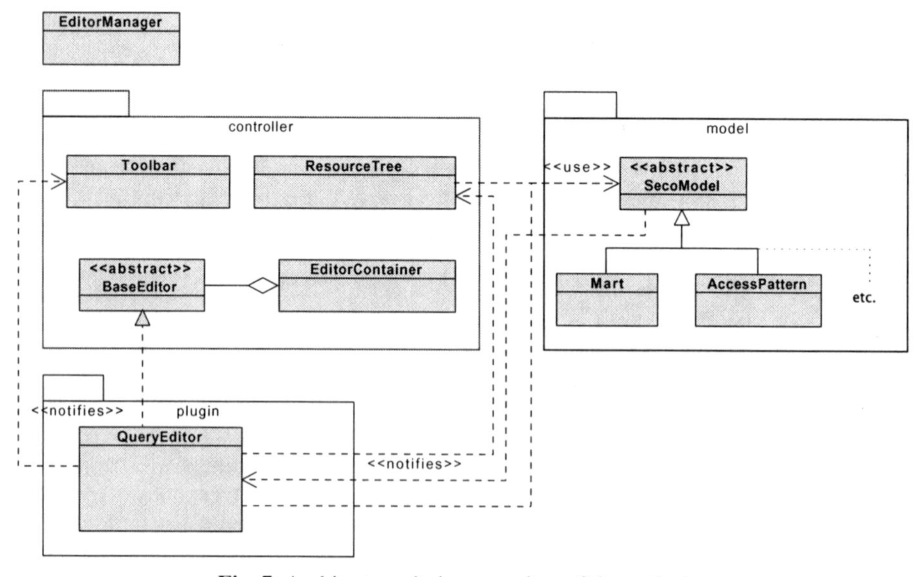

Fig. 7. Architecture design overview of the toolsuite

inform the plugin of using actions or other events). This separation is also visible in the UI: the toolbar and resource tree are provided by the framework (and are controlled via a notification API), while the main design area is controlled by the plugin.

7 Conclusions

This chapter described the set of tools that have been devised for supporting the design and configuration of search computing applications. Besides describing the tools features, the chapter also highlighted the operations allowed on the resources managed by the tools and described and overview of the architectural aspects.

References

[1] Bozzon, A., Brambilla, M., Ceri, S., Corcoglioniti, F., Gatti, N.: Building Search Computing Applications. In: Ceri, S., Brambilla, M. (eds.) Search Computing. LNCS, vol. 5950, pp. 268–290. Springer, Heidelberg (2010)
[2] Bozzon, A., Brambilla, M., Ceri, S., Fraternali, P.: Exploring the Web with Search Computing. In: Ceri, S., Brambilla, M. (eds.) Search Computing II. LNCS, vol. 6585, pp. 10–25. Springer, Heidelberg (2011)
[3] Bozzon, A., Brambilla, M., Corcoglioniti, F., Vadacca, S.: A service-based Architecture for Multi-domain Search on the Web. In: Maglio, P., Weske, M., Yang, J., Fantinato, M. (eds.) ICSOC 2010. LNCS, vol. 6470, pp. 663–669. Springer, Heidelberg (2010)
[4] Braga, D., Corcoglioniti, F., Grossniklaus, M., Vadacca, S.: Efficient Computation of Search Computing Queries. In: Ceri, S., Brambilla, M. (eds.) Search Computing II. LNCS, vol. 6585, pp. 141–155. Springer, Heidelberg (2011)
[5] Brambilla, M., Campi, A., Ceri, S., Quarteroni, S.: Semantic Resource Framework. In: Ceri, S., Brambilla, M. (eds.) Search Computing II. LNCS, vol. 6585, pp. 73–84. Springer, Heidelberg (2011)
[6] Brambilla, M., Ceri, S., Tisi, M.: Search computing: A model-driven perspective. In: Tratt, L., Gogolla, M. (eds.) ICMT 2010. LNCS, vol. 6142, pp. 1–15. Springer, Heidelberg (2010)
[7] Daniel, F., Soi, S., Casati, F.: From Mashup Technologies to Universal Integration: Search Computing the Imperative Way. In: Ceri, S., Brambilla, M. (eds.) Search Computing. LNCS, vol. 5950, pp. 72–93. Springer, Heidelberg (2010)
[8] IBM. IBM Mashup Center, `http://www-01.ibm.com/software/info/mashup-center/`
[9] IBM. IBM WebSphere, `http://www-01.ibm.com/software/websphere/`
[10] Jackbe. Presto, `http://www.jackbe.com/Products/`
[11] Open Mashup Alliance. EMML, `http://www.openmashup.org/omadocs/v1.0/index.html`
[12] Yahoo. Yahoo Pipes, `http://pipes.yahoo.com`
[13] Yu, J., Benatallah, B., Casati, F., Daniel, F.: Understanding Mashup Development. IEEE Internet Computing 12(5) (2008)

Distributed User Interface Orchestration: On the Composition of Multi-User (Search) Applications

Florian Daniel, Stefano Soi, and Fabio Casati

University of Trento, 38123 Povo (TN), Italy
{daniel,soi,casati}@disi.unitn.it

Abstract. While mashups may integrate into a new web application data, application logic, and user interfaces sourced from the Web – a highly intricate and complex task – they typically come in the form of simple applications (e.g., composed of only one web page) for individual users. In this chapter, we introduce the idea of *distributed user interface orchestration*, a mashup-like development paradigm that, in addition to the above features, also provides support for the coordination of *multiple users* inside one shared application or process. We describe the concepts and models underlying the approach and introduce the *MarcoFlow* system, a platform for the assisted development of distributed user interface orchestrations. As a concrete development example, we show how the system can be profitably used for the development of an advanced, *collaborative search application*.

1 Introduction

After workflow management (which supports the automation of business processes and human tasks) and service orchestration (which focuses on web services at the application layer), web mashups [1] feature a significant innovation: **integration at the UI level**. Besides web services or data feeds, mashups indeed reuse pieces of UIs (e.g., content extracted from web pages or JavaScript UI widgets) and integrate them into new web pages or applications. While mashups therefore manifest the need for reuse in UI development and for suitable UI component technologies, so far they produced rather simple applications consisting of one web page only.

We argue that there is a huge spectrum of applications that demand for development approaches that are similar to those of mashups but that go far beyond single page applications and, in fact, support multiple pages, multiple actors, complex navigation structures, and – more importantly – process-based application logic or navigation flows. We call this type of applications **distributed UI orchestrations** [2], as (i) both components and the application itself may be distributed over the Web and operated by different actors, (ii) in addition to traditional web services we also integrate novel JavaScript UI components, and (iii) services and UIs are orchestrated in a homogeneous fashion.

Developing distributed UI orchestrations therefore raises the need for the coordination of individual actors and the development of a distributed user interface *and* service orchestration logic. Doing so requires:

S. Ceri and M. Brambilla (Eds.): Search Computing II, LNCS 6585, pp. 182–191, 2011.

- Understanding how to *componentize UIs and compose them*;
- Defining a logic that is able to *orchestrate both UIs and web services*;
- Providing a language and tool for *specifying distributed UI compositions*; and
- Developing a runtime environment that is able to *execute distributed UI and service compositions*.

In this chapter, we describe how the above challenges have been solved in the context of the MarcoFlow project [2] and how the resulting approach can be leveraged for the development of a distributed search computing application that requires the coordination of services, UIs, and people.

This chapter is organized as follows: Next, we describe the search application that raises the need for distributed UI orchestration. In Section 3, we look at how existing techniques and technologies may support the development of such kind of application, while in Section 4 we introduce the distributed UI orchestration approach in order to fill the gaps. In Section 5, we describe our current development prototype, and in Section 6 we conclude the paper.

2 A Search Scenario

Let us consider the following collaborative search scenario illustrated in Figure 1, an extension of the single-user scenario discussed in [3].

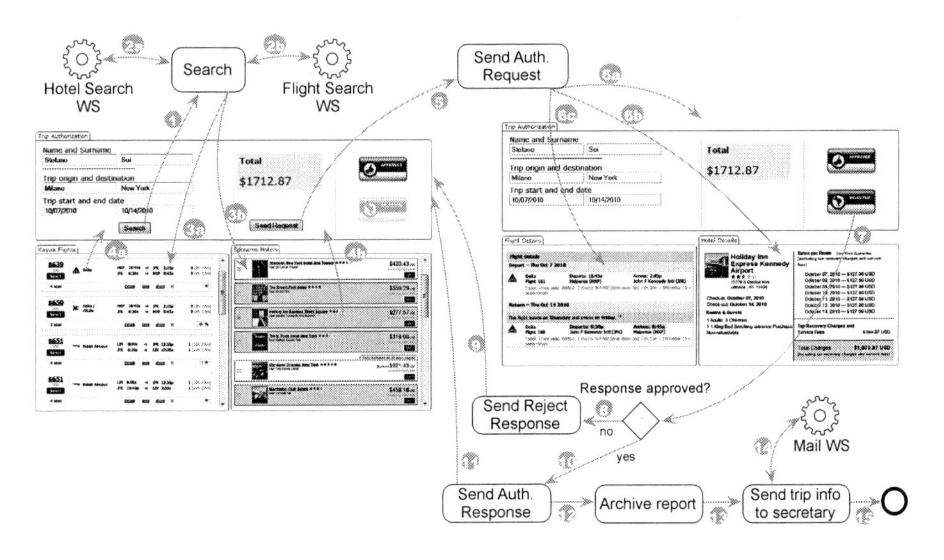

Fig. 1. Trip authorization distributed application. The gray arrows indicate synchronization or orchestration points; the number labels indicate their order in time.

This time we want to assist an employee that needs to request the authorization of a business trip to his superior. The employee can enter the relevant information about the trip (the origin and destination, and the start and end dates) and search for related flights and accommodations. He/she can select his/her preferred choices from the list of flights and hotels and send the request to the superior. The superior can inspect the request, along with its details, and send a response (accept or redo) to the employee, together with some optional comment. If the request is approved, the response will be sent to the employee and the procedure will end with archiving and mailing operations. If the request is rejected the previous steps (all or a part of them) can be repeated until the authorization request is approved.

If we analyze the scenario, we see that the envisioned application (as a whole) is *distributed* over the Web. The application includes, besides the process logic, two mashup-like, web-based control consoles for the employee and the superior that are themselves part of the orchestration and need to interact with the underlying process logic. The UIs for the actors participating in the application are composed of UI components, which can be components developed in-house (like the *Trip Authorization* component) or sourced from the Web (like the *Hotel Search* and the *Kayak Flights Search* component); service orchestrations are based on web services. In our case, the latter two UI components serve only to render content, which still needs to be queried from the Web via dedicated web services. The *Trip Authorization* component, instead, collects data from the user (via its input fields) and from the other two UI components (via suitable synchronization operations). Finally, the two applications for the employee and the superior are instantiated in different web browsers, contributing to the distribution of the overall UI and raising the need for synchronization.

3 Distributed UI Orchestration

The **key idea** to approach the coordination of (i) UI components inside web pages, (ii) web services providing data or application logic, and (iii) individual pages (as well as the people interacting with them) is to split the coordination problem into two layers: *intra-page UI synchronization* and *distributed UI synchronization and web service orchestration*.

UIs are typically event-based (e.g., user clicks or key strokes), while service invocations are coordinated via control flows. In [2] we show how to describe UI components (as also introduced in [4]) in terms of standard WSDL descriptors, how to bind them to JavaScript, and how to extend the standard BPEL language in order to support the two above composition layers. We call this extended language **BPEL4UI**. Fig. 2 shows the simplified meta-model of the language and details all the new modeling constructs necessary to specify UI orchestrations (gray-shaded), omitting details of the standard BPEL language, which are reused as is by BPEL4UI. The model in Fig. 2 exclusively focuses on the composition aspects, while the events and operations of UI components are defined in their WSDL descriptors [2].

In terms of standard BPEL [5], a UI orchestration is a *process* that is composed of a set of associated *activities* (e.g., sequence, flow, if, assign, validate, or similar), *variables* (to store intermediate processing results), *message exchanges*, *correlation sets* (to correlate messages in conversations), and *fault handlers*. The services or UI components integrated by a process are declared by means of so-called *partner links*, while *partner link types* define the roles played by each of the services or UI components in the conversation and the *port types* specifying the operations and messages supported by each service or component.

Modeling UI-specific aspects requires instead introducing a set of **new constructs** that are not yet supported by BPEL. The constructs, illustrated in Fig. 2, are: *UI type* (the partner link type for UI components), *page* (the web pages over which we distribute the UI of the application), *place holder* (the name of the place holders in which we can render UI components), *UI component* (the partner link for UI components), *property* (the constructor parameters of UI components), and *actor* (the human actors we associate with web pages).

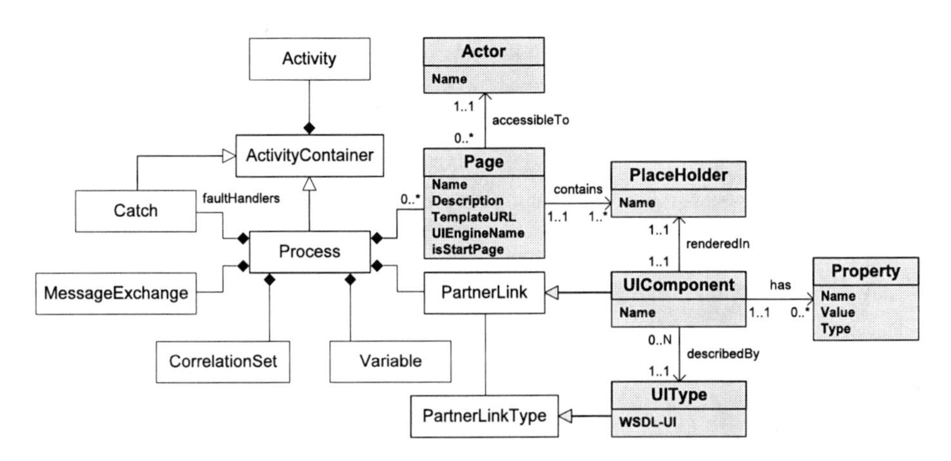

Fig. 2. Simplified BPEL4UI meta-model in UML. White classes correspond to standard BPEL constructs; gray classes correspond to constructs for UI and user management.

It is important to note that although syntactically there is no difference between web services and UI components (the new JavaScript binding introduced into WSDL to map abstract operations to concrete JavaScript functions comes into play only at runtime), it is important to distinguish between services and UI components as their *semantics* and, hence, their usage in the model will be different. A detailed description of the new constructs and their usage can be found in [2], while in Figure 3 we illustrate the BPEL4UI model of our example search application (shown in Figure Fig. 1) as modeled in our extended Eclipse BPEL editor.

The BPEL4UI model is structured into four main blocks: one *repeat-until* and three *sequences* (sub-types of the *Activity* entity in Fig. 2). The repeat-until block at the left manages the flights and hotels search operations. The processing of the block starts upon the reception of the relevant data about the trip from the employee's console, then it invokes the external search web services and, finally, sends the respective results to the UI components rendering the flight and hotel offers. This block of operations can be repeated an arbitrary number of times (e.g., in case the employee want to input new search criteria or a trip request has been rejected and needs to be redone), until the authorization is accepted. Once the search results are rendered in their UI components, the employee can choose a flight and a hotel combination by clicking on the respective choices. This allows the employee to compose his trip request summarized in the *Trip Authorization* component. The two sequence blocks (*Flight Selection* and *Hotel Selection*) in the middle of the model implement the operations that are necessary to synchronize the *Trip Authorization* UI-component, which is then in charge of storing the combination and computing the total cost of the trip. These communications, involving only UI-components belonging to the same page, are completely managed inside the employee's web browser. Once all the trip data are available, the *Send Request* button in the employee console is activated and can be used to forward the authorization request to the superior. Receiving the authorization request starts the right block in the model (*Authorization Request and Response*), which waits for the trip request data and then forwards them to the *Trip Authorization*, *Hotel* and *Flight* UI-components of the superior's console. Now the superior can inspect the request and send a response that is forwarded to the employee's console. If the superior approves the request, two web services are invoked, respectively for archiving and mailing, and, finally, the process is terminated. If the response is a reject, the whole block of operations can be repeated, allowing the employee to modify his request. The right block of service orchestration hence requires the coordination of the two actors, i.e., employee and superior, and the distributed orchestration of UI components and web services. Doing so requires the help from the BPEL engine and the setting of a suitable BPEL correlation set.

As for the **layout** of distributed UI orchestrations, defining web pages and associating UI partner links with placeholders requires implementing suitable HTML

Fig. 3. BPEL4UI modeling example for the Trip Authorization application

templates that are able to host the UI components of the orchestration at runtime. For the design of layout templates, we do not propose any new development instrument and rather allow the developer to use his/her preferred development tool (from simple text editors to model-driven design tools). The only requirement the templates must satisfy is that they provide place holders in form of HTML DIV elements that can be indexed via standard HTML identifiers following a predefined naming convention, i.e., *<div id="marcoflow-left">...</div>*. For instance, all the activities with a "[UI]" prefix in Figure 3 are associated to placeholders, in order to fill the two pages composing our reference scenario.

As this discussion shows, the main **methodological goals** in implementing our UI orchestration approach were (i) relying as much as possible on existing *standards*, (ii) providing the developer with only *few and simple new concepts*, and (iii) implementing a runtime architecture that associates each concern to the *right level of abstraction and software tool* (e.g., UI synchronization is handled in the browser, while service orchestration is delegated to the BPEL engine). These decisions, for instance, allow us to reuse BPEL's internal exception handling mechanisms to manage also exceptions in distributed UI orchestrations.

4 The MarcoFlow Environment

Fig. 4 shows the (simplified) architecture of the MarcoFlow environment, which aids the development and execution of distributed UI orchestrations. The architecture is partitioned into design time, deployment time, and runtime components, according to the three phases of the software development lifecycle supported by MarcoFlow.

The **design** part comprises the*BPEL4UI editor* that supports the full BPEL4UI language as defined in [2]. The editor is an extended Eclipse BPEL editor with (i) a panel for the specification of the pages in which UI components can be rendered and (ii) a property panel that allows the developer to configure the web pages, to set the properties of UI partner links, and to associate them to place holders in the layout.

The **deployment** of a UI orchestration requires translating the BPEL4UI specification into executable components: (i) a set of *communication channels* that mediate between the UI components in the client browser and the BPEL engine; (ii) a *standard BPEL specification* containing the distributed UI synchronization and web service orchestration logic; and (iii) a set of *UI compositions* (one for each page of the application) containing the intra-page UI synchronizations. This task is achieved by the *BPEL4UI compiler*, which also manages the deployment of the generated artifacts in the respective runtime environments.

The **execution** of a UI orchestration requires the setup and coordination of three independent runtime environments: (i) the interaction with users and intra-page UI synchronization is managed in the client browser by an *event-based JavaScript runtime framework*; (ii) a so-called *UI engine server* runs the web services implementing the communication channels; and (iii) a *standard BPEL engine* manages the distributed UI synchronization and web service orchestration.

In order for the superior and the employee to manage their trip authorizations, MarcoFlow also comes with a simple **task manager** (not detailed in Fig. 4), which allows them to start new trip authorizations (the employee) and to participate in running instances of the application (the manager). Each new request requires a new

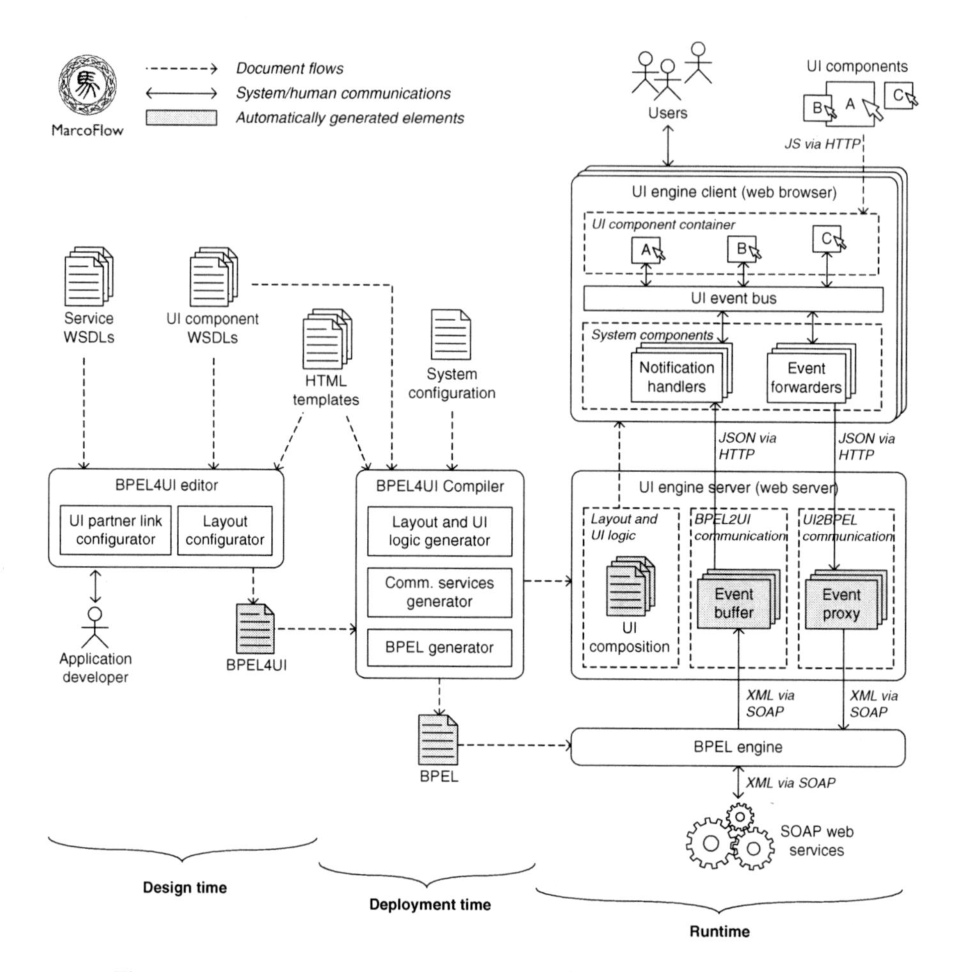

Fig. 4. From design time to runtime: overall system architecture of MarcoFlow

instantiation of the process. All running instances are shown to both actors in their personalized lists. An instance terminates upon successful approval of the trip.

The MarcoFlow system shown in Fig. 4 is fully implemented and running. A patent application for parts of the system has been filed. A detailed demonstration of how MarcoFlow can be used for the development of distributed UI orchestration is available at *http://mashart.org/marcoflow/demo.htm*.

5 Related Work

In most **service orchestration** approaches, such as BPEL [5], there is no support for UI design. Many variations of BPEL have been developed, e.g., aiming at the invocation of REST services [6] or at exposing BPEL processes as REST services [7]. IBM's Sharable Code platform [8] follows a slightly different strategy in the composition of REST and SOAP services and also allows the integration of user interfaces for the Web; UIs are however not provided as components but as ad-hoc Ruby on Rails HTML templates.

BPEL4People [9] is an extension of BPEL that introduces the concept of people task as first-class citizen into the orchestration of web services. The extension is tightly coupled with the **WS-HumanTask** [10] specification, which focuses on the definition of human tasks, including their properties, behavior and operations used to manipulate them. BPEL4People supports people activities in form of inline tasks (defined in BPEL4People) or standalone human tasks accessible as web services. In order to control the life cycle of service-enabled human tasks in an interoperable manner, WS-HumanTask also comes with a suitable coordination protocol for human tasks, which is supported by BPEL4People. The two specifications focus on the coordination logic only and do not support the design of the UIs for task execution.

The systematic development of web interfaces and applications has typically been addressed by the web engineering community by means of **model-driven web design approaches**. Among the most notable and advanced model-driven web engineering tools we find, for instance, WebRatio [11] and VisualWade [12]. The former is based on a web-specific visual modeling language (WebML), the latter on an object-oriented modeling notation (OO-H). Similar, but less advanced, modeling tools are also available for web modeling languages/methods like Hera, OOHDM, and UWE. These tools provide expert web programmers with modeling abstractions and automated code generation capabilities for complex web applications based on a hyperlink-based navigation paradigm. WebML has also been extended toward web services [13] and process-based web applications [14]; reuse is however limited to web services and UIs are generated out of HTML templates for individual components.

A first approach to component-based UI development is represented by **portals and portlets** [15], which explicitly distinguish between UI components (the portlets) and composite applications (the portals). Portlets are full-fledged, pluggable Web application components that generate document markup fragments (e.g., (X)HTML) that can however only be reached through the URL of the portal page. A portal server typically allows users to customize composite pages (e.g., to rearrange or show/hide portlets) and provides single sign-on and role-based personalization, but there is no possibility to specify process flows or web service interactions (the new WSRP [16] specification only provides support for accessing remote portlets as web services). Also **JavaServer Faces** [17] feature a component model for reusable UI components and support the definition of navigation flows; the technology is however hardly reusable in non-Java based web applications, navigation flows do not support flow controls, and there is no support for service orchestration and UI distribution.

Finally, the web mashup [1] phenomenon produced a set of so-called **mashup tools**, which aim at assisting mashup development by means of easy-to-use graphical user interfaces targeted also at non-professional programmers. For instance, Yahoo! Pipes (http://pipes.yahoo.com) focuses on data integration via RSS or Atom feeds via a data-flow composition language; UI integration is not supported. Microsoft Popfly (http://www.popfly.ms; discontinued since August 2009) provided a graphical user interface for the composition of both data access applications and UI components; service orchestration was not supported. JackBe Presto (http://www.jackbe.com) adopts a Pipes-like approach for data mashups and allows a portal-like aggregation of UI widgets (so-called mashlets) visualizing the output of such mashups; there is no synchronization of UI widgets or process logic.IBM QEDWiki (http://services.alphaworks.ibm.com/qedwiki) provides a wiki-based (collaborative) mechanism to glue together JavaScript or PHP-based widgets; service composition is not supported. Intel Mash Maker (http://mashmaker.intel.com) features a browser plug-in which interprets annotations inside web pages allowing the personalization of web pages with UI widgets; service composition is outside the scope of Mash Maker.

In the mashArt [4] project, we worked on a so-called universal integration approach for UI components and data and application logic services. MashArt comes with a simple editor and a lightweight runtime environment running in the client browser and targets skilled web users. MashArt aims at simplicity: orchestration of distributed (i.e., multi-browser) applications, multiple actors, and complex features like transactions or exception handling are outside its scope. The CRUISe project [18] has similarities with mashArt, especially regarding the componentization of UIs. Yet, is does not support the seamless integration of UI components with service orchestration, i.e., there is no support for complex process logic. CRUISe rather focuses on adaptivity and context-awareness. Finally, the ServFace project [19] aims at supporting even unskilled web users in composing web services that come with an annotated WSDL description. Annotations are used to automatically generate form-like interfaces for the services, which can be placed onto one or more web pages and used to graphically specify data flows among the form fields. The result is a simple, user-driven web service orchestration. None of these projects, however, supports the coordination of multiple different actors inside a same process, and none of the approaches discussed supports the distribution of UIs over multiple browsers.

6 Conclusion and Future Works

In this chapter, we addressed the problem of designing and orchestrating *component-based web applications* that are distributed over *multiple web browsers* and that involve *multiple different actors*. We particularly discussed the case of a search computing application that leverages on a collaborative search and browsing approach, an application feature whose development with traditional techniques would be everything but trivial. In fact, while the integration of UIs and web services is, for instance, also supported by current mashup platforms, the coordination of the actors involved in the application and the synchronization of their respective UIs would still require manual intervention. The MarcoFlow platform introduced in this chapter, instead, supports the seamless integration of services, UIs, and people in one and the same development environment, sensibly speeding up the development of process-based, mashup-like web applications.

The basic idea of MarcoFlow, i.e., the component-based development of applications is inspired by current web mashup practices, which in many cases aim at enabling also the *less skilled developer* (or even unskilled end users) to compose own applications. Given the complexity of the applications supported by MarcoFlow, it is however important to note that MarcoFlow rather targets skilled developers (e.g., developers that are familiar with composite web service development in BPEL).

One of the challenges to be addressed in our future work is therefore lowering the complexity of the design environment for distributed UI orchestrations, hiding BPEL4UI behind an easier to learn, graphical modeling language. Also, we would like to extend the approach toward streaming web services, for example to support the design of continuous queries over sensor networks.

References

1. Yu, J., Benatallah, B., Casati, F., Daniel, F.: Understanding Mashup Development and its Differences with Traditional Integration. IEEE Internet Computing 12(5), 44–52 (2008)
2. Daniel, F., Soi, S., Tranquillini, S., Casati, F., Heng, C., Yan, L.: From People to Services to UI: Distributed Orchestration of User Interfaces. In: Hull, R., Mendling, J., Tai, S. (eds.) BPM 2010. LNCS, vol. 6336, pp. 145–161. Springer, Heidelberg (2010)
3. Daniel, F., Soi, S., Casati, F.: From Mashup Technologies to Universal Integration: Search Computing the Imperative Way. In: Ceri, S., Brambilla, M. (eds.) Search Computing. LNCS, vol. 5950, pp. 72–93. Springer, Heidelberg (2010)
4. Daniel, F., Casati, F., Benatallah, B., Shan, M.-C.: Hosted Universal Composition: Models, Languages and Infrastructure in mashArt. In: Laender, A.H.F., Castano, S., Dayal, U., Casati, F., de Oliveira, J.P.M. (eds.) ER 2009. LNCS, vol. 5829, pp. 428–443. Springer, Heidelberg (2009)
5. OASIS. Web Services Business Process Execution Language Version 2.0 (April 2007), http://docs.oasis-open.org/wsbpel/2.0/OS/wsbpel-v2.0-OS.html
6. Pautasso, C.: BPEL for REST. In: Dumas, M., Reichert, M., Shan, M.-C. (eds.) BPM 2008. LNCS, vol. 5240, pp. 278–293. Springer, Heidelberg (2008)
7. van Lessen, T., Leymann, F., Mietzner, R., Nitzsche, J., Schleicher, D.: A Management Framework for WS-BPEL. In: ECoWS 2008, Dublin, pp. 187–196 (2008)
8. Maximilien, E.M., Ranabahu, A., Gomadam, K.: An Online Platform for Web APIs and Service Mashups. Internet Computing 12(5), 32–43 (2008)
9. Active Endpoints, Adobe, BEA, IBM, Oracle, SAP. WS-BPEL Extension for People (BPEL4People), Version 1.0 (June 2007)
10. Active Endpoints, Adobe, BEA, IBM, Oracle, SAP. Web Services Human Task (WS-HumanTask), Version 1.0 (June 2007)
11. Acerbis, R., Bongio, A., Brambilla, M., Butti, S., Ceri, S., Fraternali, P.: Web Applications Design and Development with WebML and WebRatio 5.0. In: TOOLS 2008, pp. 392–411 (2008)
12. Gómez, J., Bia, A., Parraga, A.: Tool Support for Model-Driven Development of Web Applications. In: Ngu, A.H.H., Kitsuregawa, M., Neuhold, E.J., Chung, J.-Y., Sheng, Q.Z. (eds.) WISE 2005. LNCS, vol. 3806, pp. 721–730. Springer, Heidelberg (2005)
13. Manolescu, I., Brambilla, M., Ceri, S., Comai, S., Fraternali, P.: Model-Driven Design and Deployment of Service-Enabled Web Applications. ACM Trans. Internet Technol. 5(3), 439–479 (2005)
14. Brambilla, M., Ceri, S., Fraternali, P., Manolescu, I.: Process Modeling in Web Applications. ACM Trans. Softw. Eng. Methodol. 15(4), 360–409 (2006)
15. Sun Microsystems. JSR-000168 Portlet Specification (October 2003), http://jcp.org/aboutJava/communityprocess/final/jsr168/
16. OASIS. Web Services for Remote Portlets (August 2003), http://www.oasis-open.org/committees/wsrp
17. Oracle. JavaServer Faces Technology, http://java.sun.com/javaee/javaserverfaces/
18. Pietschmann, S., Voigt, M., Rümpel, A., Meißner, K.: CRUISe: Composition of Rich User Interface Services. In: Gaedke, M., Grossniklaus, M., Díaz, O. (eds.) ICWE 2009. LNCS, vol. 5648, pp. 473–476. Springer, Heidelberg (2009)
19. Feldmann, M., Nestler, T., Jugel, U., Muthmann, K., Hübsch, G., Schill, A.: Overview of an end user enabled model-driven development approach for interactive applications based on annotated services. In: WEWST 2009, pp. 19–28 (2009)

On Development Practices for End Users

Alessandro Bozzon[1], Marco Brambilla[1],
Muhammad Imran[2], Florian Daniel[2], and Fabio Casati[2]

[1] Politecnico di Milano, Dipartimento di Elettronica e Informazione, 20133 Milano, Italy
{bozzon,mbrambil}@elet.polimi.it
[2] University of Trento, Via Sommarive 14, 38123 Povo (TN), Italy
{imran,daniel,casati}@disi.unitn.it

Abstract. The paper discusses some trends in end user programming (EUP) and takes inspiration from the discussions in a panel and in a vertical session on research evaluation within the second Search Computing workshop. We discuss the controversial successes and failures in this field and we elaborate on which facilities could foster adoption of end user programming. We discuss various dimensions of end user programming, including vertical versus horizontal language definition, declarative versus imperative approaches. We exemplify our discussion in the realistic scenario of research evaluation by comparing the Search Computing and ResEval approaches.

Keywords: crowd programming, end user development, mashup, conceptual modeling, declarative programming.

1 Introduction

In recent years, several research projects such as Search Computing, ResEval[1], and FAST[2] spent substantial effort towards empowering end users (sometimes called expert users, to distinguish them from generic, completely unskilled users), with tools and methods for software development. However, the success of **end user development** (EUD) and its potential adoption are still controversial. In this chapter we look at this field from a *search* perspective and we elaborate on which paradigms and ingredients best aid end users in performing development tasks, and most notably formulating complex search queries. We also discuss various dimensions of end user programming, including vertical versus horizontal language definition, declarative versus imperative approaches.

The chapter is organized as follows: Section 2 discusses the problem of identifying the developer classes that could be addressed by EUD approaches; Section 3 presents some practices that could foster the adoption of EUD and Section 4 discusses some

---

[1] http://reseval.org - ResEval is a Web-based tool for evaluating the research impact of individual researchers and groups by integrating multiple scholarly data sources.

[2] http://fast-fp7project.morfeo-project.org - FAST aims at the development of a new visual programming environment for business processes based on semantic Web services.

S. Ceri and M. Brambilla (Eds.): Search Computing II, LNCS 6585, pp. 192–200, 2011.

relevant dimensions of Domain Specific Languages within EUD approaches. Section 5 shows a declarative approach (namely, Search Computing) and an imperative approach (namely, ResEval) at work on a realistic scenario; and Section 6 draws some conclusions.

2 Target User Classes

End user development comprises several alternative approaches, spanning from mashup development, to software configuration, to simple programming tasks and search. These approaches are often antithetic, but sometimes they can be combined together to exploit the respective strength points.

For instance, while users are getting more and more used to *configuring* applications, also thanks to the pervasiveness of mobile and gaming software, *mashup platforms* for the development of simple Web applications are also gaining popularity. Yet, mashups were actually born as a hacking phenomenon, where very expert developers build applications by integrating reusable content and functionality sourced from the Web (for instance, see www.programmableweb.com), and – despite the numerous attempts – mashup development is still for skilled programmers only.

Actually, mashup tools initially targeting end users slowly moved towards the expert user, then to the developer, and finally to the expert developer. In fact, our experience on both model-driven web engineering [13] and mashup development [8] has shown that there are basically only two target users in the real world:

- **Developers**, who want to see the source code and to write imperative code. These users do not trust model-driven approaches, because they feel this can reduce their freedom in application development;
- **Non-developers**, who want to ignore all the technical issues and have simple, possibly visual or parameter-based configuration environments for setting up their applications.

The rest of the stratification of users into expert users, entry-level developers, developer/designer that can be theoretically defined does actually not exist. Recognizing the distinction of only two major user classes, empowering non-developers becomes more focused, but also non-trivial.

3 Enabling Practices and Techniques

Enabling end users to develop own applications or compose simple mashups or queries means simplifying current development practices. A variety of options may help simplifying the user development; we discuss the most important ones in the following, in order to use them in the next section to analyze two approaches that partly aim at supporting end users in composing complex queries.

Simple programming models. The first issue is to understand which programming paradigms are best suited for end user programming. The solution to this issue can take inspiration from existing experiences in orchestration and mashup languages which are targeted at process automation and at relatively inexperienced users (although they have not been that successful in reaching out to non-IT experts, as yet). The aim is to find programming abstractions that are simple enough to appeal to domain experts and at the same time complex enough to implement enterprise procedures and Web application logic.

For instance, some mashup approaches heavily rely on connections between components (this is the case of Yahoo! Pipes[3] and IBM Damia [1], for instance), and therefore are inherently imperative; other solutions completely disregard this aspect and only focus on the components and their pre- and post-conditions for automatically matching them, according to a declarative philosophy like the one adopted in choreographies (for instance, see the proposal of the FAST European project [9]).

Domain-specific languages (DSLs). Simple programming models are not enough. Typically, end users simply don't understand what they can do with a given development tool, a problem that is basically due to the fact that the development tools does not speak the language of the user and, hence, programming constructs don't have any meaning to the user. Domain-specific languages aim at adding domain terminology to the programming model, in order to give constructs domain meaning.

In some fields, such as database design, domain-specific languages are a consolidated practice: declarative visual languages like the ER model are well accepted in the field. Other, more imperative approaches, like WebML, address developers that are willing to embrace conceptual modeling. Business people, on the other hand, are well aware of workflow modeling practices and are able to work with formalisms like BPMN, completely ignoring what happens behind the scenes both in terms of technological platform and of transformations applied to get to a running application. Another example in this category is Taverna[4], a workflow management system well known in the biosciences field. A more precise classification of DSLs is provided in Section 4.

Intuitive interaction paradigms. User interfaces of development tools may not be a complex theoretical issue, but acceptance of programming paradigms can be highly influenced by this aspect too. The user interface comprises, for instance, the selection of the right graphical or textual development metaphor so as to provide users with intelligible constructs and instruments. It is worth investigating and abstracting the different kinds of actions and interactions the user can have with a development environment (e.g., selecting a component, writing an instruction, connecting two components), to then identify the best mix of interactions that should be provided to the developer.

Reuse of development knowledge. Finally, even if a tool speaks the language of the user, it may still happen that the user doesn't speak the language of the tool, meaning that he/she still lacks the necessary basic development knowledge in order to use the tool profitably. Such a problem is typically solved by asking more expert users (e.g., colleagues or developers) for help – if such are available. The challenge is how to reuse or support the reuse of development knowledge from more expert users in an automated fashion inside a tool, e.g., via recommendations of knowledge [12].

---

[3] http://pipes.yahoo.com/pipes/

[4] http://www.taverna.org.uk/

Recommendations can be provided based on several kinds of information, including components, program specifications, program execution data, test cases, simulation data, and possibly mockup versions of components and program fragments used for rapid prototyping. Information may or may not be tagged with semantic annotations. When present, the annotations can be used to provide better/more accurate measures of similarity and relevance. In a general sense, the approach we envision is an alternative to design patterns for exploiting the expertise of good developers, thus allowing reuse of significant designs.

Programming, testing, and prototyping experiences of peers or of more experienced developers may support the entire development lifecycle. If knowledge is harvested and summarized from peers (e.g., by analyzing their mashup definitions), this opens the door to what we can call "implicit collaborative programming" or "crowd programming", where users, while going through a software engineering lifecycle for implementing procedures of their own interest, create knowledge that can be shared and leveraged by other domain experts for their own work.

4 Domain-Specific Languages: Assessment Dimensions

We have seen that **Domain-Specific Languages** (DSLs), i.e., design and/or development languages that are designed to address the needs of a specific application domain, are important to provide the end user with familiar concepts, terminology and metaphors. That is, DSLs are particularly useful because they are tailored to the requirements of the domain, both in terms of semantics and expressive power (and thus do not enforce end users to study more comprehensive general-purpose languages) and of notation and syntax (and thus provide appropriate abstractions and primitives based on the domain).

While a broad discussion on DSLs is outside the scope of this paper[5], we wish to highlight a few possible classifications of these languages, which can become handy for EUD. In particular, we describe the dimensions of focus, style and notation.

The **focus** of a DSL can be either vertical or horizontal. *Vertical DSLs* aim at a specific industry or field. Examples of vertical DSLs may include: configuration languages for home automation systems, modeling languages for biological experiments, analysis languages for financial applications, and so on. On the other side, *horizontal DSLs* have a broader applicability and their technical and broad nature allows for concepts that apply across a large group of applications. Examples of horizontal DSLs include SQL, Flex[6], WebML[7], and many others.

The **style** of a DSL can be either declarative or imperative. *Declarative DSLs* adopt a specification paradigm that expresses the logic of a computation without describing its control flow. In other words, the language defines what the program should accomplish, rather than describing how to accomplishing it. *Imperative DSLs* instead specifically require defining an executable algorithm that states the steps and control flow that needs to be followed to successfully complete a job.

[5] See a thorough discussion on this topic here:
http://lostintentions.com/2009/08/15/a-look-into-domain-specific-languages/

[6] http://www.adobe.com/products/flex/

[7] http://www.webml.org/

The **notation** of a DSL can be either graphical or textual. The *graphical DSLs* (also known as Domain Specific Modeling Languages, DSML) imply that the outcomes of the development are visual models and the development primitives are graphical items such as blocks, arrows and edges, containers, symbols, and so on. The *textual DSLs* comprise several categories, including XML-based notations, structured text notations, textual configuration files, and so on.

Despite the various experiences in DSL design and application, there is no general assessment on the preferences of the developers for one or the other kind of language depending on the user profile. However, typically languages oriented to the end users tend to be more visual and declarative, while the ones for developers are often textual and imperative.

5 EUD in Practice: Two Examples for Research Evaluation

To exemplify how EUD can be supported in practice, we describe two approaches to a domain-specific problem, namely **research evaluation**. Research evaluation has received a lot of interest in the last years since everybody is producing research artifacts at his best, and in this race everyone wants to lead. Assessing the impact of researchers and publications is highly demanded and important [1]. Yet, the very problem of finding experts or high-profile people in some specific area is still a challenging endeavor: simply imagine you want to assess the *independence of young researchers* or to evaluate the *quality of a supervisor*: different metrics should be defined and used.

There are attempts of *applications* that aim to help assessment problems like the above. Typically, they support citation-based metrics for the evaluation of research impact. Examples are Web of Science[8], Scopus[9], Publish or Perish[10], and similar. However, all the currently available tools lack some key features, such as completeness of data, data cleaning options, or comparison features, which imply that their outputs are not always satisfying and reliable.

In the rest of this section we discuss the imperative approach proposed in ResEval and the declarative approach proposed in Search Computing for addressing the issue, and, finally, we provide a possible combination of the two for leveraging on the respective strength points.

5.1 Research Evaluation in ResEval

ResEval is a research evaluation platform that is currently being developed at the University of Trento. ResEval is a tool that is based on citation-based indicators like h-index, g-index, noise ratio, and citation count. ResEval provides self-citation analysis, the possibility to find top co-authors, and top citers of a researcher.

ResEval is mainly based on *citation analysis*, which is today's de-facto standard practice. Despite its widespread use, citation analysis does however not come without controversy [6]. Everyone has his own, sometimes very *subjective logic* for assessing

[8] http://scientific.thomson.com/products/wos/
[9] http://www.scopus.com/home.url
[10] http://www.harzing.com/

research impact. Also ResEval comes short if we want to assist sophisticated, user-defined assessment metrics, since – as in all the other tools – the supported features are pre-defined and custom metrics cannot be expressed. In order to support users in defining their own research evaluation logic, we introduce a new ingredient that we think will allow us to further lower the complexity of the mashup process: we propose a *domain-specific information mashup approach* [7] through which users (possibly with no programming skills) can define, execute, and visualize a metric's combination logic and its result.

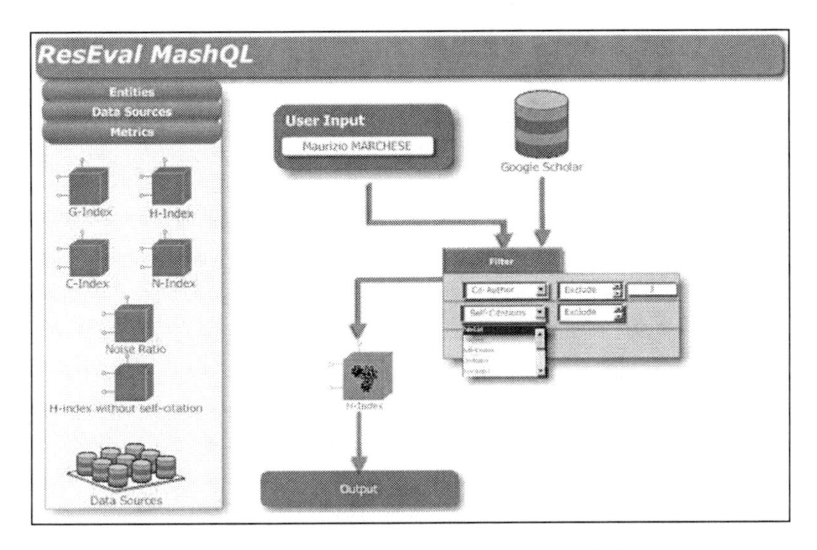

Fig. 1. Mockup of a research evaluation mashup in ResEval

Specifically, ResEval is currently being extended with a ***mashup platform*** (an adaptation and extension of the mashArt platform [8]), allowing users to combine both UI components (widgets that can show charts and trends) and information from a variety of (Web) sources. ResEval features a ***domain-specific language*** that constrains the concepts and the functionality of the platform for the sake of ease of use [10], in order to have a mashup platform that can be used by non-programmers. The focus of the proposed DSL is ***vertical*** (research evaluation), while its style is ***imperative***, and its notation is ***graphical*** so as to fine-tune the ***development metaphors*** to the domain. ***Reuse*** of development knowledge mainly comes in the form of ready components that can be composed into value-adding manners, while keeping large part of the complexity inside the components.

For instance, Fig. 1 shows a simple mashup example where data sources, operators, filters, and UI components are used to compute a metric. Specifically, the mashup fetches all data from Google Scholar (other sources like DBLP and Microsoft Academic and unions thereof are also supported), filters out only those publications by a given author, and applies a set of filter conditions (e.g., we exclude self-citations). Then we compute the conventional h-index metric over to so cleaned list of publications and render the result in a UI component so that the user can inspect the output of the computation.

5.2 Research Evaluation with Search Computing

Search Computing complies with all the EUD practices described in Section 3. In particular, it provides ***simple programming models*** (based on visual registration of search services and configuration of queries expressed on such services) defined on a set of domain-specific languages that cover service registration, query design, and query plan refinement (expressed using the Panta Rhei notation).

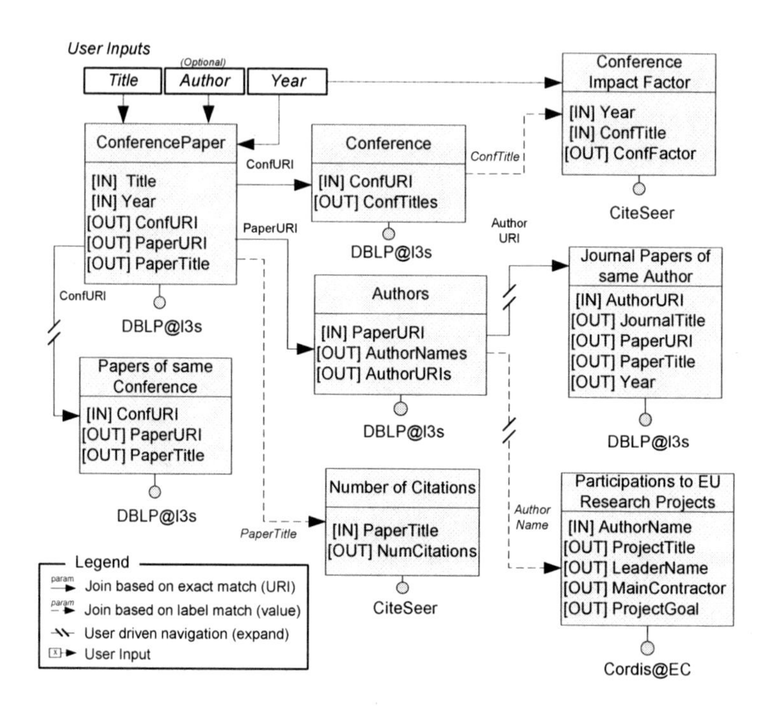

Fig. 2. A liquid query template for a research products and venues search application

A set of ***graphical design tools*** has been devised to support developers in their work. These tools also comprise support to ***design reuse*** (in terms of registered services and queries), together with additional facilities for design validation (e.g., checking the correctness of the query against the properties of the registered services). At the moment no recommendation features are provided, although some basic ones are scheduled as future work. Generally speaking, Search Computing provides a set of ***horizontal*** DSLs, being focused on search applications but applicable to any industrial field. Most of them are ***declarative***, except for Panta Rhei, which specifies the query plans as an executable orchestration of search services and therefore is ***imperative***. The notations are graphical, except for SeCoQL and the Liquid Query configuration language [3], which are textual notations.

Being a general-purpose approach (i.e., a horizontal DSL), search computing can be easily applied to the research productivity field. To provide a better understanding of our approach, consider a scenario in which a research evaluation application is built at design time [4] and then consumed at runtime by a user through the Liquid Query interface [1].

For instance, we assume the final application expects the user to search for a conference paper and the systems to provide him the list of matching papers, the number of citations for each paper, the authors, the and the details of conference where it was presented, including its impact factor. Subsequently, the user can decide to extend the results by navigating toward other papers published in the same conference, the journal papers of each author, and the European research projects the authors have been responsible for. The declarative specification of such a query template can be displayed visually, as shown in Fig. 2, while the results are shown in the Atom View interface [3] represented in Fig. 3[11].

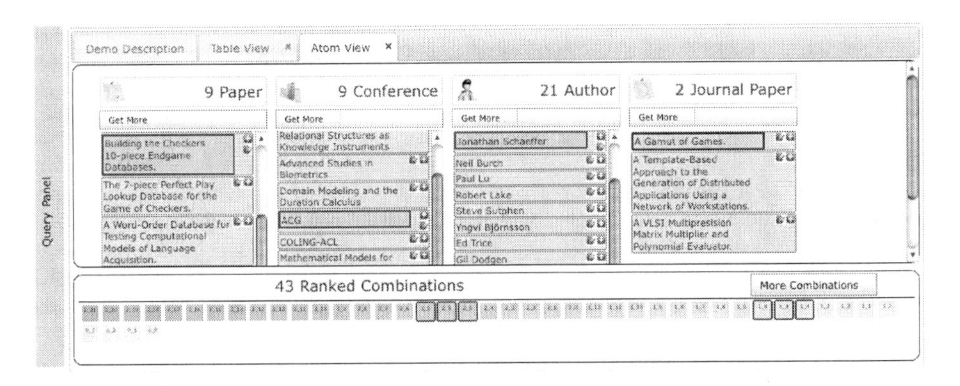

Fig. 3. Structure of the query upon research products and venues

5.3 On the Combined Potentials of ResEval and Search Computing

The previous two sections show that advanced research evaluation scenarios can be approached from at least two different perspectives. ResEval proposes an *imperative* paradigm for the specification of custom *data processing logic*; it is particularly strong in the *flexibility* with which evaluation logic can be expressed. Search Computing proposes a *declarative* paradigm for the specification of *data integration logic*; it is particularly strong for its *ease of use*. Although the two instruments approach the same problem from a different perspective, it is important to note that, rather than representing alternatives, the two instruments complement each other and that, hence, an integration of the two may be beneficial to both.

In particular, Search Computing allows for easy data integration, by natively providing join mechanisms on heterogeneous sources, and efficient data retrieval by means of parallel and asynchronous invocation of services, with non-blocking behavior with respect to the overall query plan.

On the other side, ResEval can contribute pre-defined complex ranking indexes for the specific domain of research evaluation, and composition flexibility thanks to the mashup-based development.

[11] See the prototype and a video at: http://www.search-computing.org/demo/ui

6 Conclusions

This chapter discussed the problem of end user development and investigated the most important enabling features and classification dimensions. The Search Computing and ResEval approaches have been considered as representative examples. While the future of EUD is still uncertain, we strongly believe that combinations of declarative and imperative approaches can bring value to end users, who can be empowered with basic development capabilities, at least in limited domain scenarios.

References

1. Altinel, M., Brown, P., Cline, S., Kartha, R., Louie, E., Markl, V., Mau, L., Ng, Y.-H., Simmen, D., Singh, A.: Damia – A Data Mashup Fabric for Intranet Applications. In: Proceedings of VLDB 2007, Vienna, Austria (2007)
2. Bozzon, A., Brambilla, M., Ceri, S., Fraternali, P.: Liquid query: multi-domain exploratory search on the Web. In: Proceedings of WWW 2010, pp. 161–170. ACM, New York (2010)
3. Bozzon, A., Brambilla, M., Ceri, S., Fraternali, P.: Exploring the Web with Search Computing. In: Ceri, S., Brambilla, M. (eds.) Search Computing II. LNCS, vol. 6585, pp. 10–25. Springer, Heidelberg (2011)
4. Brambilla, M., Tettamanti, L.: Search computing processes and tools. In: Ceri, S., Brambilla, M. (eds.) Search Computing II. LNCS, vol. 6585, pp. 169–181. Springer, Heidelberg (2011)
5. Informatics Europe Report, Research evaluation for computer science (2008), `http://`
6. `www.informatics-europe.org/docs/research_evaluation.pdf`
7. Chapman, A.J.: Assessing research: citation count shortcomings. The Psychologist, 336–344 (1989)
8. Imran, M., Daniel, F., Casati, F., Marchese, M.: ResEval: A Mashup Platform for Research Evaluation. In: Proceedings of ECSS 2010, Prague, Czech Republic (2010)
9. Daniel, F., Casati, F., Benatallah, B., Shan, M.-C.: Hosted Universal Composition: Models, Languages and Infrastructure in mashArt. In: Laender, A.H.F., Castano, S., Dayal, U., Casati, F., de Oliveira, J.P.M. (eds.) ER 2009. LNCS, vol. 5829, pp. 428–443. Springer, Heidelberg (2009)
10. Hoyer, V., Janner, T., Delchev, I., Fuchsloch, A., López, J., Ortega, S., Fernández, R., Möller, K.H., Rivera, I., Reyes, M., Fradinho, M.: The FAST Platform: An Open and Semantically-Enriched Platform for Designing Multi-channel and Enterprise-Class Gadgets. In: Baresi, L., Chi, C.-H., Suzuki, J. (eds.) ICSOC-ServiceWave 2009. LNCS, vol. 5900, pp. 316–330. Springer, Heidelberg (2009)
11. Soi, S., Baez, M.: Domain-specific Mashups: from all to all you need. In: Daniel, F., Facca, F.M. (eds.) ICWE 2010. LNCS, vol. 6385, pp. 384–395. Springer, Heidelberg (2010)
12. Baez, M., Birukou, A., Casati, F., Marchese, M.: Addressing Information Overload in the Scientific Community. IEEE Internet Computing 99 (2010) (prePrints)
13. Roy Chowdhury, S., Rodríguez, C., Daniel, F., Casati, F.: Wisdom-Aware Computing: On the Interactive Recommendation of Composition Knowledge. In: Proceedings of WESOA 2010. Springer, Heidelberg (December 2010)
14. Ceri, S., Fraternali, P., Bongio, A., Brambilla, M., Comai, S., Matera, M.: Designing Data-Intensive Web Applications. Morgan Kaufmann, San Francisco (2002)

Part 7
Bio-SeCo

In this session, the main search computing assets, namely data integration from distributed resources and computation of their global ranking, are discussed in the bio-medical context. This context is characterized by numerous heterogeneous data and algorithmic resources, which often provide ordered data; their integration can be instrumental to the answer of complex bio-medical questions.

The first chapter presents Bio-SeCo as a case study of the use of search computing for describing well known bioinformatics resources as search services, and for carrying out integrated analyses over the resulting services. In particular, data from sequence comparisons and from gene expression results are integrated in a way that takes account of the ranked results from the different types of data. In so doing, the use of ranking as a first class citizen for data integration in the life sciences is illustrated and open issues are identified for further investigation.

The second chapter focuses on the use of workflows in the Life Sciences to integrate and analyze dispersed information, and their connection with search computing. The chapter describes Taverna, a well established platform for designing and executing workflows, which allows the automation of experimental methods through the use and integration of a number of different life science services (such as Web services). The benefits of making ranked data and partial/incremental results as first class citizens in life science workflows, as well as the use of domain-specific service collections and provenance traces as complements to the search computing paradigm, are investigated and discussed.

Finally, the third chapter discusses the user needs concerning complex bio-medical searches and bio-molecular knowledge discovery, highlighting potential issues and benefits for search computing applications in the bio-medical field.

Bio-SeCo: Integration and Global Ranking of Biomedical Search Results

Marco Masseroli and Giorgio Ghisalberti

Dipartimento di Elettronica e Informazione, Politecnico di Milano,
Piazza Leonardo da Vinci 32, 20133 Milano, Italy
{masseroli,ghisalberti}@elet.polimi.it

Abstract. This chapter presents how well known bioinformatics resources can be described as search services in the search computing framework and how integrated analyses over such services can be carried out. An initial set of bioinformatics services has been described and registered in the search computing framework and a bioinformatics search computing (Bio-SeCo) application using these services has been created. This current prototype application, the available services which it uses, the queries which are supported, the kind of interaction which is therefore made available to the users, and the future scenarios are here described and discussed.

Keywords: search, bioinformatics, data integration, ranked data.

1 Introduction and Motivation

In the life sciences, questions are often complex and simultaneously regard several different functional and structural aspects of an organism and its biomolecular entities (e.g. the genes expressed in certain conditions, their mutations and their involvement in pathological phenotypes or diseases, the proteins with their protein domains and 3D structure, their participation in different biochemical pathways and biological processes, etc.). An example is the following: "Which genes encode proteins in different organisms with high sequence similarity to a given protein and are significantly expressed in the same given tissue or condition?" Such questions can be addressed only by exploring, comprehensively searching and globally evaluating the numerous available data and their relationships, which are of different types and often inherently ordered or associated with ranked confidence values. Access to these data is being increasingly provided, and more relevantly than in other fields, by web services, which offer both generic and domain-specific *search services*, i.e. bioinformatics services that provide results (often ranked) of user defined searches within data repositories. These services provide users with rapid and selective access to biomedical data from potentially huge repositories. However, individual search tools are often ineffective for use in applications in which the answer to a request involves combining results from more than one search engine. In particular, available search services typically provide *vertical* search capabilities [1], in which they are focused on a single domain. They seek individual items that meet the criteria specified in a request, whereas in practice information relevant to a biomedical requirement may be spread

S. Ceri and M. Brambilla (Eds.): Search Computing II, LNCS 6585, pp. 203–214, 2011.

over several resources. Furthermore, it is often essential to combine multiple vertical search services to create multi-domain searches, where the different domain searches either refine or augment previous results. For example, if the user is interested in knowing which genes both encode proteins with high sequence similarity to a given protein and are significantly expressed in the same given biological condition or tissue, current practice typically involves the integration of results from three different searches (for similar proteins, protein encoding genes and gene expressions), where the individual search results are themselves likely to be ranked by some criteria [2]. Such an integration task, taking account of the rankings, is termed a *multi-domain search*, and may be carried out manually or by a custom program, but has not typically been supported directly by data integration platforms.

Search computing [3] [4] and its information exploration paradigm based on semantic resource framework (see this book, Part 3 – Chapter 1: Semantic Resource Framework and Part 1 – Chapter 2: Information Exploration in Search Computing), provide a platform for expressing requests over multiple search services, such that the results of the integrated requests take account of the rankings of individual search results. The several different types of biomedical data and their relationships, as well as the numerous bioinformatics web services available, can be semantically represented through a resource framework such that depicted in Figure 1. Thus, by using available web services for searching bioinformatics data and taking advantage of the attributes they define for providing a ranking, search computing techniques can be applied to efficiently explore available data and to search for globally ranked answers to complex biomedical questions.

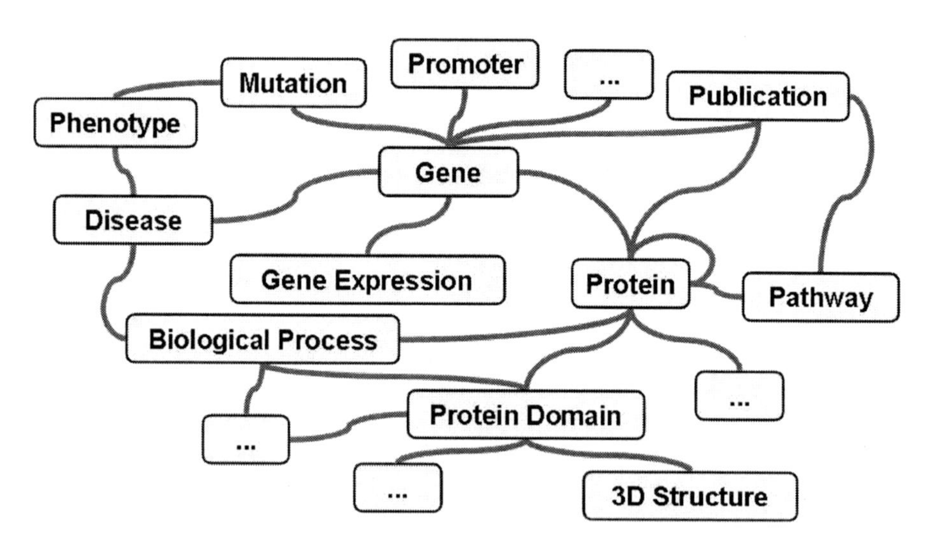

Fig. 1. Biomedical Semantic Resource Framework

For example, let us consider the above mentioned multi-domain case study question: "Which genes encode proteins in different organisms with high sequence similarity to a given protein and are significantly expressed in the same given tissue or condition?" It can be addressed by first looking for proteins with high sequence similarity to a given protein. Then, the results of this first search can be expanded by searching for the genes that encode the similar proteins found. Finally, the encoding genes found are filtered by looking for only those genes that are significantly expressed in the same given biological tissue or condition (Figure 2).

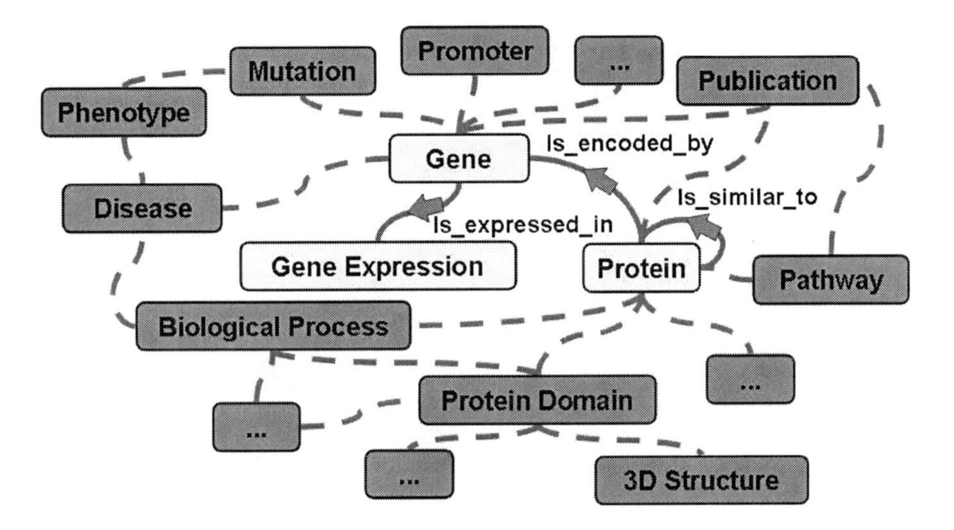

Fig. 2. Exploring biomedical information by moving in the Semantic Resource Framework through selected semantic connections

This chapter complements a previous exploration of the envisaged relevance of search computing to the life science domain [2] by illustrating the application of the implemented search computing platform and semantic resource framework in a biomedical use case.

The remainder of the chapter is as follows. Section 2 presents both the description of bioinformatics search services that makes them usable for search computing and the definition of requests that span multiple search services. Section 3 illustrates the types of bioinformatics web services currently described and registered in the search computing framework, where they are available to be used for search computing applications. Section 4 describes a prototype bioinformatics search computing (Bio-SeCo) application able to answer an example of a biomedical question, which requires integrating biological sequence and gene expression service results; we created such Bio-SeCo application as a demonstrator of the capabilities of search computing technology to answer complex multi-domain biomedical questions. Section 5 discusses future scenarios in the development of the prototypical Bio-SeCo application and in the use of source computing in the life sciences. Some conclusions, in particular on this last aspect, are presented in Section 6.

2 Bioinformatics Resource Representation for Search Computing

In order to be used in a search computing platform, a resource must be *registered* as a search service in the search computing framework. As described in Chapter 9 of [4], this is done by describing and making the resource available to search computing through a standard format, called *service mart*. The latter describes that type of resource, and defines the binding between the service mart and the operation to be invoked on the service that provides access to the resource, with its input and output parameters. A *service mart* is a conceptual abstraction that masks the different implementation styles of services and is tailored to a specific need to expose *search services* results. Such results are produced by interacting with concrete data sources, which are made available through service interfaces, wrappers, or direct access to extensional data collections (e.g. databases, excel files, and so on). A service mart models a specific type of service by describing it and its properties; each service mart definition includes a name (the service type name) and a collection of attributes (the typical input and output attributes exposed by the services of that type). Service marts have atomic attributes and repeating groups consisting of a non-empty set of sub-attributes that collectively define a property of the service mart. Atomic attributes are single-valued, while repeating groups are multi-valued.

Each service mart is associated with one or more specific *access patterns*, which abstract and logically describe the way in which data access can be effectively performed. An access pattern is a signature of the service mart in which each attribute or sub-attribute is characterized as either input (I) or output (O), depending on the role that the attribute plays in the service call. Moreover, an output attribute is designated as ranked (R) if the service produces its results in an order that depends on the value of the attribute. Access patterns can include a subset of the associated service mart attributes that are relevant for the specific data access; they can also have additional specific attributes (i.e. external attributes, see Part 3 – Chapter 1) not included in their service mart since they are not typical for the majority of services described by the service mart.

Each service mart is associated with one or more *service interfaces*; each of them maps an access pattern to a specific implementation and is represented as a triple including a name, a given access pattern and a service.

Pair-wise coupling of service marts is defined through *connection patterns*, which completely specify the connection semantics. Every pattern has a *conceptual name* and a *logical specification*, consisting of a sequence of simple comparison predicates between pairs of attributes or sub-attributes of the two connected services; such predicates are interpreted as a conjunctive Boolean expression, and can therefore be implemented by joining the results returned by the calling service implementations.

Through service marts, access patterns and connection patterns, existing resources can hence be represented in a standard format; this enables to register, use and combine existing bioinformatics resources in the search computing framework to perform multi-domain searches that provide results globally ordered according to the ranking of the retrieved single domain results.

3 Bioinformatics Search Services Registered in the Search Computing Framework

By using the above described standard formats for representing resources for search computing, an initial set of bioinformatics search services has been described and registered in the search computing framework. Three of the most common types of search services in bioinformatics, i.e. for biomolecular sequence alignment and search (in a databank of nucleotide or amino acid sequences), for protein ID look up (in a gene or protein databank) and for gene expression result search (in a databank of experimental gene expression results) have been considered. Their service marts and some of their access patterns have been defined as follow, together with some service interfaces for a few specific bioinformatics services of such types.

3.1 Services for Biomolecular Sequence Alignment and Search

The far most used algorithms for biomolecular sequence alignment and search in a repository of biomolecular sequences are BLAST (Basic Local Alignment and Search Tool) [5] and FASTA (FAST All) [6]. Numerous implementations of each of these algorithms, in some cases optimized for specific purposes, exist and are publicly available as search services. They usually have many input and output attributes: the most important can be described by the following *sequenceAlignmentSearch* service mart:

sequenceAlignmentSearch(sequenceAlignmentProgram, searchedDatabase, querySequence, querySequenceID, querySequenceIDName, foundSequenceSymbol, foundSequenceID, foundSequenceIDName, foundSequenceDescription, foundSequenceOrganism, alignments(score, expectation, probability, matchQuerySequence, matchFoundSequence, matchPattern))

A couple of different access patterns have been defined to logically describe the ways to access these biomolecular sequence alignment and search services and the data they provide. They are *sequenceAlignmentSearch_bySequence* and *sequenceAlignmentSearch_byID*, that describe the two possible ways of expressing the input biomolecular sequence, which is used as query sequence for searching similar sequences: by its nucleotide or amino acid sequence, or by its ID in the searched databank (specified by the two input attributes *querySequenceID*^{I}, and *querySequenceIDName*^{I}).

sequenceAlignmentSearch_bySequence(sequenceAlignmentProgram^I*, searchedDatabase*^{I}*, querySequence*^{I}*, foundSequenceSymbol*^{O}*, foundSequenceID*^{O}*, foundSequenceIDName*^{O}*, foundSequenceDescription*^{O}*, foundSequenceOrganism*^{O}*, alignments.score*^{R}*, alignments.expectation*^{R}*, alignments.probability*^{R}*, alignments.matchQuerySequence*^{O}*, alignments.matchFoundSequence*^{O}*, alignments.matchPattern*^{O}*)*

sequenceAlignmentSearch_byID(sequenceAlignmentProgram^I*, searchedDatabase*^{I}*, querySequenceID*^{I}*, querySequenceIDName*^{I}*, foundSequenceSymbol*^{O}*, foundSequenceID*^{O}*, foundSequenceIDName*^{O}*, foundSequenceDescription*^{O}*, foundSequenceOrganism*^{O}*, alignments.score*^{R}*, alignments.expectation*^{R}*, alignments.probability*^{R}*, alignments.matchQuerySequence*^{O}*, alignments.matchFoundSequence*^{O}*, alignments.matchPattern*^{O}*)*

The *sequenceAlignmentProgram* is the input attribute used to specify the sequence alignment program (e.g. *BLASTN*, *BLASTP*) to use in order to search, in the *searchedDatabase* database (e.g. *UniProtKB*), for the sequences similar to a specific query sequence; the retrieved sequences are described through the *foundSequenceSymbol*, *foundSequenceID*, *foundSequenceIDName*, *foundSequenceDescription* and *foundSequenceOrganism* output attributes. In the first access pattern, the query sequence is specified by providing as input its actual sequence (through the *querySequence* input attribute); in the second access pattern, the query sequence is specified by providing its ID as input (through the two *querySequenceID* and *querySequenceIDName* input attributes) in the database in which the search is performed (specified through the *searchedDatabase* input attribute). In all cases, *alignments.score*, *alignments.expectation* and *alignments.probability* are the output attributes that can be used for providing three different rankings of the retrieved sequences and their local alignments with the query sequence (*alignments.matchQuerySequence, alignments.matchFoundSequence, alignments.matchPattern*), according to their similarity with the query sequence.

Service interfaces for the BLAST implementations of the Washington University (WU BLAST) (http://www.ebi.ac.uk/Tools/blast2/) and the US National Center for Biotechnology Information (NCBI BLAST) (http://blast.ncbi.nlm.nih.gov/) have been created as follow:

WU_BLAST_bySequence("Washington University BLAST",
sequenceAlignmentSearch_bySequence,
http://www.ebi.ac.uk/Tools/webservices/wsdl/WSWUBlast.wsdl)

NCBI_BLAST_bySequence("National Center for Biotechnology Information BLAST", sequenceAlignmentSearch_bySequence,
http://www.ncbi.nlm.nih.gov/blast/Blast.cgi?CMD=Put&QUERY=
<querySequence>&DATABASE=<searchedDatabase>&PROGRAM=
<sequenceAlignmentProgram>)

3.2 Services for Protein ID Look Up

Several bioinformatics web services are available to retrieve a variety of protein information from different databanks; in particular some of them provide the ID and the symbol of the genes a given protein is associated with, either since the protein interacts with the gene, or because the gene encodes for the protein. The access to this type of information can be described for search computing by the following *protein2gene* service mart and the *protein2gene_byID* access pattern, which represents a way of defining the input protein whose associated genes are looked for: by its ID, or by its symbol and organism.

protein2gene(proteinID, proteinIDName, geneID, geneIDName, geneSymbol, organism, associationType(type), associationProvenance(database)) *protein2gene_byID(proteinID^{I}, proteinIDName^{I}, geneID^{O}, geneIDName^{O}, geneSymbol^{O}, organism^{O}, taxonomyID^{O}, associationType.type^{O}, associationProvenance.database^{O})*

A few service interfaces to access our Genome Function INtegrated Discoverer (GFINDer) (http://www.bioinformatics.polimi.it/GFINDer/) integrative Genomic and Proteomic Data Warehouse (GPDW) [7] [8] have been created; an example is as follow.

GPDW_byID("Genomic and Proteomic Data Warehouse", protein2gene_byID, http://www.bioinformatics.polimi.it/GFINDer/)

3.3 Services for Gene Expression Result Search

A few repositories of gene expression experimental data exist, some of which are publicly accessible through web interfaces and services. Among such repositories, Array Express Gene Expression Atlas (http://www.ebi.ac.uk/gxa/) [9] and Gene Expression Omnibus (http://www.ncbi.nlm.nih.gov/geo/) [10] are the most important ones; their data access can be described for search computing with the following *geneExpressionSearch* service mart and *geneExpressionSearch_byGeneID* access pattern, which represent the inputs used to describe the gene and its expression data that are looked for in the repository.

geneExpressionSearch(queryProperty, queryPropertyValue, queryOrganism, queryRegulation, queryFactorTerm, queryFactorValue, foundGeneSymbol, foundEnsemblGeneID, foundGeneSynonyms(geneSymbol), expressionFactorTerm, expressionFactorValue, expressionFactorOntologyID, expressionRegulation, experimentNumber, bestExperimentPvalue, bestExperimentID)

geneExpressionSearch_byGeneProperty(queryProperty^{I}, queryPropertyValue^{I}, queryOrganism^{I}, queryRegulation^{I}, queryFactorTerm^{I}, queryFactorValue^{I}, foundGeneSymbol^{O}, foundEnsemblGeneID^{O}, foundGeneSynonyms.geneSymbol^{O}, expressionFactorTerm^{O}, expressionFactorValue^{O}, expressionFactorOntologyID^{O}, expressionRegulation^{O}, experimentNumber^{R}, bestExperimentPvalue^{R}, bestExperimentID^{O})

Some service interfaces to access the Array Express repository have been created; an example is:

Array_Express_byGeneID("Array Express Gene Expression Atlas", geneExpressionSearch_byGeneID, http://www.ebi.ac.uk/gxa/api?geneIs=<queryEnsemblGeneID>&format=xml&indent)

4 A Bioinformatics Search Computing (Bio-SeCo) Application

To demonstrate the effectiveness of Search Computing in addressing complex biomedical questions and searching for their globally ranked answers, we considered the multi-domain case study question mentioned above ("Which genes encode proteins in different organisms with high sequence similarity to a given protein and are significantly expressed in the same given tissue or condition?"). We created a bioinformatics search computing (Bio-SeCo) application that enables users to run online such multi-domain biomedical query for different user selected proteins, gene expression regulation types and biological tissues or conditions, and obtain globally ranked ordered results.

The case study question can be decomposed into the following three single domain sub-queries: "Which proteins in different organisms have high sequence similarity to a given protein?"; "Which genes encode which proteins?"; and "Which genes are significantly expressed in the same given tissue or condition?". Each of these sub-queries can be mapped to an available specific search service, i.e. a sequence similarity search program such as BLAST, in one of its many implementations (e.g. WU BLAST), a query service in a database of genomic and proteomic data such as our GFINDer GPDW, and a search engine over a repository of gene expression data such as Array Express Gene Expression Atlas, respectively. As above described, these bioinformatics services have been registered and are now available in the search computing framework; thus, in the search computing framework they can be composed to automatically perform the multi-domain searches required to answer the considered example question. As described in Chapter 9 of [4], the composition of the bioinformatics services useful for computing the answer to the considered question can be done by defining, between the service marts that model those services, the following two pair-wise coupling connection patterns:

existsCodingGene_byProteinID(sequenceAlignmentSearch, protein2gene):
[(sequenceAlignmentSearch.foundSequenceID = protein2gene.proteinID
AND sequenceAlignmentSearch.foundSequenceIDName =
protein2gene.proteinIDName)]

existsExpressedGene_byGeneSymbol(protein2gene, geneExpressionSearch):
[("Gene" = geneExpressionSearch.queryProperty
AND protein2gene.geneSymbol = geneExpressionSearch.queryPropertyValue
AND protein2gene.taxonomyID = geneExpressionSearch.queryOrganism)]

4.1 Query Submission

The example query can hence be expressed in the search computing framework according to the semantic resource framework in Figure 1 and executed using a user interface such as the one described in Part 1 – Chapter 2. To achieve this aim, in the search computing platform we specified the three single domain sub-queries, in which the example query can be decomposed as follows, and created an execution plan implementing the two connection patterns above.

similarProteins(queryProteinIDName, queryProteinID, list_of(A, B, C, D))

codingGene(A, B, E, F)

expressedGene("Gene", E, F, queryExpressionRegulation, queryFactorValue, G, H) with *A*: *similarProteinID*, *B*: *similarProteinIDName*, *C*: *similarProteinSymbol*, *D*: *similarityExpectation*, *E*: *codingGeneSymbol*, *F*: *organism*, *G*: *experimentNumber*, *H*: *p-value*.

Furthermore, we defined a suitable ranking composition function (see Part 6 of this book – Rank Join) to aggregate the ranked results from the single domain searches produced by the composed services and generate a unique global ranking. While ranking attributes of the two composed services that provide ranked results (i.e. the *sequenceAlignmentSearch* and *geneExpressionSearch* services), we respectively considered the attributes *bestAlignmentExpectation* and *bestExperimentPvalue*. Given the dimensionless nature of the values in such attributes, we decided to define the ranking composition function as the product of these ranking attributes. In addition, since in a single search a *geneExpressionSearch* service can provide results for different values of the *expressionFactorValue* attribute (e.g. for different types of a given biological tissue or condition), we decided to generate such ranking composition for each value retrieved for the *expressionFactorValue* attribute and order the obtained global ranking in decreasing order of relevance. Finally, we created a web user interface to enter search constraints as query parameter values, submit the query execution (Figure 3) and present the global-ranked multi-domain search results obtained (Figure 4). Through such interface freely available on line (http://www.search-computing.it/UIDemoBio/) the user can interact with the prototypical Bio-SeCo application created. He/she can both submit searches for his/her chosen search constrains and browse the retrieved results, with the possibility to hide or to show retrieved attributes and/or change the result visualization order according to any set of retrieved attributes.

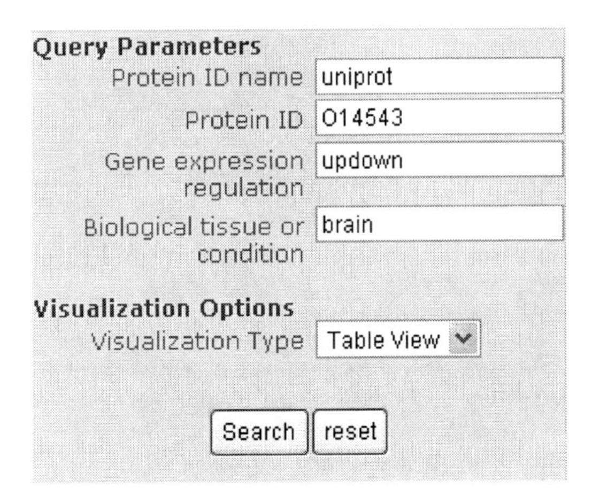

Fig. 3. Bio-SeCo application user interface to enter search constraints as query parameter values, choose the result visualization type and submit the query execution.

4.2 Query Results

In Figure 3 an example set of search constraints is shown: Protein ID name (*sequenceAlignmentSearch querySequenceIDName*) = "*uniprot*", Protein ID (*sequenceAlignmentSearch querySequenceID*) = "*O14543*", Gene expression regulation (*geneExpressionSearch queryRegulation*) = "*updown*" and Biological tissue or condition (*geneExpressionSearch queryFactorValue*) = "*brain*". In Figure 4 an excerpt of the global ranked results obtained for the user specified search constraints in Figure 3 is depicted. The resulting genes (*Socs3* in mouse, human and rat, *Socs2* in human and rat, and *socs8* in zebrafish) represent the ordered list of genes that encode proteins with high sequence similarity to the input *O14543* protein (human *Suppressor of cytokine signaling 3*) and are significantly differentially over or under expressed (*up regulated* or *down regulated*) in the *brain*. Hence, according to the partial ranked results provided on August 5th, 2010 by the WU BLAST, GPDW and Array Express services registered in the search computing framework, they constitute the global ranked answer to the considered example question, with the search constraints specified in Figure 3. The search computing framework automatically builds such results by integrating the partial ranked results provided by each considered service, as shown in Chapter 11 of [4]. As expected, the resulting genes include the gene that encodes the input protein.

Combination	Sequence Alignment			
Rank	Protein ID	Protein Name	Protein Symbol	Expectation
3.365e-121	O35718	Suppressor of cytokine signaling 3	SOCS3_MOUSE	2.9999999999999993e-98
3.365e-121	B1AQL6	Suppressor of cytokine signaling 3	B1AQL6_MOUSE	2.9999999999999993e-98
6.533e-108	O14543	Suppressor of cytokine signaling 3	SOCS3_HUMAN	2.5999999999999996e-99
6.533e-108	Q6FI39	SOCS3 protein	Q6FI39_HUMAN	2.5999999999999996e-99
1.140e-106	O88583	Suppressor of cytokine signaling 3	SOCS3_RAT	2.0999999999999999e-97
9.259e-29	O14508	Suppressor of cytokine signaling 2	SOCS2_HUMAN	3.1e-20
6.701e-23	A9JRX2	Socs8 protein	A9JRX2_DANRE	3.6e-21
1.322e-22	O88582	Suppressor of cytokine signaling 2	SOCS2_RAT	2.5e-20

Gene Protein Association		Gene Expression			
Gene Symbol	Organism	Factor	Regulation	Experiment Number	P-value
Socs3	Mus musculus	brain	UP	24	1.1218185040451748e-23
Socs3	Mus musculus	brain	UP	24	1.1218185040451748e-23
SOCS3	Homo sapiens	brain	UP	11	2.512857477645065e-9
SOCS3	Homo sapiens	brain	UP	11	2.512857477645065e-9
Socs3	Rattus norvegicus	brain	DOWN	6	5.427190918894098e-10
SOCS2	Homo sapiens	brain	DOWN	12	2.9868274520339355e-9
socs8	Danio rerio	brain	DOWN	1	0.0186142735183239
Socs2	Rattus norvegicus	brain	DOWN	5	0.005287489853799343

Fig. 4. Global ranked (Rank) results provided by the search computing to the example question for the user input Protein ID = "*O14543*", Protein ID name = "*uniprot*", Gene expression regulation = "*updown*" and Biological tissue = "*brain*". Expectation: similarity expectation value of the best BLAST alignment of the input protein sequence with the sequence of the protein with the Protein ID, Protein Name and Protein Symbol shown; P-value: most statistical significance p-value of the differential expression (Regulation) of the gene with the indicated Gene Symbol in the Organism and biological tissue (Factor) shown, according to the experiments (Experiment Number) considered.

5 Future Scenarios

The created Bio-SeCo application fully enables the user to run the example multi-domain biomedical query and demonstrates the capabilities of the search computing technologies to be effectively applied to efficiently search for globally ranked answers to complex biomedical questions. Yet, the described application is just a first prototype that can and will be improved in the near future. Among others, in accordance with planned developments of the search computing technology, future improvements will concern ranking composition function, query expansion and web user interface. Different types of ranking composition function will be defined, which will include weight coefficients for each composed service. Values for such coefficients will be definable and interactively modifiable by the user at query time, in order to allow customizing global ranking calculation also in accordance with the retrieved results by each specific individual search. According to the information exploration paradigm based on semantic resource framework, expansion of global search results will be made possible in order to both refine search results and explore additional search domains, not included in the initial search but related according to the semantic resource network of domain services registered in the search computing framework. For instance, for the considered example query, interesting expansions could concern the search for similarity among the promoters of the genes found significantly expressed in the same given biological tissue or condition, or the search for co-occurrence in the same biomolecular pathways of the proteins found with amino acid sequence similar to the one of a given protein. A more advanced result visualization interface will allow interactively browsing and expanding individual search results, highlighting global combinations of results with particular relevance. Finally, the increasing number of bioinformatics services that will be registered in the search computing framework will enable more possible combinations of service compositions, with the consequent increasing capability of answering more and even more complex biomedical questions.

6 Conclusions

This chapter has shown that mainstream bioinformatics resources can be described and composed using search computing constructs in order to automatically answer complex multi-domain biomedical queries, where global ranking of the integrated retrieved results is automatically computed by the search computing platform based on the rankings of the individual searches. The shown example and the implemented application take into account only three bioinformatics services and just one composition of them. When more services are registered in the search computing platform, they can be composed in different ways to answer a broader variety of complex biomedical queries and refine or augment query results. In so doing, search computing can support exploratory search and curiosity driven browsing of life science data that are difficult to perform otherwise, thus enabling ambitious data driven biological knowledge discovery and verification. Further work, required to enable more than a single mechanism for aggregating ordered data sets, will allow the use of multiple global ranking mechanisms. This will enable users to customize the global rankings, to reflect individual preferences on a search-by-search basis, and allow meeting the variety of requirements of biomedical users.

References

1. Goble, C.A., Belhajjame, K., Tanoh, F., Bhagat, J., Wolstencroft, K., Stevens, R., Pettifer, S., Nzuobontane, E., McWilliam, H., Laurent, T., Lopez, R.: BioCatalogue: a curated Web Service registry for the Life Science community. ISMB/ECCB 2009. Technology Track: TT40 (2009)
2. Masseroli, M., Paton, N.W., Spasić, I.: Search Computing and the Life Sciences. In: Ceri, S., Brambilla, M. (eds.) Search Computing. LNCS, vol. 5950, pp. 291–306. Springer, Heidelberg (2010)
3. Braga, D., Ceri, S., Daniel, F., Martinenghi, D.: Mashing up search services. IEEE Internet Comput. 12(5), 16–23 (2008)
4. Ceri, S., Brambilla, M. (eds.): Search Computing - Challenges and Directions. LNCS, vol. 5950. Springer, Heidelberg (2010)
5. Altschul, S.F., Gish, W., Miller, W., Myers, E.W., Lipman, D.J.: Basic Local Alignment Search Tool. J. Mol. Biol. 215(3), 403–410 (1990)
6. Pearson, W.R.: Using the FASTA program to search protein and DNA sequence databases. Methods Mol. Biol. 24, 307–331 (1994)
7. Masseroli, M., Ceri, S., Campi, A.: Integration and mining of genomic annotations: Experiences and perspectives in GFINDer data warehousing. In: Paton, N.W., Missier, P., Hedeler, C. (eds.) DILS 2009. LNCS, vol. 5647, pp. 88–95. Springer, Heidelberg (2009)
8. Masseroli, M., Ceri, S., Tettamanti, L., Campi, A., Sormani, S.: Integration of distributed heterogeneous biomolecular data to support biological discovery. In: BITS 2009: Sixth Annual Meeting Bioinformatics Italian Society, Liberodiscrivere edizioni, Genova, IT, pp. 113–114 (2009)
9. Parkinson, H., Sarkans, U., Shojatalab, M., Abeygunawardena, N., Contrino, S., Coulson, R., Farne, A., Lara, G.G., Holloway, E., Kapushesky, M., Lilja, P., Mukherjee, G., Oezcimen, A., Rayner, T., Rocca-Serra, P., Sharma, A., Sansone, S., Brazma, A.: ArrayExpress - a public repository for microarray gene expression data at the EBI. Nucleic Acids Res. 33(Database issue), D553–D555 (2005)
10. Barrett, T., Troup, D.B., Wilhite, S.E., Ledoux, P., Rudnev, D., Evangelista, C., Kim, I.F., Soboleva, A., Tomashevsky, M., Marshall, K.A., Phillippy, K.H., Sherman, P.M., Muertter, R.N., Edgar, R.: NCBI GEO: archive for high-throughput functional genomic data. Nucleic Acids Res. 37(Database issue), D885–D890 (2009)

Workflows for Information Integration in the Life Sciences

Paolo Missier, Norman Paton, and Peter Li

School of Computer Science, University of Manchester
Oxford rd., Manchester, UK
{firstname.lastname}@cs.manchester.ac.uk

Abstract. The increasingly computationally- and data-intensive nature of experimental science motivates recent interest in workflows, as a way to specify complex data processing and integration pipelines in a fairly intuitive way. Such workflows orchestrate the invocation of data retrieval services in a way that resembles, to some extent, Search Computing query plans. While the former are manually specified, however, the latter are the result of an automated translation process. Using lessons learnt from experience in workflow design, in this chapter we discuss some of the requirements on service curation that make automated, on-demand data integration processes possible and realistic.

1 Workflows for Computational Science and Information Integration

In many disciplines of natural science, research advances increasingly rely upon the automated acquisition, transformation and analysis of large-scale data. In this chapter, we exemplify and discuss the use of workflow technology as a way to address the needs of data analysis automation in science [1]. Our examples refer to two emerging areas in the life sciences which has been the focus of data-intensive research, namely next generation DNA sequencing (NGS) and systems biology.

NGS is having a profound impact on the expectations and the methods of genomics research. First introduced around 2005, NGS makes it possible to sequence entire genomes in weeks, spurring ambitious new efforts like the 1000 Genomes Project [2]. While these projects underpin the study of the genetic causes of human diseases, they come with new challenges at multiple levels. Firstly, they push the limits of current data repositories. For example, the Short Read Archive, the European repository that accepts data submissions from NGS machines at the EMBL[1], received 30TB of data in the first six months of operation, making data submission rate the new bottleneck for advances in genomics[2] [3]. At the same time, a secondary effect of these new whole-genome sequencing studies is the exponential growth in the number of submissions to

[1] European Molecular Biology Lab: http://www.ebi.ac.uk/ena/

[2] The EMBL-Bank grows in size at the rate of 200% per annum.

S. Ceri and M. Brambilla (Eds.): Search Computing II, LNCS 6585, pp. 215–225, 2011.

SNP databases[3] [4]. In turn, advances in data production drive the need for the development of highly automated pipelines for the analysis of NGS data, both primary (sequence) and "downstream" (SNP analysis, for example).

While such experimental processes predictably involve a combination of data-centric (data retrieval, format mappings) as well as compute-intensive tasks, their exact nature and composition into a complete process tend to change rapidly, following data availability and other technological advances. In practice, the experimental nature of the projects extends from data generation technology, to the development of novel techniques for data analysis. In this setting, workflow technology addresses the scientists' needs for rapid prototyping of innovative applications. Workflows embody high level programming models that let users specify the coordinated execution, known as *orchestration*, of various types of executable software components, or *tasks*, often implemented as Web services. Workflow languages tend to be higher level than traditional scripting languages, such as Perl, resulting in more manageable specifications of complex data processing pipelines. At the same time, their computational models are more understandable by domain experts with a limited knowledge of general-purpose programming. For these experts, workflows are a way to maintain control over all phases of their computational experiment, from design, to execution, to analysis of the results. Workflow systems offer additional advantages over general scripting environments, including managing the scheduling of tasks and their deployment on HPC infrastructures, such as clouds.

A variety of workflow systems for science have emerged over the past few years, in response to these scientists' needs. Their commonalities and differences have been described at length in the literature [5,6]. In most cases, however, the focus is on workflows that accomplish compute-intensive tasks, such as large-scale simulations [7]. Less emphasis is placed on a class of workflows whose main purpose is to retrieve and integrate data from multiple sources, usually in order to enable some more complex processing downstream. The importance and impact of these *resource-oriented* workflows on the e-science data infrastructure is growing with the number and size of the available databases, as mentioned earlier. A recent EBI statistic [3], for example, compares interactive Web page accesses to programmatic (i.e., Web service-based) access to its 63 databases, reporting about 1 million automated data retrieval jobs / month in 2009 from services.

Resource-oriented workflows resemble less a scientific experiment, and more a distributed query plan in which the nodes are service invocations, a characterisation that makes them particularly interesting in the context of Search Computing. In the rest of the chapter we use an example from the bioinformatics area of systems biology, implemented using the Taverna workflow system [8], to discuss opportunities and limitations of using workflows as a form of on-the-fly data integration.

[3] Single Nucleotide Polymorphisms, or SNPs, are single-base mutations on a chromosome. About .5k new SNPs are detected for each genome that is sequenced, leading to over 100 million submissions, by early 2010, to dbSNP, the SNP database at the NCBI: `http://www.ncbi.nlm.nih.gov/snp/`.

2 Example: Automating a Systems Biology Pipeline

The following example of a workflow-based systems biology study demonstrates how workflow modelling can be used effectively for data integration. In the next section we will discuss the opportunities and the limitations of this approach. In bottom-up system biology studies, genome-scale models of biological systems, found in existing databases of pathways, are used as a starting point for the creation of quantitative models that support various types of simulations. A typical study involves the following steps: (i) identify the pathway or portion of a network that is to be modeled; (ii) associate the model with functions and parameter values that represent its dynamic behavior, either from databases or experimentation; and (iii) analyze and/or simulate the resulting model to understand its properties [9]. Model construction is typically a manual process, involving the association of a model with experimental data. While such an approach can produce good quality models, it is hardly scaleable. The goal of workflow-based modelling in this case is to automate the entire process, from network data retrieval, to the result of a simulation. At a conceptual level, the entire process consists of a sequence of the following four workflows, shown in abstract form in Fig. 1 along with the databases involved.

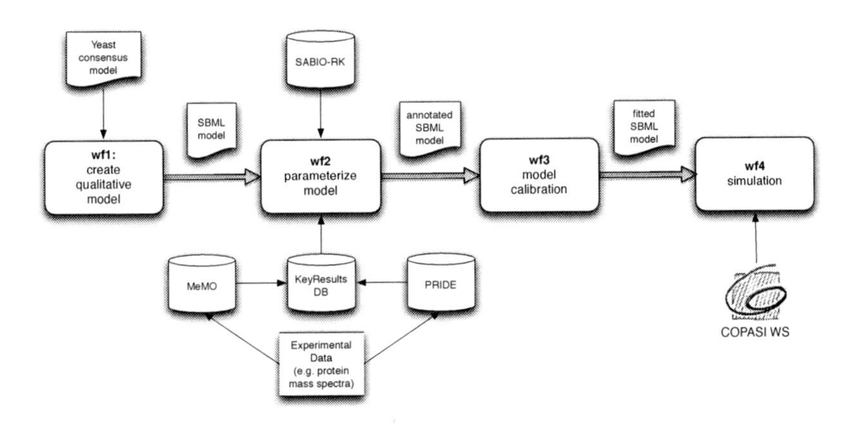

Fig. 1. Conceptual workflow view of systems biology modelling

1. The first workflow produces a small and manageable qualitative model that is focused on a specific metabolic pathway using data from a consensus model of yeast metabolism [10]. This model represents the components and their relationships in a biological system. In a metabolic pathway, for example, the nodes of this system represent metabolites and enzymes, whilst the edges between these components represent biochemical reactions. Importantly, the model is expressed using the standard System Biology Markup Language (SBML) syntax[4]. SBML provides a common and interoperable schema for

[4] `http://sbml.org`

model representation that is shared through all phases of the study, and makes its component workflows reusable within the systems biology community.
2. The qualitative SBML model is fed to a second *parameterisation* workflow, which uses information on reaction kinetics, found in the SABIO-RK database,[5] as well as primary experimental data stored in the Key Results Database (KRDB)[6], to produce a quantitative model, i.e., an annotated version of the input model with additional details.
3. The quantitative model is then fitted to experimental data, and finally the resulting *calibrated* model is analysed using a simulation algorithm offered by the COPASI service [11] to predict how the concentrations of metabolites vary over time.

3 Workflows and Search Computing

We now discuss the use of resource-oriented workflows with a view on Search Computing (SeCo), using a specific workflow model, namely Taverna [8].

3.1 Workflow-Based Service Orchestration

Taverna is a language and computational model designed to support the automation of complex, service-based and data-intensive scientific processes. It is best known for its application to the life sciences, where it has been used to support experimental investigation into a variety of research areas. In the example above, we have used Taverna workflows to automate the otherwise time consuming, partially documented process of manual model design and refinement. These workflows can be viewed as procedural specifications of data and service integration pipelines. Integration of data is made possible by the adoption of SBML, a *de facto* standard language for model description, as well as of the MIRIAM guidelines for curating quantitative models of biological systems [12]. Integration of services is provided by the workflow programming model, whereby service invocations, represented by tasks, are specified along with the intended flow of data amongst them.

Fig. 2 shows a detail of the second workflow in the example pipeline[7]. The data retrieval tasks are either invocations of Web service operations, i.e., on the SABIO-RK service, or are scripts that encode *ad hoc* SQL queries, as suggested in the figure. Data flows are specified as data dependencies from the output of a task, to the input of another task. Workflow execution is entirely data-driven[8]. Users can easily extend the collection of Web services that can be used in a workflow. Specifically, given a generic WSDL interface (reachable through

[5] `http://sabio.villa-bosch.de/`

[6] `http://code.google.com/p/keyresultsdb/`

[7] The complete workflow is reproduced in [9] as Fig.1, and the entire suite of workflows is available from the myExperiment repository: `http://www.myexperiment.org/users/221/workflows`

[8] Additional details on the Taverna model and architecture can be found in [8].

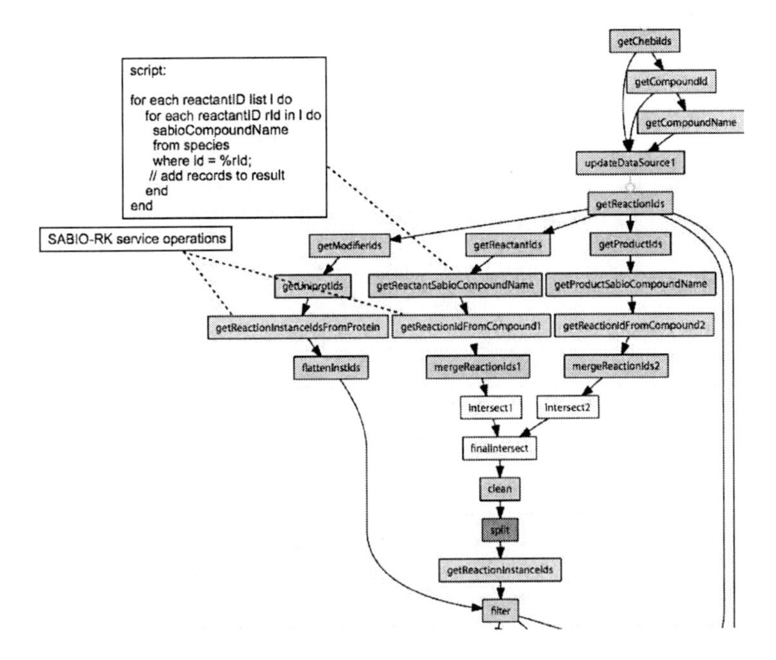

Fig. 2. Detail of model parameterisation Taverna workflow

its URL), Taverna generates a set of tasks, each corresponding to one interface operation, with input and output ports that correspond to the operation parameters[9]. These tasks are immediately available for use in any workflow.

3.2 Providing Support for Integration Workflow Design

The service orchestration model just presented suggests that integration workflows like the ones used in our example can be viewed as *specifications of distributed query plans*, in which the elementary steps correspond to Web service invocations. As such, they may be relevant in the context of Search Computing (SeCo). There are, however, major obvious differences between SeCo query plans and workflows; namely, SeCo logical query plans incorporate notions of *chunking* and can deal with *ranked* results returned by services, neither of which is part of the Taverna model of computation. Nevertheless, one can still gain insight into the potential complexity of automated SeCo query plan generation, by looking at the typical manual workflow design process, where the primary concern is to ensure consistency of the integration. As in the case of traditional schema mapping, the design of an integration workflow too relies on a number of elements:

[9] A plugin for importing REST services is also avaiable. Tasks that represent REST service invocations have an input port for each configurable parameter in the URL template that represents the main access path to the service.

- knowledge of the data types and formats produced and consumed by each of the involved services, as well as of the services' functionality (intended behaviour);
- knowledge of the schema and data access mechanisms (SQL, for example), as in the task that retrieves SABIO compound names in Fig. 2;
- a collection of additional adapter tasks, required whenever format or other types of transformations are needed to achieve integration.

Our running example, and specifically the second workflow, is a fortunate case of integration in which the services' signatures and behaviour are well specified, and furthermore, they are designed by the service provider to work together, thus simplifying their composition and removing the need for adapters. Additional ad hoc local scripts were still needed, however, to supplement the service-based data access operations. In the example, the designers use their knowledge of each service's functionality and of the type of each service port, to identify the connections that are consistent with the data values and formats.

In general, however, workflow-based integration efforts are not immune from the curse of heterogeneity that affects traditional data integration projects. This results in complex integration workflows that often contain more adapters, which are custom-made and hardly reusable scripts (colloquially known as "shims"), than they do actual data retrieval steps. This is typically the case when the composition involves arbitrary services[10]. Indeed, the ability of the Taverna model to turn any public, WSDL-based service into a set of tasks is both its strength and its weakness, as this generality comes at the price of an increased burden on the workflow designer to ensure the consistency of the service composition.

3.3 Restricting the Space of Target Services: Examples

These considerations suggest, unsurprisingly, that composition within a limited and controlled space of services may be easier to accomplish, automate, and verify than when we are faced with an open space of arbitrary third party services that exist "in the wild". Assisted, or even automated workflow design comes with a price, however, by either requiring a systematic curation effort to provide a knowledge-rich description of the services and datasets involved, or by putting additional constraints onto the space of services and components that are available to the users.

Notable examples of workflow systems that take these approaches include Wings/Pegasus, Galaxy, Triana, and Kepler, amongst many others. We briefly recall their main features in the rest of this section. These systems are all geared primarily towards the design of complex, compute-intensive data pipelines (as opposed to data retrieval and integration tasks) for the benefit of a specific user community. As the space of available components is both limited and controlled

[10] Examples of phenomenon can be easily found on the myExperiment web site. For a rather extreme case, see workflow `http://www.myexperiment.org/workflows/1212.html`, where 2 computational tasks (one script and one database lookup) are surrounded by 33 adapters.

as a consequence of this focus, better guidance (in the form of components recommendations and type-checking, for example) can be provided to workflow designers.

Wings/Pegasus. The Wings project [13] is an exploration into the trade-off between component curation and automated workflow generation. Wings implements a suite of AI-type algorithms that operate on a knowledge base of annotated workflow templates, as well as of a curated catalog of components and datasets, to generate complete Pegasus workflows [14] from high-level user requirements. User requirements are expressed in the form of constraints over the types and features of the components, and logic reasoning is used to incrementally refine an initial workflow template into a ground (i.e., fully specified and executable) workflow along with a set of configuration parameters. One of the key to the success of this approach is its application to a well-curated and confined area of computational science, namely in a library of components for machine learning and data mining.

Galaxy. The Galaxy system[11] is a rich, interactive environment for genomics and metagenomics data analysis [15]. Its primary goal is to make *in silico* research accessible to bioinformaticians, by letting them build complex data analysis pipelines interactively, with the option to save them as workflows that can be shared and reused, possibly with different inputs and parameters. With a strong emphasis on *reproducibility* of scientific data analysis, Galaxy is characterised by a controlled space of component tools which are exposed to the environment through special configuration directives. Reproducibility is achieved by ensuring that the tools remain available after execution, as well as by keeping a detailed execution log that can be used for partial re-run of a workflow. As we have observed earlier, control over the set of tools available to scientists is also beneficial in reducing the need for ad hoc *shims*, which indeed are not part of Galaxy workflows.

An analysis of Galaxy features that are missing from Taverna can be found in [16]. Such analysis has been used to justify a recent integration effort, consisting of a hybrid environment code-named *Tavaxy*, whereby Taverna and Galaxy workflows can call each other. While this has remained a largely isolated effort (possibly because the authors are not affiliated with either project), a more recent interoperability effort has been initiated jointly by the Taverna development group in Manchester, UK, and the Netherland Bioinformatics Centre (NBIC), where both Galaxy and Taverna are routinely used. The goal of this project, which at press time is at too early a stage too have been documented, is to enable two-way, programmatic invocation of workflows, thereby offering users the advantages of both systems. These efforts suggest not only that these workflow models share some portion of the bioinformatics research space, but also, in a more technical sense, that building bridges across them is feasible and, hopefully, will remove the need for organisations to choose one camp over another.

[11] `http://galaxy.psu.edu/`

Triana. Originally designed to cater to a particular branch of physics (gravitational wave [17]), the Triana workflow system [18] comes with a rich toolkit of over 400 "native" components, which was subsequently expanded image manipulation, data mining[12], and other types of specialised application areas.

As anticipated, Triana offers a rich graphical workflow design environment complete with type checking and other assistive features like on-the-fly creation of tools, along with a facility for deploying tasks onto a variety of Grid computing environments, and thus distributing the entire workflow computation.

Kepler. Designed for e-science applications, the Kepler workflow system [19] is characterized by a decoupling of the workflow language syntax, from its semantics. The same workflow can be interpreted differently by different *directors*, each of which implements a different Model of Computation (MoC) for the language. The strong structural constraints imposed by some of these directors on the workflow topology actually facilitates their composition, as well as analysis (behaviour prediction) and optimisation. The COMAD MoD (Collection Oriented Modelling and Design), for example, operates of data pipelines where each node, or *actor*, can consume and produce exactly one input message [20]. Such messages are nested data collections, expressed as XML documents, and an actor's configuration includes path expressions (similar to XPath) that specify the portions of the input document that the actor will update. Such update semantics for actors can be captured formally, and in addition, the model lends itself well to assistive workflow design. Indeed, once again we see an example of a controlled space of workflow tasks, as actors are dedicated components that are specifically developed for use in Kepler workflows.

3.4 Taverna in a Controlled and Annotated Service Space

In contrast to the approaches just surveyed, the generality of the Taverna approach to composition leaves the burden of workflow design entirely with its author. To alleviate the problem, the strategy for supporting the designers is centred around a home-grown service registry, called BioCatalogue[13]. The registry accepts direct contributions from the community, as well as ingesting service descriptions from a variety of other registries, including BioMoby Central[14], DAS[15], SeekDa[16], and EMBRACE[17]. Services in BioCatalogue are described in a variety of ways, reflecting the heterogeneity of annotations supplied natively by the different providers. These include structured data, i.e., WSDL interface specifications, free text, user-supplied tags, as well as more formal ontology terms. A recent account of BioCatalogue [21] describes four main annotation categories, covering: (i) functional properties, which attempt to capture both the overall

[12] `http://www.dataminingrid.org/`
[13] `http://www.biocatalogue.org.`
[14] `http://www.biomoby.org/`
[15] `www.biodas.org`
[16] `www.seekda.com`
[17] `www.embraceregistry.net`

scientific purpose of the service operations, as well as describing how operations are to be combined into specific patterns in order to deliver some high-level service functionality; (ii) operational properties, which specify for example conditions on the service usage; (iii) service performance and health profiling obtained through monitoring, and (iv) the provenance of the description, including who provided the entry and its annotations.

While basic service description lookup functionality has been integrated into the Taverna workflow design enironment, service annotations are not yet, however, exploited to their full potential to actively support the design process. Our ongoing research, specifically in the area of functional characterisation of services [22], starts from the observation that the common approach of hand-crafting ontology-based annotations may be at the same time too expensive, as it requires support from skilled curators, and inadequate to drive the semi-automated composition of services into a workflow. OWL-based annotations in particular are proving difficult to "get right", to understand by non-experts, and to use effectively. The main problem appears to be one of level of abstraction in service description. Simple models for the semantic annotatation of WSDL specifications like SAWSDL, a W3C recommendation[18], are limited to individual interface elements, and do not provide any facilities for describing functions that are only delivered through the grouping of operations into specific patterns (the System Biology SABIO-RK service mentioned in Sec. 2 provides examples of such patterns). Providing simple functional descriptions that are easy to maintain and can drive the workflow design process is one of the priorities for BioCatalogue.

The functional annotations approach is being tested on a small number of restricted application domains with a "closed world" of services. These include, amongst others, (i) the ongoing *ChemTaverna* effort[19] to provide a set of generic and chemistry-specific workflow components, that can be coupled together for data analyses without the need for shims, and (ii) the EU-funded e-Lico project,[20] which involves the automated generation of Taverna workflows that use a suite of data mining services, starting from a process specification in the OWL Semantic Web language.

4 Summary

As experimental science becomes increasingly data-intensive [1], computational scientists are coming to appreciate process models that can be used to specify complex data pipelines in a simple and intuitive way. In particular, workflow systems that are based on a dataflow programming model have proven successful in describing integration of data as an orchestration of data-retrieval services. Such orchestrations are interesting in the context of Search Computing, as they resemble, at least superficially, a distributed query plan whose nodes consist of Web service invocations.

[18] `www.w3.org/2002/ws/sawsdl/`

[19] `http://www.taverna.org.uk/introduction/related-projects/chemtaverna/`

[20] `http://www.e-lico.eu/`

In this chapter we have shown an example of such orchestration, from the area of System Biology, that has been implemented using the Taverna workflow system. Motivated by the parallels between such workflows and SeCo queries, we have then briefly discussed how the ability to automatically generate workflows from higher-level specifications essentially hinges upon the curation effort that can be afforded on the service collections involved, with "services in the wild" at one extreme of the spectrum, and well-curated services with rich metadata descriptions, at the other.

References

1. Hey, T., Tansley, S., Tolle, K. (eds.): The Fourth Paradigm: Data-Intensive Scientific Discovery. Microsoft Research (2009)
2. Via, M., Gignoux, C., Burchard, E.G.: The 1000 Genomes Project: new opportunities for research and social challenges. Genome medicine 2(1), 3 (2010)
3. Southan, C., Cameron, G.: Beyond the Tsunami: Developing the Infrastructure to Deal with Life Sciences Data, Microsoft Corp., pp. 117–123
4. Koboldt, D.C., Ding, L., Mardis, E., Wilson, R.: Challenges of sequencing human genomes. Briefings in bioinformatics (Epub ahead of print) (June 2010)
5. Deelman, E., Gannon, D., Shields, M., Taylor, I.: Workflows and e-Science: An overview of workflow system features and capabilities. Future Generation Computer Systems 25(5), 528–540 (2009)
6. Taylor, I.J., Deelman, E., Gannon, D., Shields, M. (eds.): Workflows for e-science, Scientific workflows for Grids. Springer, Heidelberg (2006)
7. Ludascher, B., Altintas, I., Bowers, S., Cummings, J.: Scientific Process Automation and Workflow Management. In: Computational Science. Chapman & Hall, Boca Raton (2010)
8. Missier, P., Soiland-Reyes, S., Owen, S., Tan, W., Nenadic, A., Dunlop, I., Williams, A., Oinn, T., Goble, C.: Taverna, reloaded. In: Gertz, M., Hey, T., Ludaescher, B. (eds.) SSDBM 2010. LNCS, vol. 6187, pp. 471–481. Springer, Heidelberg (2010)
9. Swainston, N., Jameson, D., Li, P., Spasic, I., Mendes, P., Paton, N.: Integrative Information Management for Systems Biology. In: Lambrix, P., Kemp, G. (eds.) DILS 2010. LNCS, vol. 6254, pp. 164–178. Springer, Heidelberg (2010)
10. Herrgård, M.J., Swainston, N., Dobson, P.: A consensus yeast metabolic network reconstruction obtained from a community approach to systems biology. Nature Biotechnology 26(10), 1155–1160 (2008)
11. Dada, J.O., Mendes, P.: Design and Architecture of Web Services for Simulation of Biochemical Systems. In: Paton, N.W., Missier, P., Hedeler, C. (eds.) DILS 2009. LNCS, vol. 5647, pp. 182–195. Springer, Heidelberg (2009)
12. Novère, N.L., Finney, A., Hucka, M., Bhalla, U.S., Campagne, F., Collado-Vides, J., Crampin, E.J., Halstead, M., Klipp, E., Mendes, P., Nielsen, P., Sauro, H., Shapiro, B., Snoep, J.L., Spence, H.D., Wanner, B.L.: Minimum information requested in the annotation of biochemical models (MIRIAM). Nature Biotechnology 23(12), 1509–1515 (2005)
13. Gil, Y., Gonzalez-Calero, P., Kim, J., Moody, J., Ratnakar, V.: A Semantic Framework for Automatic Generation of Computational Workflows Using Distributed Data and Component Catalogs. Journal of Experimental and Theoretical Artificial Intelligence (to appear, 2010)

14. Deelman, E., Singh, G., Su, M.-H., Blythe, J., Gil, Y., Kesselman, C., Mehta, G., Vahi, K., Berriman, B.G., Good, J., Laity, A.C., Jacob, J.C., Katz, D.S.: Pegasus: A framework for mapping complex scientific workflows onto distributed systems. Scientific Programming 13(3), 219–237 (2005)
15. Nekrutenko, A.: Galaxy: a comprehensive approach for supporting accessible, reproducible, and transparent computational research in the life sciences. Genome Biology 11(8), R86 (2010)
16. Abouelhoda, M., Alaa, S., Ghanem, M.: Meta-workflows: pattern-based interoperability between Galaxy and Taverna. In: Proceedings of the 1st International Workshop on Workflow Approaches to New Data-centric Science, Wands 2010, pp. 1–8. ACM, New York (2010)
17. Taylor, I.: Triana Generations. e-Science, 143 (2006)
18. Churches, D., Gombas, G., Harrison, A., Maassen, J., Robinson, C., Shields, M., Taylor, I., Wang, I.: Programming Scientific and Distributed Workflow with Triana Services. Concurrency and Computation: Practice and Experience (Special Issue: Workflow in Grid Systems) 18, 1021–1037 (2006)
19. Ludäscher, B., Altintas, I., Berkley, C.: Scientific Workflow Management and the Kepler System. Concurrency and Computation: Practice and Experience 18, 1039–1065 (2005)
20. Altintas, I., Barney, O., Jaeger-Frank, E.: Provenance Collection Support in the Kepler Scientific Workflow System. In: Moreau, L., Foster, I. (eds.) IPAW 2006. LNCS, vol. 4145, pp. 118–132. Springer, Heidelberg (2006)
21. Bhagat, J., Tanoh, F., Nzuobontane, E., Laurent, T., Orlowski, J., Roos, M., Wolstencroft, K., Aleksejevs, S., Stevens, R., Pettifer, S., Lopez, R., Goble, C.: BioCatalogue: a universal catalogue of web services for the life sciences. Nucleic Acids Research (May 2010)
22. Missier, P., Wolstencroft, K., Tanoh, F., Li, P., Bechhofer, S., Belhajjame, K., Goble, C.: Functional Units: Abstractions for Web Service Annotations. In: Procs. IEEE 2010 Fourth International Workshop on Scientific Workflows (SWF 2010), Miami, FL (2010)

Complex Search, Ranks, and Biological Discovery: A User's Perspective

Paolo Romano[1] and Luciano Milanesi[2]

[1] National Cancer Research Institute, c/o Centro Biotecnologie Avanzate, Largo Rosanna Benzi, 10, I-16132 Genova, Italy
paolo.romano@istge.it
[2] Institute for Biomedical Technologies, National Research Council, Via Fratelli Cervi 93, I-20090 Segrate (MI), Italy
luciano.milanesi@itb.cnr.it

Abstract. This chapter presents a users perspective regarding the potential applications of the Search Computing technology for biomedical discovery. Recent research on human inherited diseases has increased the number of information resources useful to bridge medicine and biology and to associate genotype and phenotype. The application of the Search Computing technology is discussed in the frame of a number of techniques that can be applied in Life Sciences for managing distributed biomedical data: Federated databases, Grids, Cloud computing, Web Services, Workflow. Particular attention is then devoted to challenges and opportunities deriving from the application of ranking and the management of missing information. Finally, the definition of a standard score function, that could be adopted by all service providers in order to merge all the collected scores for the Search Computing, and the combined use of workflow management systems and Search Computing, are discussed.

Keywords: Search Computing, Grid computing, workflow, web services, Bioinformatics.

1 Introduction and Motivation

The recent advent of Next-Generation Sequencing technologies has produced an impressive increase of data and information for Life Sciences research. Moreover, the introduction of the new and rapidly expanding Systems Biology methodology has created a scope for integrating and making sense of '-omics' data, by relating it to high-level physiological data and by using it to analyze and simulate pathways, cells, tissues, organs and disease mechanisms. These data create an enormous amount of different genomes information, including genes, proteins, and all related functional properties and characteristics. At the same time, a number of new web tools and databases is increasing the potentiality of the distributed management, processing, analysis, and visualization of large quantities of bioinformatics data from genomics, proteomics, transcriptomics, and system biology studies.

S. Ceri and M. Brambilla (Eds.): Search Computing II, LNCS 6585, pp. 226–235, 2011.

This huge amount of biomedical data, that is currently available through Internet, makes the task of searching, retrieving and integrating information very difficult and time-consuming. Data is often manually analyzed by accessing several bioinformatics web servers and databases which are available on the Internet. The large amount of data and the exponential accumulation of new information imply that biologists and scientists need to deal with a growing stream of information generated by genomics, proteomics and other data-intensive technologies.

Due to this need, it has become urgent to provide the bioinformatics research communities with tools allowing them to manage large, complex, multimedia datasets and to navigate through an increasingly intricate and potentially confusing information landscape. A number of techniques can be applied to solve this urgent problem, e.g. Federated databases, Grids, Cloud computing, Web Services, Workflows, and Search Computing. In the following, some of the main distributed systems used in Bioinformatics are introduced in order to discuss some challenges and opportunities related to query processing for Search Computing.

2 Web Services and Workflow Management Systems in Life Sciences

Among current ICT technologies, Workflow Management Systems (WMS) in connection with Web Services (WS), seem to be the most promising ones. A workflow is defined as "a computerized facilitation or automation of a business process, in whole or part" (Workflow Management Coalition). The goal of workflows is the implementation of data analysis processes in standardized environments; their main advantages relate to effectiveness, reproducibility, reusability of intermediate results, and traceability. Web Services are software-oriented network services communicating by the use of SOAP (Simple Object Architecture Protocol, a framework for the distribution of XML structured information) over HTTP. They offer a good solution for automated retrieval of information and many WS have already been set up for the biomedical domain [1].

Some workflows management systems have already been proposed and are increasingly being adopted in the biomedical domain. Their utilization of graphical and user-friendly interfaces simplifies the access to, and use of, public in-silico data analysis tools. Among available systems, Taverna Workbench is emerging as one of the most known and appreciated ones by Bioinformatics communities [2]. It was jointly developed at the University of Manchester and at the European Bioinformatics Institute (EBI) in the frame of the myGrid project. It is able to build complex analysis workflows by leveraging on access to both remote and local services. Workflows can be enacted and results can be displayed in various formats. Its only requirement is the availability of a Java Run-time Environment (JRE).

Taverna is complemented by myExperiment [3], a repository of workflows for biomedical research that applies social network techniques, and by BioCatalogue, a user-curated catalog of Web Services for biomedical research [4]. By using these tools, it is possible to build new effective automated data analysis processes. The Search Computing work should take that into account and try to co-operate by offering its added value to this development framework.

As it is said, Taverna is maybe the best known and supported WMS for Life Sciences, but various alternative systems also exist [2]. From the architectural point of view, it is possible to distinguish libraries, standalone software, client/server systems, and web servers. Libraries (or modules) are add-on for programming languages and tools, allowing to implement workflows' features into new software. Standalone systems usually accommodate in a single software all features that are needed to design, implement and execute workflows on a workstation. Client/server systems divide all needed functionalities between the client side, which is usually devoted to the workflow design and the visualization of results, and the server, where the actual enactment of the workflow occurs.

What can make a WMS more interesting and appealing for a biomedical researcher is its ability to cope with the great variety of Web Services that are available for biological information sources and analysis tools. From this point of view, capacity of dealing with Web Services implemented by using, e.g., BioMart, SoapLab, and NCBI EUtils, may be a great advantage. In this sense, Taverna appears to be more performing than its alternatives.

New methods based on WMS and WS may be useful for several distributed communities, especially because of the advent of high-bandwidth computer networks. This could support a new approach to distributed biomedical research as to increase productivity and scientific quality of research carried out by minor laboratories. Nevertheless the effectiveness of major bioinformatics service provision centers, like the National Center for Biotechnology Information (NCBI) and the European Bioinformatics Institute (EBI), would not be reduced.

3 Grid and Bioinformatics

Computation resources are fundamental to process Life Sciences data, yet many problems are associated to them. As a matter of fact, since the completion of the Genome Project, the number of sequencies available for analysis is so vast that problems have dramatically increased. Moreover, the amount of data continues to increase at lightning speed because high throughput expression analysis technologies provide researches with a continuous flow of new information. In the meantime, the study of comparative genomics and genetic variation by means of modern analysis methods, aimed to identify in details the different sets of genes involved in diseases, amplifies the computational load problem.

Grid technology is a very important step forward from the Web, which simply allows the sharing of information on the Internet. This new distributed computing paradigm aims at promoting the development and advancement of technologies that provide seamless and scalable access to wide-area distributed resources. Computational Grids enable the sharing, selection, and aggregation of a wide variety of geographically distributed computational resources, such as supercomputers, compute clusters, storage systems, data sources, and instruments. Indeed, the Grid presents itself as a single, unified resource as to solve large-scale and data intensive computing applications. Grid computing is an emerging solution to establish a flexible environment for dealing with data produced using new high throughput technology, giving suitable solutions to researchers as to perform genome scale analysis.

In this context, the European Grid Infrastructure (EGI), that was developed and deployed within the European 7th Framework Program, has implemented a "pan-European" distributed computing model, where easy access to geographical computing and data management resources may be provided to large multi/inter-disciplinary Virtual Organizations (VO) made of both developers and users.

Grid Computing is another technological and societal revolution in high-performance distributed computing, much as the World Wide Web has been for the last ten years as far as the availability and interpretation of global information is concerned. The aim is to operate this widely distributed computing environment as a uniform service which looks after resource management, exploitation, and security independently from individual technological choices.

Bioinformatics is a discipline that aims to perform analysis of complex biological systems producing huge amounts of data through a computer science approach. Several applications in different fields of Bioinformatics have been developed for the Grid infrastructure and tested in the framework of the BioinfoGRID project. These applications can be seen as example cases which should drive the development of Grid technology in Life Sciences.

The EGI fulfills both the computational and the management requirements, even if this technology is still complex to use, in particular for non expert users, who encounter many problems during the computations of the challenges. In particular, problems are related to the dynamic nature of the Grid, in which the status of resources changes constantly and, therefore, transient problems can happen and provoke the failure of jobs.

Although the monitoring and logging systems of the Grid put actually a lot of effort to limit these problems, consistency between the effective status of the system and the provided information is very difficult to accomplish, due to the size of the infrastructure and its geographical dispersed structure. Moreover, the virtual file system implemented over the Grid, through the combination of the file catalogue and the storage resources, has still a lot of inconsistencies because of temporary unavailable resources which discontinue the accessibility to files.

The Grid is an effective system to cope with the increasing demand of computational power by bioinformatics, in particular in the field of proteomics. In case of independent computations, the system scalability is very high, even when taking into account the time needed for scheduling jobs and transferring data, which can be supposed to be higher in the context of bigger challenges.

The computing power that has been made available on demand thanks to the recent introduction of Cloud computing, shows a new concrete methodology to face challenges, which were thought to be impossible, in Life Sciences and in Medicine up to a few years ago.

4 Some Issues with Scores and Rankings in Life Sciences

4.1 Missing Information

Life Sciences data currently available in biomedical databases are only a fraction of all data that could be useful for new knowledge discoveries in biology. Although apparently trivial, this information may radically change the perspective of the application of Search Computing techniques in biology.

Many factors may affect this situation. As experimental research is often extremely focused, it ends with deeply analyzing a limited phenomenon, instead of more complex systems. As a consequence, it may produce highly detailed data on limited knowledge domains, instead of coherent, uniformly spread information on wide knowledge domains. There are cases when technological limitations do not allow to collect all possible interesting information: this is the case, as an example, of the three-dimensional structure of some proteins that are difficult to crystallize. It is also clear that knowledge discovery in biology may often lead to new hypothesis that, in turn, may lead to new information needs in an apparently never-ending cycle: an exhaustive dataset for biological information is, presently, impossible to achieve or, even, to imagine.

It is also known that biological databases are sparse, i.e. they miss a great part of data. Although complete genomes have been more and more sequenced and made available, the vast majority of them is still undetermined. Moreover, available genomes are virtual, since they do not relate to any actual living organism, but are derived from cells taken from many individuals. This is extremely relevant in the case of the human genome, for which only a very limited fraction of all polymorphisms and gene variations have been identified. It has been estimated that the human body may contain over two million proteins and some 20 to 30 thousands genes, but the UniProtKnowledgeBase only includes less than 100,000 records for human proteins, some 20,000 of which annotated (UniProt release 2010_10, October 5, 2010, see http://www.uniprot.org/).

This makes the Life Sciences domain very different from other knowledge domains where one can assume that all needed information is available (e.g., as far as a tourism reservation system is concerned, one can assume that all available hotels, as well as all possible flights, museums, etc., and all desired information, are likely to be listed).

Life science is, in other words, an "open world". The "open world assumption" (see http://en.wikipedia.org/wiki/Open_world_assumption) must therefore be taken into account. It states that "the truth-value of a statement is independent of whether or not it is known by any single observer or agent to be true". Being the opposite of the "closed world assumption", holding that any statement that is not known to be true is false, it leads us to the conclusion that we cannot assume that what is missing, and therefore unknown, is also false. The clear consequence is that any kind of in-silico analysis may not be exhaustive and definitive, since it can only relates to a subset of data, e.g. an homology search may find the small fraction of all homologues that is already recorded in current databases.

In a computer science context, this problem can be compared to making a left join instead of an inner join in a SQL statement. With reference to the Search Computing approach, its consequences are quite evident. If one has only a few information to merge, it may be simple to overcome this problem; on the contrary, when one has many information, coming from different tools and referring to various data sources, each of which is limited in a different way, it may become very difficult, and somehow misleading, to merge them, since only those few data, for which results can be retrieved from all sources, will be highlighted.

It is therefore essential to find strategies for merging rankings when the vast majority of information is missing. Some examples are the search of similarities of proteins at different structure levels, since the three-dimensional information is rarely available; comparison of gene co-expression, due the very limited number of gene expression experiments that are currently available; protein-protein interactions, again due to the few data available.

4.2 Composition of Scores

Scores provided by bioinformatics tools usually are the result of some mathematical computations and are therefore defined either on the basis of special numerical ranges, that may greatly differ from one algorithm to another, or as a statistical measure, like, a p value. Whichever those values are, they are neither defined to be compared among them, nor merged anyway, their only aim being internal to the algorithm and tool. Their comparison, or merging, may unavoidably lead to over- or under-estimate some of them, unless they are first somehow normalized.

Scores may also have inverted rankings: a clear distinction must be made between rankings giving higher values for best results and those that, vice versa, assign low values for best results. The aggregation function must of course be different.

However, the assessment of scores' relevance is often left to researchers, to whom all computed results are shown. So it is essential to define, for each type and range of scores, a threshold in order to separate those values that may be admitted to further automated elaborations, e.g. score merging, and those that should be removed and not taken into any further consideration.

In order to cope with heterogeneous ranges, a standard ranking metrics could be usefully defined and adopted. As an example, some scores may be quite near (the absolute value of the difference is small) but they may also have a very different meaning; otherwise values may also be very far (the difference is great), but with a similar meaning.

Various measures of similarity between two protein structures with different tertiary structures exist and they can be used for demonstration purposes in this context. The Root Mean Square Deviation (RMSD) [5,6], that is the average distance between the backbones of superimposed proteins; the Longest Continuous Segment (LCS); the Global Distance Test (GDT) [7], a measure of similarity between two protein structures with identical amino acid sequences, but different tertiary structures; and the Template Modeling Score (TM-score), that is intended as a more accurate measure of the quality of full-length protein structures than previous ones and is independent from protein lengths: these are all measures of similarity between three dimensional protein structures.

Given this multiplicity of possible measures, provided by distinct servers (see e.g. http://zhanglab.ccmb.med.umich.edu/TM-score/), one can imagine to scan a database of structures for identifying the protein that best matches the structure of a given protein by merging and aligning results. In this context, the Search Computing could find a useful application.

However, an RMSD value is expressed in length units, whereas the most commonly used unit in structural biology is the Angstrom (Å) which is equal to 10^{-10} m, while the TM-score is in the range [0,1], where 1 indicates a perfect match [8]. Scores below 0.20 correspond to unrelated proteins and should be discarded from any further evaluation. Scores higher than 0.5 relate to proteins having a very similar fold [9]. A score equal to 0.5 identifies proteins having a good probability of being in the same structural family (37% in the same CATH Topology family, 13% in the same SCOP Fold family) [10].

From the above example, it is clear that normalization of ranking may be a problem in Life Sciences. Rankings may be linear or not, they may represent different information, with different scales and to some extent different meaning.

5 Complex Searches and Bio-molecular Knowledge Discovery

A new type of software development environment for distributed data management in Bioinformatics is evolving rapidly. It only needs a short developing time for achieving requirements of today's most demanding applications. The advantages of technological improvements are usually followed by data management drawbacks, as it happened in other research fields. In Life Sciences, the increase of experimental analytical capacities corresponds to an explosion of data throughput and to a real need for new systems for data access, extraction, integration, and management. These requirements need new architectures with improved reliability, stability, scalability and security characteristics. In order to achieve high-level performances in Bioinformatics, Search Computing needs to interact with analytical tools that are based on high performance environments and versatile data types.

New data are continuously produced by many existing methods developed to solve genetic problems, from genome annotation to systems biology. In this case, one of the main difficulties is represented by the analysis of a big amount of data that need to be integrated with publicly available data.

Data Mining techniques [11,12,13] may be used to extract new knowledge from existing data. However, in order to use the most modern ICT technologies, biological databases must be well engineered, consistent and harmonized with high data quality standards. Without these premises, data mining and Search Computing procedures could lead to inconsistent results when applied to biological databases. However, many Life Sciences resources available in Internet, which are often heterogeneous and fragmented, do not follow well defined standards for data sharing and accessing.

Typically, data warehouses are used in Bioinformatics in order to integrate heterogeneous databases by providing both uniform conceptual schemas, that resolve representational heterogeneities, and querying capabilities, that aggregate and integrate distributed data [14,15,16]. Research in this area has applied a variety of database and knowledge-base techniques, including semantic data modeling, ontology definition, query translation and query optimization.

Of course, users would like to retrieve as much information as possible, but, on the other side, results of their searches must be very specific and well standardized, ready to be used for further analysis. In particular, the need to filter results has become very important due to big availability of data that can easily create the risk of an information overflow. In this regard, search computing should offer a possibility to merge and focus the main resulting information.

Another important open issue is the use of the experimental data that is generated by the user directly from his local laboratory and its integration with the great amount of available knowledge that, although retrievable through databases and in-silico analysis tools, may not be ranked [17]. User needs, requirements, preferences, and ways of reaching data may change depending both on preliminary and intermediate results and on the quality of rankings. Driving prioritization is a very important part of the searching procedure and typical quality indexes (like evidence and provenance) are not always optimal for the actual needs in Bioinformatics and Life Sciences[18].

6 Conclusions

These conclusions have partially been derived from some points raised and discussed at the Second Search Computing Workshop: Challenges and Directions, held in Como on May 26-28, 2010.

6.1 A Standard Scoring System for Bioinformatics

Due to extreme variability of score systems and scores in Life Sciences, a proper implementation of Search Computing techniques can only be achieved by an improved scoring system, widely adopted and standardized, that can be easily used to merge search results.

Such a scoring system should be aimed at merging scores, and, as such, it should be normalized and made linear. It could be related to probability and its range should therefore be from 0 to 1. It should also include a threshold value, under which the data should be removed from any further elaboration. A minimum level for the inclusion of data in the analysis should then be defined by researchers, separately for each data set.

In this context, the role of negative results should carefully be taken into account. This is extremely important since in Life Sciences we often have contradictory data from distinct databases, e.g. over- and under-expression of the same gene in the same situation from distinct experiments.

6.2 Integration with Workflow Management Systems

As already stated, automated data analysis workflows promise to be among the most powerful tools for tomorrow Life Science research, due to their flexibility, ability to cope with huge amount of information and information sources, and user-friendliness. During next years, the great majority of databases and analysis tools will be made available through APIs and this will allow researchers to create analysis workflows on almost every topic of interest in any domain. The development of specialized integration tools will further support this trend.

From this point of view, the Search Computing techniques could also be implemented as APIs and support automated data elaboration, e.g. merging of ranked results from other elaborations.

SeCo and workflow management systems could effectively be interconnected: the simplest way would be to offer some features from the SeCo environment as Web Services. Taverna and other WMS are quite sound, stable and effective. It could be easy for users to benefit from SeCo while working with these tools: one simple way could be allowing researchers to provide the rankings to SeCo through Web Services and ask it to return merged rankings and scores.

6.3 Search Computing as a Consensus Generation Tool

As already stated, there is a need for focalization of in-silico research, so that these huge and complex data can be integrated and reduced to the actual needs of researchers. Also, the automation of analysis processes in the form of workflows can be the basis for the achievement of the flexibility that is needed in current biomedical research. From these considerations, it is possible to delineate one useful target for Search Computing in Life Sciences: it could act as a specialized and effective data filter and focuser.

Indeed, this role could be achieved in a relatively simple and effective way by developing specialized merging tools able to create consensus among alternative elaborations in the same analysis domain. This is the case of the above examples of measures of similarity between protein structures with different tertiary structures. One could imagine to develop an interface aimed both at comparing methods and measures for similarity among structures and at creating a consensus on the most similar protein, from the point of view of its three dimensional structure. This could be implemented by Web Services as well, so that workflow management systems can access and exploit it.

Acknowledgements

This work has been supported by the FP7 SHIWA, INBIOMEDVision and by MIUR FIRB: ITALBIONET (RBPR05ZK2Z), BIOPOPGEN (RBIN064YAT) projects.

References

1. Romano, P.: Automation of in-silico data analysis processes through workflow management systems. Briefings in Bioinformatics 9, 57–68 (2008)
2. Hull, D., Wolstencroft, K., Stevens, R., Goble, C., Pocock, M., Li, P., Oinn, T.: Taverna: a tool for building and running workflows of services. Nucleic Acids Research 34 (Web Server issue), W729–W732 (2006)
3. Bhagat, J., Tanoh, F., Nzuobontane, E., Laurent, T., Orlowski, J., Roos, M., Wolstencroft, K., Aleksejevs, S., Stevens, R., Pettifer, S., Lopez, R., Goble, C.A.: BioCatalogue: a universal catalogue of web services for the life sciences. Nucleic Acids Research 38 (Web Server issue), W689–W694 (2010)
4. Goble, C.A., Bhagat, J., Aleksejevs, S., Cruickshank, D., Michaelides, D., Newman, D., Borkum, M., Bechhofer, S., Roos, M., Li, P., De Roure, D.: myExperiment: a repository and social network for the sharing of bioinformatics workflows. Nucleic Acids Research 38 (Web Server issue), W677–W682 (2010)
5. Armougom, F., Moretti, S., Keduas, V., Notredame, C.: The iRMSD: a local measure of sequence alignment accuracy using structural information. Bioinformatics 22, e35–e39 (2006)

6. Kabsch, W.: A solution for the best rotation to relate two sets of vectors. Acta Crystallographica 32, 922–923 (1976)
7. Zemla, A.: LGA: A method for finding 3D similarities in protein structures. Nucleic Acids Research 31, 3370–3374 (2003)
8. Zhang, Y., Skolnick, J.: Scoring function for automated assessment of protein structure template quality. Proteins 57, 702–710 (2004)
9. Zhang, Y., Skolnick, J.: TM-align: a protein structure alignment algorithm based on the TM-score. Nucleic Acids Research 33, 2302–2309 (2005)
10. Xu, J., Zhang, Y.: How significant is a protein structure similarity with TM-score=0.5? Bioinformatics 26, 889–895 (2010)
11. Hastie, T., Tibshirani, R., Friedman, J., Franklin, J.: The elements of statistical learning: data mining, inference and prediction. The Mathematical Intelligencer 27, 83–85 (2005)
12. Han, J., Kamber, M.: Data mining: concepts and techniques. The Morgan Kaufmann Series in Data Management Systems. Morgan Kaufmann Publishers, San Francisco (2006)
13. Witten, I.H., Frank, E.: Data Mining: Practical Machine Learning Tools and Techniques, Second Edition, 2nd edn. The Morgan Kaufmann Series in Data Management Systems. Morgan Kaufmann Publishers, San Francisco (2005)
14. Mosca, E., Alfieri, R., Merelli, I., Viti, F., Calabria, A., Milanesi, L.: A multilevel data integration resource for breast cancer study. BMC Systems Biology 4, 76 (2010)
15. Mosca, E., Bertoli, G., Piscitelli, E., Vilardo, L., Reinbold, R.A., Zucchi, I., Milanesi, L.: Identication of functionally related genes using data mining and data integration: a breast cancer case study. BMC Bioinformatics 10(Suppl 12), 8 (2009)
16. D'Ursi, P., Chiappori, F., Merelli, I., Cozzi, P., Rovida, E., Milanesi, L.: Virtual screening pipeline and ligand modelling for H5N1 neuraminidase. Biochem. Biophys. Res. Commun. 383(4), 445–449 (2009)
17. Milanesi, L., Petrillo, M., Sepe, L., Boccia, A., D'Agostino, N., Passamano, M., Di Nardo, S., Tasco, G., Casadio, R., Paolella, G.: Systematic analysis of human kinase genes: a large number of genes and alternative splicing events result in functional and structural diversity. BMC Bioinformatics 6(Suppl 4), S20 (2005)
18. Milanesi, L., Romano, P., Castellani, G., Remondini, D., Liò, P.: Trends in Biomedical Complex Systems. BMC Bioinformatics 10(Suppl. 12), I1 (2009)

Part 8

Towards a Sustainable Exploitation

To what extent should a research project be driven by market considerations? Or in other words, should basic research take the market into account? Different studies in the innovation management literature debate on the origin of innovation, whether it is technology push or demand pull, coming to the conclusion that both trajectories are possible. While it is normally acknowledged that technology-pushed innovation has a higher probability to be radical and disruptive, it is equally well known that the progress in technology by itself does not guarantee economic success, when it does not match with market needs. After two years of project, many technological and technical decisions still have to be taken. Most of these decisions have a technological answer and need technological research, but these answers will heavily affect the future market potential of the system. In order to take the right decisions from now on, it is necessary to shade some light on the potential use of technology.

Two main problems may affect brilliant technological solutions when they reach the market: on one side, the risk of remaining unused because the path to solution development is too hard. On the other side, the risk that despite being used by the market, these services or products are not really exploited by those who introduced the technology, but rather by someone else who is more able to capture the value of the innovation. The two chapters of this part aim at shedding some light on these two risks, trying to foresee a potential market exploitation of the SeCo project, in order to drive some insights about how taking the very next decisions in term of research and technology.

The first chapter introduces the basic idea of user centered design and shapes a process for designing potential killer application services based on search computing technologies. This process has been tested and validated in two different environments and allowed to obtain a list of service requirements that the future search computing technology must be able to fulfill, in order to be able to answer to real market needs.

The second chapter is aimed at understanding the various ways in which the value created with search computing applications can be appropriated by market players. The value chain of search computing application development is shaped and the main roles of involved characters are identified. Subsequently, the hypothesis of having some of these roles managed by other companies in the market is taken into account. Depending on how many and which roles are externalized, it is possible to draw some archetypal business models, each one with specific strengths and weaknesses. These last considerations are used to derive some recommendations for the continuation of the project.

An Experience in Applying User Centered Design to Search Computing

Tommaso Buganza[1], Marta Corubolo[2],
Emanuele Della Valle[3], and Elena Pellizzoni[1]

[1] Dipartimento di Ingegneria Gestionale, Politecnico, 20133 Milano, Italy
[2] Dip. di Ind. Design, Art, Communication and Fashion, Politecnico, 20133 Milano, Italy
[3] Dipartimento di Elettronica e Informazione, Politecnico, 20133 Milano, Italy
{name.surname}@polimi.it

Abstract. This chapter presents our experience in applying User Centered Design methods to identify several design opportunities for future Search Computing applications. The chapter starts introducing the definition of User Centered design, its principles and methodologies. It presents the five steps process we followed in order to get user behavioral insights and design opportunities. It illustrates each phase that composes the research process focusing on the visualization of the user experience and on the design questions we were able to identify by analyzing it. The core contribute of the chapter is in proposing answers to the design questions in the conference travel context for two user profiles: the Fast&Furious and the Discover&Learn.

1 Introduction

Search Computing (SeCo) brings about a technology able to support users in accomplishing **complex search**, i.e., a combination of multiple domain data primarily based on ranking and a sophisticated manipulation of the results that includes their expansion through "liquid" reformulation of the original query.

In a technology-based project as SeCo, technology is usually the main driver of innovation. However, it cannot be sufficient to assure the success on the market. As Bill Moggridge states in his famous book on "Designing Interaction" [2], "In the past, those who built interactive systems tended to focus on the technology that makes them possible rather than on the interfaces that allow people to use them. But a system isn't complete without the people who use it. Like it or not, people – irritable, demanding and often distracted people like ourselves – and their goals are the point of our system, and we must design for them".

The value and meaning that the end-users will attribute to SeCo technology are the result of a combination between their expectations and their expertise, attitudes and background. Trusting [1] that the best-designed products and services result from understanding the needs of the people who will use them, we analyzed the innovative context that SeCo enables with a "User Centered design" approach.

The User Centered design process involves the end-user in an active and collaborative way. Whenever possible it recommends [3] to put demonstrators of the product/service under design in the hands of the user, and to proceed conducting

S. Ceri and M. Brambilla (Eds.): Search Computing II, LNCS 6585, pp. 239–255, 2011.

interviews, and observing the user in the natural habitat (at work, at home, etc.). The goal is to extract the essence of what people do (or need to do), what they think (attitudes) and what they feel (emotional aspects).

The major challenge we faced in applying a User Centered design approach to SeCo was that, at the time this research was conducted, SeCo applications were proof-of-concept and could not demonstrate the full potential of the SeCo technology. We could not design experiments to observe new actions, new habits, and new praxis in the resolution of problems enabled by SeCo. However, we were able to evoke problems that SeCo technology can solve and analyze how people try to solve them without it. As a result, we mapped the user experiences and identified unexpressed motivations and behavior insights. This can be considered a starting level for SeCo innovation from a user point of view, and a potential entry point for SeCo applications to penetrate into the market.

The rest of the chapter is organized as follows. Section 2 describes the research process we followed from a methodological point of view. Section 3, 4, and 5 report the results obtained by performing the intermediate activities of our research process. Section 6 is fully dedicated to the illustration of the central findings of our investigation, i.e., the map of the end-user experience. Finally, in Section 7 the potential implications of our findings for designing future SeCo applications are discussed.

2 Our Research Process

When embracing a user centered design research for SeCo enabled applications, we refined and contextualized one of the possible methodological processes typical of this discipline to SeCo. We constructed a 5 steps process (Fig. 1.) that starts from the selection of a context of application, gathers data from the direct observation of the end-user experience and leads to the identification of open gaps and design opportunities useful for future SeCo applications.

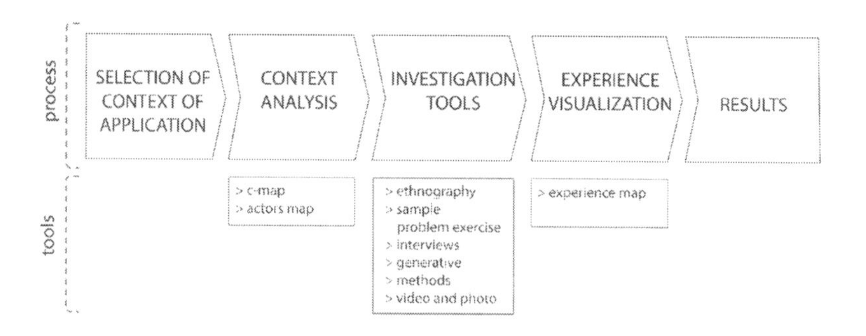

Fig. 1. The research process we followed

The main aim of the first step of our process was **identifying specific contexts of application** suitable to involve the end-user and, at the same time, characterized by SeCo complex search. As anticipated in Section 1, a user centered analysis requires an active role of the users and a close observation of them in their daily life: the selection of a specific context of application allows to specify a sample group of users and avoids the gathering of generalized data which are not useful in the design phase.

Once we defined a specific context of application, the second step was to **map an overview of the context** we wanted to observe. We conducted this step on two parallel levels. On the one hand, we investigated the demand of information expressed by the end-user (*"what are the users looking for?"*). On the other hand, we focused on the actors operating on the system as sources of information (*"which are the answers of the system?"*), on their roles and relations.

For capturing the information on the first level, we used a Conceptual Map [4], a graphical tool for organizing and representing knowledge. A Conceptual Map includes concepts, usually expressed as keywords enclosed in circles and relationships between them represented by a line linking two concepts. We used them to categorize and build a graph that represents all the topics searched by the user and their characteristics.

We studied the system of stakeholders with their mutual relations; we used an Actors Map [5] to represent the results. An Actor Map provides a systemic view of the service and its context. The map is built through the observation of the service from a specific point of view that becomes the center of the whole representation. In this research, we selected the point of view of the users of a SeCo application and the map visualizes all the stakeholders starting from their relations with them. The focus is on roles, grouping and relations. The grouping aspect of the technique is used to organize the actors by their function.

The third step focuses on the selection of the investigation tools. As anticipated in Section 1, a User Centered approach influences the identification of investigation tools; it recommends focusing on those that allow for both a closer observation and a collaborative involvement of a user. The diagram (Fig. 2.) lays out **different kinds of research tools.** The horizontal scale characterizes design opportunities and user needs, from explicit (left) to latent (right). The vertical scale indicates the difference in techniques from macro (top) to micro (bottom). Since our goal is the design of a SeCo application that has not even been thought yet, we could not place the application in the hands of the user. We had to use methods that enable to discover latent needs and desires that will help SeCo team to define potential opportunities. As we detail in Section 5, we adopted: video ethnography techniques on the macro scale, where stop frame video is set up to watch a space or task to reveal patterns of use; then, on the micro scale, we opted for observational techniques, and went to wherever the design context exist to see what people really do, as opposed to what they say they do [2].

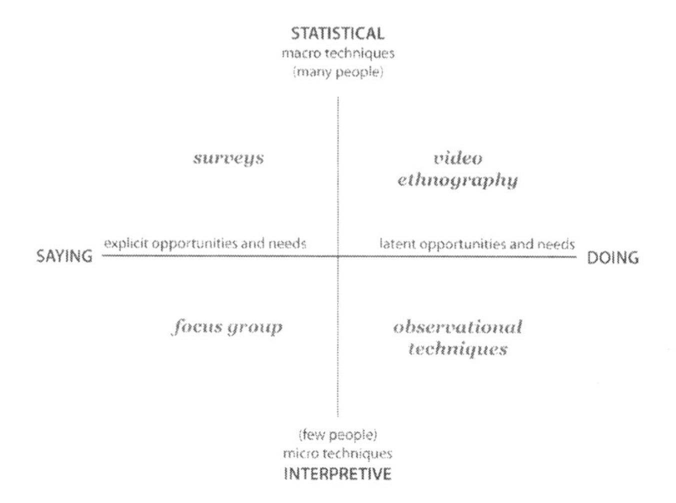

Fig. 2. The space of all possible research tools

In the last step we ordered the quantitative and qualitative information gathered in the previous step in a way that allows it to be shared with those that will take the future design decision for SeCo applications.

This step's crucial point concerns the interpretation of the gathered information, connecting the dots of experience, finding implicit or innovative connections between the different steps of the process using a tool able to **visualize the experience** with a holistic approach, and maintaining at the same time the user's point of view. We chose to use an "Experience Map" [5], a graphical tool for capturing the service journey of a user. An Experience Map shows the users' perspective from the beginning, middle and end as they engage a service to achieve their goals. It shows the range of tangible and quantitative interactions as well as the intangible ones, and qualitative motivations, frustrations and meanings.

The creation of an experience map led us to observe the search process from the user's point of view. The value of the experience map is in the connection that the researcher can do between resources, feelings and motivations, actions and tools, emphasizing points of pleasure and points of pains/barriers that a user may encounter during the process. Finally, this led us to the **definition of open gaps** in the users' experience, that can be assumed **as design opportunities** and that could be exploited in the future phases of SeCo application design.

3 Selection of the Contexts of Application

As far as this research is concerned, we selected two different contexts of application:

- Conference Travel: facing the organization of a business journey in combination with a weekend of tourism and vacation.

- Real Estate: the search of an house that responds both to the end-user desires of comfort and location/position and to his/her financial incomes.

Both of them involve the typical actions enabled by SeCo technology such as the comparison and integration of multiple results (e.g., combine travel with accommodations and local transportation in the Conference Travel context, or houses with city districts and information about local shops for the Real Estate one), the expansion of the search, and the manipulation of the results. But, at the same time, they present differences in their inner characteristics, for example:

- Time - the search of a house could last for months or years and probably it will be a unique experience if compared to the organization of a conference travel;
- Effort - the effort requested to the user is totally different: booking a hotel or a flight can not be compared to buying a house, in terms of importance, personal significance, and money;
- Resources - internet is usually the primary resource used to organize a travel, while in the real estate context professional advice and guidance have still a central role.

The selection of two or more contexts of application highlights how different the motivations, triggers, and expectations are and, therefore, how different the experience lived by a user is, too. Due to the limited space available to illustrate this work, in this chapter we present only the findings of our research applied to the Conference travel context.

4 Context Analysis

The context analysis was conducted in two steps. Initially, the observation focused on the definition of the services offered to the users in order to construct their personal solution. For this purpose we used a Conceptual Map. Then, the focus moved to the owners of the information, to how this information is organized and presented to the users, and to the relations between the stakeholders. An Actors Map was used to capture the results of this second part of the analysis, and, in particular, to visualize the links between the stakeholders and the additional services that they present in order to complete the offer.

4.1 Conceptual Map

In regard to the Conceptual Map for the **Conference Travel** context (Fig. 3.), we started by listing/enumerating what the user could search while organizing a journey (keywords enclosed in circles). For example a user could look for hotels and places to stay, flights or trains to arrive to the destination, restaurants or pub where to have a good lunch and so on.

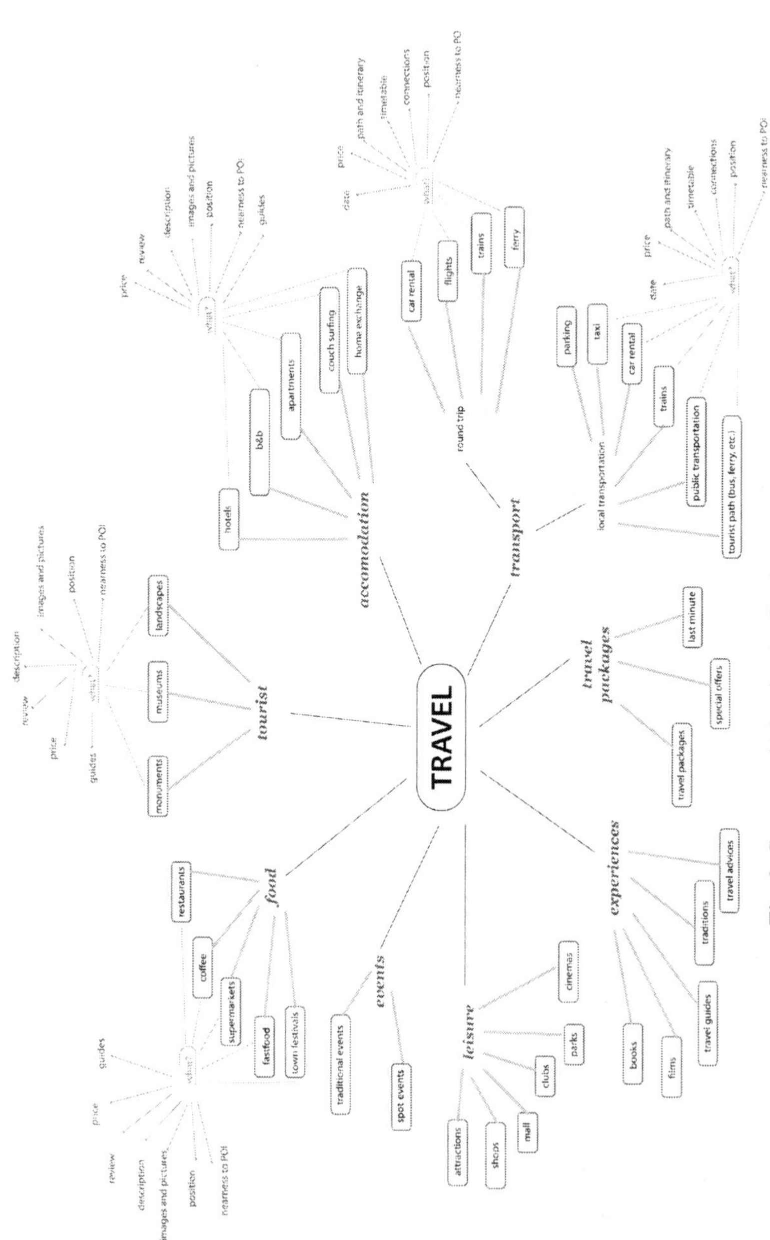

Fig. 3. Conceptual Map for the Conference Travel context

When the list was completed, we proceeded in two directions. On the one hand, we looked back in order to group this concepts in categories expressed reflecting users' needs. We obtained the bold faceted nodes directly connected to the center of the map; e.g., transport for moving around, food for eating, and leisure for having fun or knowing people. On the other hand, we went on focusing on the qualities of searched information that could be interesting for the end-user. We added the small star shaped graphs centered on a node named "what?" in the outer part of the map; e.g., when looking for a hotel users would like to have detailed information about the price (numbers), location (maps), pictures of the rooms (images), and clients reviews (rating); finally, when looking for information about transports, the end-user is once again interested in prices and locations, but the presentation of any pictures is not necessary.

4.2 Actors Map

The main stakeholders of the Conference Travel context were grouped and visualized in an Actors Map (Fig. 4.).

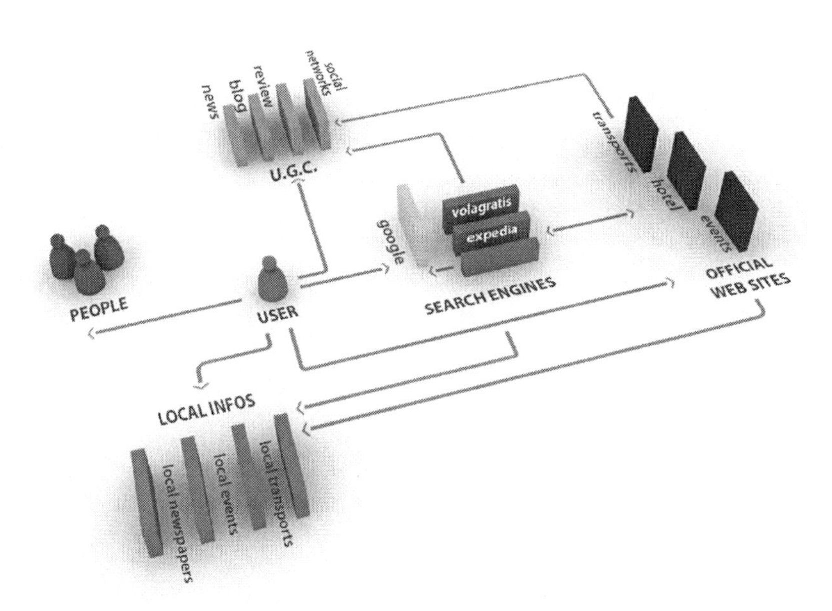

Fig. 4. Actors Map for the Conference Travel context

The content of the Actors Map can be described as follows:

- User – it is the organizer of the travel. His/her specific point of view becomes the center of the whole representation.
- Search engines - they can be divided into:
  - Generic, e.g., Google.
  - Domain-specific, e.g., Expedia, Volagratis, and Edreams.

- Official web sites - owners of specific databases and information (e.g., tourism agencies, and chambers of commerce) or service providers (rental services, hotels, and airlines).
- Local info - owners, producers or providers of local knowledge. This category groups, for example, newspapers and magazines, event organizers, transportation companies, shops, associations and other actors that work on the local environment.
- User Generated Content (UGC) - various kinds of media contents, publicly available, that are produced by end-users. This category can include users' reviews and travel experiences (like Tripadvisor), photo sharing sites (like Flickr), social networks (like Facebook or Twitter), travel blogs, and wikis.
- People - friends, colleagues or persons that can be sources of trusted information, because of their recent (travel) experiences or their deep knowledge about the topic.

The overview on the end-user's demand of information and on the stakeholders system allowed us to select and detail some areas of interest rather than others and to refer the answers of the end-user to the complete scenario of the context of application.

5 Definition of the Investigation Tools

As already said, the selection of investigation tools focused on methodologies that could answer to the needs of both a closer and direct observation of an end-user and of an analysis of SeCo complex search scenario, in order to analyze and discover latent opportunities and needs (Fig. 2.).

In regard to the capability of the investigation tools to analyze a complex search scenario, we created a **sample problem exercise** that could replicate the situations that SeCo will generate in the future. The sample problem exercise can be defined as a description of a situation that the users know since they lived it in the past and that they probably will experience again in the future. It is written in a narrative form to convey the users the right scenario environment. It is not a list of tasks to be done, but a story in which "real users" search and combine data to build a solution. Participants are asked to read it, to remember similar past situations and to show how they work in the same context. That exercise is usually followed by a feedback session, in which the researcher and the user evaluate the experience.

The aim of this tool is trying to prevent those generalizations usually produced by classic interviews or by reports of personal experiences, through focusing on a real situation conducted by a user in his/her own environment.

In the Conference Travel context (see Fig. 5.) we formulated the sample problem exercise starting from the selected areas of interest of the Conceptual Map (i.e., transportation, accommodation, event, food&leisure) and keeping in mind the peculiarities of a complex search process enabled by SeCo technology.

In the next upcoming weeks a financial conference from the EIASM[1] calendar will take place abroad, and you would like to participate. You decide to search the background of the invited speakers and their publications related to your work. Since the days when the conference will take place are next to a week-end, you will have some free days to visit the city. Many cultural events or sport matches will take place during that period and you could attend them. Due to a business commitment, you have limited flexibility in the choice of the day and the time of the departure; you are not worried about the price since your university will give you a reimbursement. The hotel where you will have to stay during the days of the conference must be selected among the partnered ones. On the contrary, for any extra day that you will spend there as a tourist, you can choose another accommodation that meets your preferences and means.

Fig. 5. Sample problem exercise for the Conference Travel context

We assumed an ethnographic approach in order to get closer to the user's experience and collect data from a firsthand source, and tried to step into users' shoes as to start gathering and define the requirements. Ethnography is based on learning about a context and people living in it, their system of values and beliefs, and the way they make sense of their experiences. We selected different kinds of tools that could operate on various levels of investigation. The aim was to build an empathic connection to the end-users, by engaging them through dialogue, discussions, games and informal conversations, in order to include also the emotional level into the analysis. As to support the sample problem exercise we used structured interviews (both guided and informal), photos, video recordings, computer screen recordings and others generative methods that bet on the engagement of the users. We asked them to play with cards, make sketches, and create a process using some tools specifically designed for the research project. Through these empathic involvements of the user, these methodologies try to explicit the motivations and the drivers that are then translated by the end-user into actions and behaviors.

6 Visualization of the User Experience

The data extracted from the interviews, the sample problem exercise, the videos, and the generative methods allowed us to build an Experience Map (Fig. 6). The "Experience Map", which we tailored to SeCo, is composed by five parts: activities, actions, data flow, sources of information, and mood path.

By **activities** we mean tasks that the majority of the users perform when trying to find a solution to the complex search problem we proposed to them in the context of the Conference Travel example problem.

[1] EIASM: The European Institute for Advanced Studies in Management (EIASM) is the international network for management research and teaching that includes more than 40,000 management scientists from all over the world. http://www.eiasm.org

The process that led us to identify the list presented in the first column of Fig. 6 is the following one:

1. We started from the four activities that described a user experience in general:
   - attraction, i.e., the identification of needs and desires,
   - preparation, i.e., the selection of tools and resources,
   - interaction, i.e., the use of the system, and
   - memory, i.e., the evaluation phase.
2. We tailored it to complex search processes by expanding:
   - preparation into the selection of the various resources needed to manually accomplish one of the complex search processes that SeCo technology can automate; and
   - interaction into several tasks typical of a complex search process such as search query formulation, results' comparison and choice, transaction, and use.
3. Finally we contextualized them in the Conference Travel obtaining the list of activities that constitute the leftmost column of the map.

Analyzing these activities carried out by the user, we can observe a recurring set of actions that a user does while trying to organize his/her journey. This sequential search is usually composed by an initial set of constraints, the selection of useful sources of information (e.g., search of vertical search engine, official web site, friends or colleagues, etc), the comparison of partial results and a definition of possible solutions. This process of search is repeated for all the interest areas (transports, accommodation, events, food and leisure). The partial results defined for each area are then compared and aggregated by hand, producing a huge amount of information that recirculate in the system.

In order to highlight these cycles, near and in relation to the list of activities, we draw a line that captures the **flow of data** in a complex search process conducted by the user. It shows and emphasizes the cycles and the recurring redefinition of the search query that often brings the user back to an initial phase.

The third column shows the recurring **actions** that the user does in relation to specific steps of the search process such as using a tool to take notes or to save the results, calling friends or colleagues or wasting time looking for the right source of information. They are *of central importance* and become meaningful actions in relation to positive (satisfaction) or negative (frustration) moments that the user experienced.

The last column presents the **source of information** employed to satisfy the users' demand of information. This column highlights:

- the typology of resource which was used to get the information (e.g., search engines, friends, colleagues, professionals, newspapers, etc,);
- the ratio of employment of every resource;
- the use of more than one resource to get the same information.

The main sources of information used in the Conference Travel context were generic and vertical search engines, official Web sites, local Web sites, friends and colleagues, Web communities, Web 2.0 review services, and people living in the city of destination.

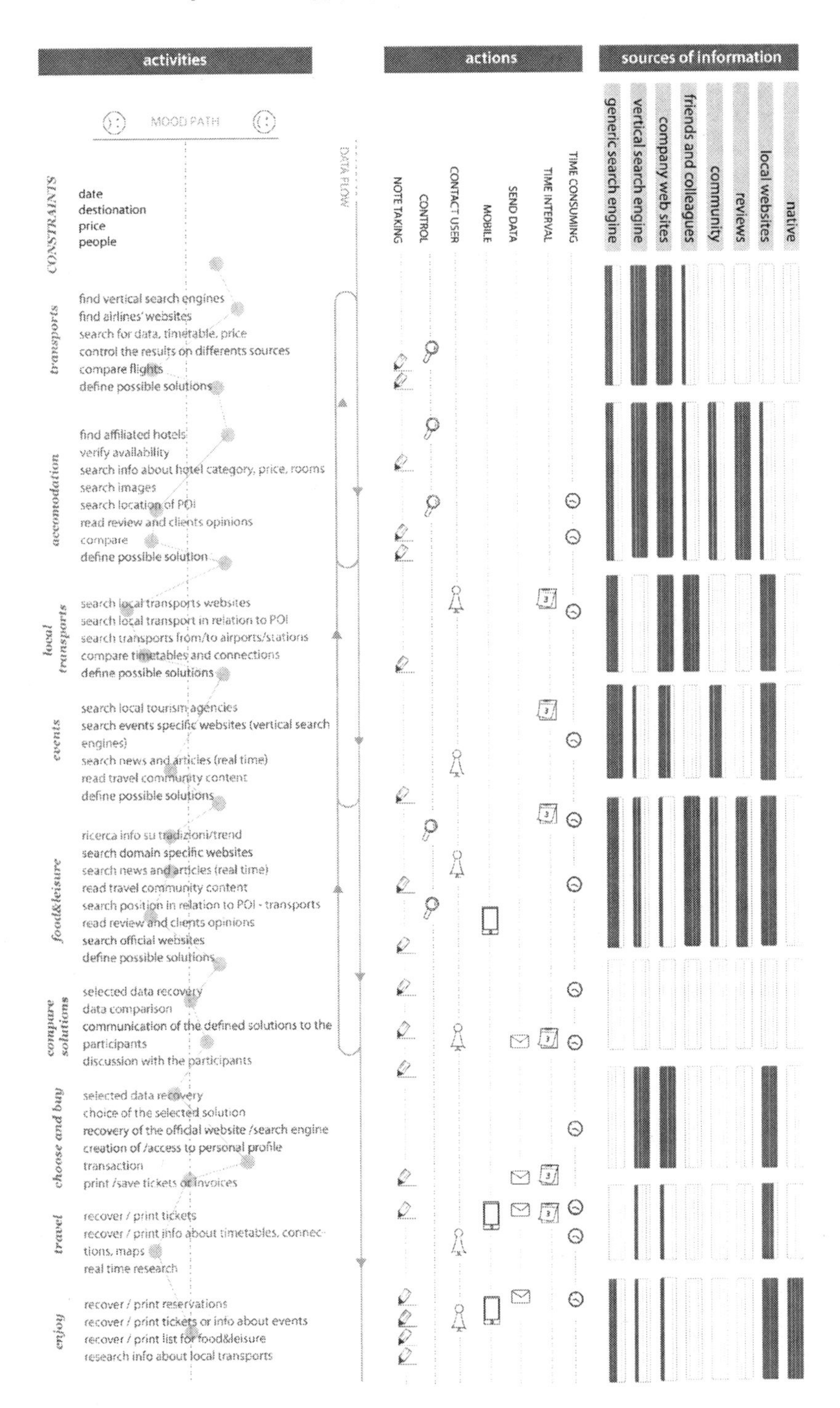

Fig. 6. Conference Travel Experience Map

The analysis on the user experience could highlight how the user makes use of different typologies of resources to get the same information. For example: even if the solution is provided by a vertical search engine like Expedia that combines, for example, flights and accommodation, the user often accesses the official Web site of the airline company or of the hotel in order to check the validity of the information.

Moreover, the internet search, which is mainly based on generic and vertical search engines, is employed to look for "technical" information like prices, timetables, and maps; on the contrary, the search of those information which are related to subjective and sentimental fields and to one's very personal experiences (preferences and attitudes) is carried on involving other users, like friends or colleagues. This investigation focuses on the Web 2.0 platforms and considers different layers of user generated content, i.e., from an impersonal system of reviews to a community.

Finally, behind the activities description, we present a **mood path** that describes the feeling perceived by the user; analyzing the information from the videos, the sample problem exercise and the generative methods, we were able to visualize the moments of satisfaction and dissatisfaction, surprise or frustration, confidence and boredom.

Analyzing the Conference Travel mood path, our attention focused more on the pains and barriers that the user found during the experience. Most of them referred to the time consumption. Therefore the user feels frustrated and confused when he/she does not know where to find the information he/she needs (e.g., looking for local news on strikes), or becomes upset and bored when spending a lot of time reformulating and refining the search (e.g., controlling the solutions, finding new offers, or combining the results in different ways).

7 Discussion of the Results

The Experience Map enables the visualisation of connections among search steps, tools/services used, and feelings/moods. It highlights both the satisfaction points and frustration moments perceived by the user. The satisfaction points can be considered strength points of the current service system and, therefore, a SeCo application in the same domain should maintain them. On the contrary, the frustration moments are the weaknesses of the analysed experience. Those weaknesses, if referred to SeCo, can be assumed as **open gaps** to be addressed in the upcoming SeCo application prototyping phase.

We grouped the opportunities emerging from the open gaps into categories that define **design questions**. We formulated these opportunities as questions, because this form helps to focus the attention on the solutions that can better answer the questions.

The **answers** to each of these design questions can be multiple, especially if, as in the case of the analysis we conducted, a unique search experience cannot be observed. The end-user goals, the effort and the time dedicated to the search, contribute and influence the perception of the experience value. They identify different **user profiles**. Consider, for example, the activities of data comparison and integration; a solution dedicated to an end-user that attributes value to the exploratory search as a learning process will be quite different from a solution designed to answer an end-user who is focused on the achievement of best combination of data in the shortest time.

The former would prefer an application that helps in managing complexity, whereas the latter would prefer an application that reduces perceived complexity offering straight away the best combination.

Let us, now, report step by step our findings for the **Conference Travel** scenario. The analysis of the experience map presented in the previous section allowed us to synthesize six open gaps that lead to the formulation of as many design questions (Table 1.).

Table 1. Open gaps in Conference Travel analysis

Open Gaps	*Design Questions*
The users are skeptic since they do not completely trust the proposed solution.	How to gain more users' confidence?
The users build a comparison for every combination of results and they often do this "by hand"	How to make the comparison easier?
The users discuss the possible solutions with other people, saving the results, exchanging info, asking for advices.	How to facilitate exchange and exportation of data and experiences?
The search is not carried on in a unique session, but is divided in different time intervals. Some new info can also modify former results, leading to a repetition of the process.	How to deal with time intervals and search query redefinition?
The users search local and global information, often with the need of their integration.	How to improve integration between local and global search engines?
It is difficult for the users to manage and integrate multiple transactions on different Web sites.	How to deal with the transaction phase?

As we explained in the beginning of this section, no generic answers to the design questions can be provided, but only answers for specific user profiles are possible. The analysis we conducted allowed us to identify two main use profiles: Fast&Furious and Discovery&Learn.

These two profiles can be considered as opposite ways of the user to construct a personal travel solution, in terms of effort, quantity and quality of the resources, time dedicated to the search, goals of the user. Table 2 shows in the first row the main characteristics of both approaches through a storyboard. The features of each kind of complex search process and the design goals that should be taken into consideration by SeCo are listed in the next two rows.

Table 2. End-users' approaches to search experience

	Fast & Furious	***Discover & Learn***
Storyboard	"Due to her work, Giorgia travels a lot during the year. In this occasion she already knows the destination and she does not have time to explore and find detailed info. She wants the faster solution that fits her needs. She is interested to the information, like round trip tickets, accommodation, position of some P.O.I. related to her work. She postpones the choices on local transports, food&leisure and events to the arrival to the destination."	"Giulia sees the search as a discovering process that can enrich herself. The value is not only in the solution that better fits her needs, but also in the construction of this process and in the selection of the sources and the results. She is interested in the culture and in the knowledge expressed by the local inhabitants, she is curious about peoples' experiences and opinions. She expands the search to topics not totally necessaries to the satisfaction of the initial needs. She prefers to plan the whole trip, looking for detailed info about traditional events, typical restaurants and leisure."
Search features	> fast search > preset best combinations	> exploratory search > learning process
Design goals	> remove the efforts > gain confidence > go straight to the solution	> decrease the efforts > enable a personalized search process > share experiences

Fast&Furious describes a user whose goal is mainly to get straight to the solution that suites her needs at best. Since she has not time or she is not interested in searching detailed information, she prefers to find a good and trustable combination of data. This kind of search process, in opposition to Discover&Learn, does not require much time and bets on presenting the best combinations of travel solutions. If we imagine, for example, to design a SeCo application for this kind of user, we should consider that reduction of effort perceived by the user and the acquisition of user's trust are the keys for a successful user experience.

Discover&Learn describes a user that perceives search as a learning process trough that she can enrich herself. Her goal is not only to find the solution to her initial needs, but to put efforts and time in constructing a search process that can lead her to the discovery of resources, local knowledge and culture and to share experiences and opinions. Differently from Fast&Furious, this approach is based on an exploratory search. If we image, for example, to design a SeCo application for this second type of user, we should decrease the effort of the user without depriving her of the control on the search process. In other words, we should enable her search process.

8 Conclusions

The experience we reported in this chapter allows us to drive some conclusions and to indicate some of the features of a SeCo application in the Conference Travel context.

We believe that, for every context of application of SeCo, an user-centered analysis should be performed since the open gaps and the design questions that can emerge from it should be a starting point for designing a SeCo application in terms of features and services that compose different end-users experiences.

As already said, in this analysis we take into consideration the whole experience of the user: from the perception of needs and desires to the process evaluation. Therefore, the open gaps, and consequently the design questions, embrace both the inner characteristics enabled by SeCo technology and those not immediately related to it. The former are the core competences and services enabled by SeCo technology (i.e., combination, integration, comparison, expansion and manipulation of data) and the design phase should concentrate on them in terms of developing the main application features. The latter are not peculiar only of SeCo technology, but they must be taken into great consideration since they are expression of end-user's needs or desires and will influence (positively or negatively) the perception of the experience (i.e., user-generated content, real time search, and transaction).

Considering the open gaps highlighted during the analysis of the experience map in the Conference Travel context, the answer to the presented questions will be quite different for Fast&Furious and Discover&Learn, because the users' goals and the values, which they assign to the experience, are different. In Table 3. we present the answers to the design questions that we recommend to integrate as features of a SeCo application in the Conference Travel context. These features can potentially build a user experience that could answer to the open gaps.

Table 3. Answer to the design questions related to Fast&Furious and Discover&Learn

Design Questions	*Fast & Furious*	*Discover & Learn*
How to gain more users confidence?	Give guarantee to users	Give control to users
How to make the comparison easier?	Reduce complexity	Manage complexity
How to facilitate exchange and exportation of data and experiences?	Save and export	Sharing platform
How to deal with time intervals and query redefinition?	Create user profiles	Organize your search experience
How to improve integration between local and global info?	Real time search	UGC and local info providers
How to deal with the transaction?	Transaction included	Included or not included

A SeCo application designed to respond to a Fast&Furious approach will be winning if it reduces the complexity perceived by the user, offering guarantees for the results. For instance, the application could manage a user profiles that remember the search preferences and the behavioral patterns of an end-user, exploiting them during future interactions of the user with the system. This design solution bets on the learning experience that a SeCo application can gather: the user gives confidence to a system that will automatically present him/her a search process that he/she recognizes as a winning one.

On the contrary, a successful SeCo application focused on the Discover&Learn approach should be able to manage the complexity of a search process in the Conference Travel context, and, at the same time, to include those features and tools that could facilitate the end-user in constructing a personal search process. This approach focuses on an exploratory search that gains value, for example, in the multiple suggestions that the system could offer as alternatives of search. The automatic generation of new ways and directions of search can help the end-user to discover unknown solutions and travel opportunities, contributing to increase the value attributed by the user to the learning process of a search. This can be translated into design features that, for example, enable an expansion of the result through the presentation of similar keywords or through the inclusion of search categories (i.e., films and recommended readings, local news, etc.) usually not presented in a traditional travel Web site.

Moreover, this futuristic SeCo application could be designed to enable different users with similar needs to exchange and share their results, widening the comparison to different combination of travel solutions to those selected by various end-users.

As far as those design questions that are not related to the core characteristics of SeCo technology are concerned, there could be two main directions of development. On the one hand, the designers of future SeCo application can focus on the market and, in particular, on those potential competitors in the travel context that are providing answers to the opens gaps trough their offers and services. For instance, the design question related to the integration of local knowledge and information could be solved as done by the potential competitor Easyviaggio.com [6]. This website attributes value to the quality and authoritativeness of the information it delivers, operating through the creation of a permanent editorial office in partnership with the AFP (Agence France Presse). This office is in charge of keeping the traveler up to date on the latest news.

On the other hand, the designers can explore solutions experimented and developed in different contexts. For example, in responding to "the need of saving and sharing the results and the chronology of search", the designer could decide to propose a solution similar to Yahoo Search Pad [7]. This service allows to take notes while searching; it assists the user in collecting, editing, organizing, saving, printing, and emailing one's own notes for immediate or future use.

To conclude, we trust that the research process, presented in this chapter, illustrates a successful experience in applying a User-Centered analysis to the potential application that can be developed using SeCo technology. We were able to map the user experience identifying unexpressed motivations and behaviour insights. The selection of a specific context of application together with the direct observation of the user experience led us to identify several design opportunities. The examples and recommendations presented in this last section are an initial set of the ideas that can drive a multidisciplinary team in designing an innovative application in the Conference Travel context using SeCo technology. The next step of this line of research should focus on prototyping the Conference Travel application and place it in the hands of the user in order to set a second and closer user-centered observation. From the results of the analysis of such observation we will be able to get more specific design requirements that can lead to a commercial product.

References

1. Design Council, UK, http://www.designcouncil.org.uk
2. Moggridge, B.: Designing interactions. MIT Press, Cambridge (2007)
3. Frog Design, http://www.frogdesign.com
4. Novak, J.D., Cañas, A.J.: The Theory Underlying Concept Maps and How to Construct and Use Them. Technical report, Florida Institute for Human and Machine Cognition, Pensacola Fl (2008)
5. Morelli, N.: New representation techniques for designing in a systemic perspective. In: Design Inquires, Stockholm (2007)
6. Easyviaggio, http://www.easyviaggio.com
7. Yahoo Search Pad, http://help.yahoo.com/l/us/yahoo/search/searchpad

Analysis of Business Models for Search Computing

Tommaso Buganza[1], Marta Corubolo[2],
Emanuele Della Valle[3], and Elena Pellizzoni[1]

[1] Dipartimento di Ingegneria Gestionale, Politecnico, 20133 Milano, Italy
[2] Dip. di Ind. Design, Art, Communication and Fashion, Politecnico, 20133 Milano, Italy
[3] Dipartimento di Elettronica e Informazione, Politecnico, 20133 Milano, Italy
{name.surname}@polimi.it

Abstract. The SeCo project is dedicated to the support of investigation-driven frontier research on search computing and its prime objective is not to develop a commercial venture. Still, not considering the possible market exploitation of such a project could be misleading also in terms of technology investigation. At the half of the project, when many technology and research decisions still must be taken, it is important to consider also the possible implications of such decisions on possible future business models. The present chapter is aimed at understanding the possible future trajectories for the exploitation options of the SeCo project. The chapter starts identifying roles and capabilities of a potential SeCo value chain. The second step is defining some archetypal Business Models investigating the option of having some of the previous roles not managed internally but delegated to the external market. Finally the business models are compared and recommendations for the following of the project are drawn.

Keywords: Search Computing, software engineering, development process, advertising models, cloud computing, software architectures.

1 Competences and Roles of the SeCo Value Chain

Each company that sells a product or a service on a market must take into account the environment in which it is operating. In particular, both the input markets and the output ones should be taken into consideration. This means focusing on two dynamics: the evolution of market needs (output) and the evolution of technologies (input). Every firm must always keep in mind that needs and technologies change in the course of time. This is especially true in high tech environments where firms are focused not only on delivering good services but also on continuously redesigning them in order to deal with frequent changes and opportunities. Successful firms tend to use iterative processes, which emphasize learning and adaptation rather than planning and execution [1-4,10]. Researchers refer to these abilities to adapt the firms' products and processes as Dynamic Capabilities: ability to adapt, integrate, and reconfigure internal and external organizational skills resources, and functional competences to match the requirements of a changing environment. Only through these capabilities, firms will be able to maintain a competitive advantage [8,9].

S. Ceri and M. Brambilla (Eds.): Search Computing II, LNCS 6585, pp. 256–271, 2011.

In order to respond to shifts on both market needs (output) and technologies (input), it is necessary to develop two kinds of dynamic capabilities: firstly, identifying reading and interpreting the evolution of the market needs (*Market Capabilities*); secondly, the firm has to be able to build specific competences as to manage and develop the technologies market (*Technological Capabilities*).

The above considerations are valid for each firm operating in high tech industries but they could not be enough for leveraging the technology developed within SeCo. Indeed, this technology combines services offered by different stand-alone publishers. These services are not just put side by side but they are combined to make complete and customized solution. This architecture therefore introduces a new capability to be developed and taken into account. To manage a SeCo based service it is necessary to create a *mart* that is a container of already existing services adapted and preworked in order to be mixable by the SeCo engine. Thus, managing the relationships with the service owners, attract them in the mart, manage the property right and legal issues and create a partnership agreement become a new and distinctive capability to be developed in order to bring the SeCo technology on the market.

Moreover this capability must be dynamic too: Web-based services owners change both the data that they publish and the interfaces they use to display them to meet changing market needs and technologies. If the new approach is to integrate these different services, it is also necessary to monitor their evolution. Thus, in order to bring the SeCo technology to the market, the dynamic capabilities needed to manage market and technology must be accompanied by "marting" capabilities to handle the relationships with services owners. (Fig. 1).

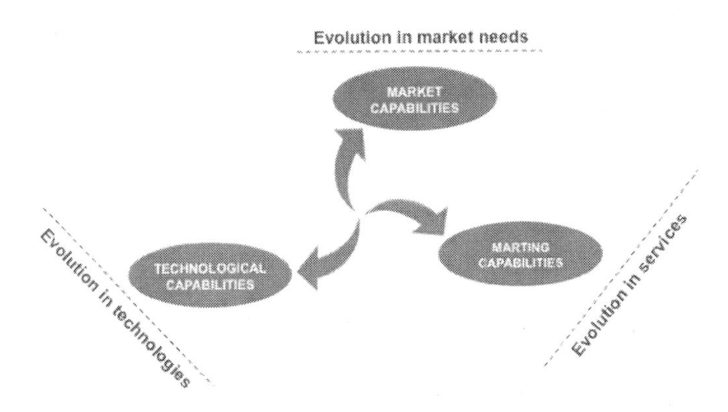

Fig. 1. Capabilities to be developed in integrated web-based services

After having considered the capabilities needed to bring the SeCo Technology to the market, we analyzed the value chain (in terms of processes and actors) needed to sell a web-based service, and according to the logical steps we followed before, we tried to understand if these ones must be different in the SeCo case.

We can summarize the value chain, from the development of the tool to the publication of the application by identifying four key roles (Fig. 2).

First of all, the *Tool Developer* who develops the tool needed to create the application on the web. Secondly, the *Mart Manager* who deals with mart data entry and manages relationships with different service providers (legal issues, contracts, privacy, etc). The Mart Manager must also monitor the continuing evolution of the services market in order to keep the mart updated and running.

Another key role is the *Application Developer*, who is responsible for taking and mixing the data in the mart to create the final application.

Finally, the *Application Publisher* who is the actual deliverer of the application on the web. This role handles the relationship with end users through a web site that can offer different services.

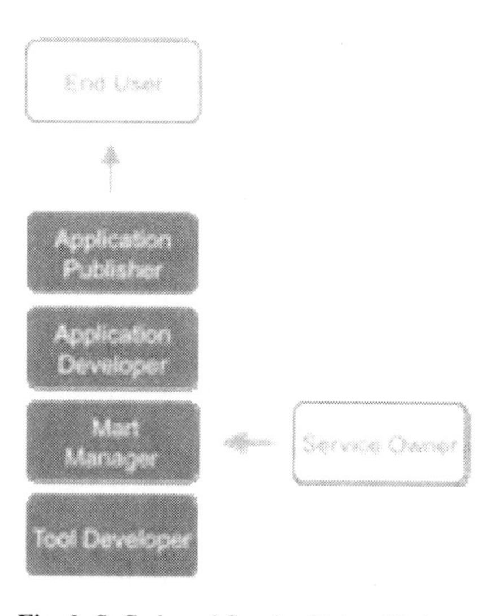

Fig. 2. SeCo based Service Value Chain. . .

Analyzing this simplified value chain, we can find two key processes: the *Run Time* and the *Developing Stage*. The two processes are different in typology and repetitiveness of the activities to be carried out. The Developing Stage is the process that goes from the creation of the developing tools to the design of the final application. This chain of activities is executed just once every application (even though upgrading and maintenance may follow). The Run Time process is completely different. It happens when the application published on a website is used by the end user. For each application there are (hopefully) several Run Times. The two processes can be considered logically sequential: the use of the application following its creation.

These two processes also involve different roles. In particularly, the Tool Developer, the Mart Manager and the Application Developer are involved in the Developing Stage, while the Run Time asks to the work of the Publisher and the Mart Manager.

It is easy to connect the different roles in the SeCo value chain with the dynamic capabilities described before. The development roles (Application and Tool Developer) will need to develop Technological Capabilities: it is impossible to create new applications and tools without knowing the evolution of already existing technologies.

In the same way the Application Publisher will need the support of Market Capabilities: it is impossible to sell a service and to concept it without knowing the targeted customer. Finally the Mart Manager, will heavily leverage on Marting capabilities to engage, control and manage the Service Owners.

Until now we referred to roles and competences without specifying whether they are embedded in a single actor or if many different actors must take care of each of them and interact to create the service. Obviously both the options are possible and many others configurations in between are feasible as well. This will lead to define alternative business models that will be discussed and compared in the following paragraphs.

2 How to Compare Alternative Business Models

Before describing and comparing different business models it is necessary to define the dimensions along which they will be compared: value creation, value appropriation, internally managed competences and support organization required.

Value creation

Value is what customers may pay for goods and services they receive. A firm is profitable only if the value it creates is greater than the costs that it involves [4]. There are two possible strategies for creating value: cost leadership and differentiation. Cost leadership is the strategy to produce and sell the same products (or deliver the same services) at a lower price than competitors. On the contrary, differentiation means focusing on providing the highest quality products even at a higher cost [6].

For each configuration that we will analyze, it is necessary to define which value creation strategy it must involve. It is possible to choose a cost leadership strategy and to focus mainly on efficiency by simplifying and making the development and the delivery of complex services cheaper. Otherwise it is possible to prioritize differentiation by creating innovative and various services for the end user.

Value Appropriation

Value creation, however, is not enough to achieve the final success [5;7]. The *value appropriation* dynamic is the second dimension along which we will compare the different configurations. A firm has to develop the ability to restrict competitive forces so as to be able to create profit. Firms that do not have the capabilities to maintain this competitive advantage are unable to appropriate the value they have created. Instead, competitors and customers claim for it.

In the value chain of web services there are two processes as we described before: Development Stage and Run Time. It is possible to make profit in both of them but in different ways. The business model configurations that we will analyze may be more suitable for one or both of them. The developing stage is performed only once for each application, thus the obtainable profit must be completely obtained in a single time shot. Furthermore, the possibility of competitors' imitations of the solution makes the value appropriation very delicate. The Run time process is completely different. In this process, a small price can be asked because the high repeatability of the process maintains the profits high. Here customers have a lower bargaining power than in developing stage. But it is always possible that competitors are able to replicate the run time process and to steal the value. Moreover the value is created during a longer time span increasing the risk of competitors.

Internally managed competences

If the value creation and appropriation strategies deal with the benefit side of each configuration, the decision of what role to play inside (outside) deals with the cost and the time needed to actually develop the configuration. Thus a further dimension we will use to compare different configurations is based on what capabilities each one of them will require to manage (and in many cases to develop) internally. It is possible to highlight three different approaches according to the different capabilities needed (Fig. 3).

In the first approach the Tool Developer would manage all the Dynamic Capabilities needed (Technological, Market and Marter). In this approach there would be the total control of the processes from the development of the tool to the run time of the application including the engagement and management of the Service Owners. In the second approach, the Tool Developer would develop some Dynamic Capabilities (Technological and Marter) leveraging on external partners to have Market Capabilities (access to market and new service concept development). Finally, in the third approach the Tool Developer would focus only on Technology Capabilities, leaving to external partner(s) the development and management of Market and Marter Capabilities.

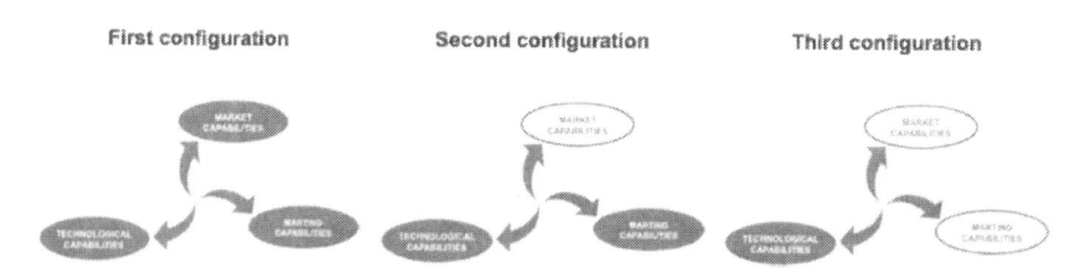

Fig. 3. Capabilities to be developed in the different configurations

Support organization required

Finally the fourth dimension along which we will compare the configurations is the (simplified) organizational chart needed to support them. In particular, we will compare the organizational charts against the case in which competences are managed internally, which would ask for the development of the following functions:

- Tool Development Function: it deals with the development and updating of the tool. It includes expert developers with technological skills.
- Marketing Function: It manages the relationship with end users and it also defines the concept of the service that they want. For these reasons, it includes mainly people with market capabilities.
- Application Development Function: It is the function in which the development of the application takes place. It includes mainly people with technological skills.

– Run Time Function: It manages the run time and the mart. This business unit is the one controlling the relationship with the Service Owners. It is necessary to develop legal and negotiation competences and also technological skills to run it.

3 First Configuration: The Service Deliverer

The first possibility analyzed is to manage inside all the three capabilities. This means that, in this configuration, the Tool Developer would directly deal with final users and Service Owners.

Table 1. The Service Deliverer

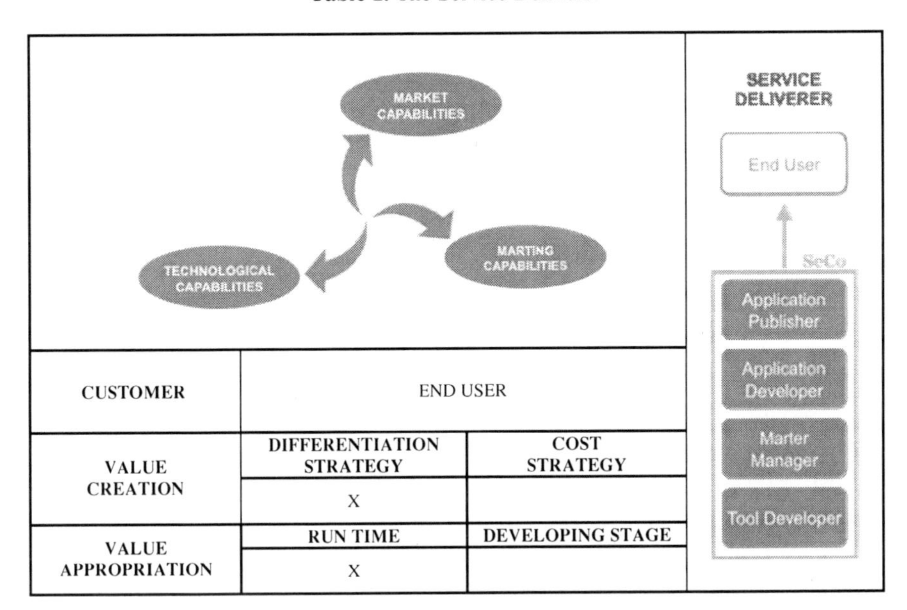

CUSTOMER	END USER	
VALUE CREATION	DIFFERENTIATION STRATEGY	COST STRATEGY
	X	
VALUE APPROPRIATION	RUN TIME	DEVELOPING STAGE
	X	

The target client is the end user. The main goal is to develop a killer application, that means pursuing a Differentiation Strategy. The end user is attracted from new web services mainly if they are better than the ones already existing on the web. The developers are focused on a specific user and design the vertical application depending on his/her needs. In this configuration, the publishing of the application is managed internally (selling and managing the runt time) too.

Managing the entire process is very complex. For this reason, it is likely to do only one application for a very specific target. We cannot assume to create a factory of services for any purpose, as wanting to manage the whole value chain implies to focus the attention on specific services. The value is done in the several cycles of run time, which are completely managed inside. Service Owners could be requested a fee for the provision of their visibility, or end-customers could be requested it too for the service benefit they receive. Besides, the role of publishers enables the management of advertising on the site.

This business model is named: Service Deliverer. The Tool Developer creates the tool and the mart and then designs and develops the vertical application. Finally it also delivers the service on the web site, manages the run time and the continuous improvement of the service and the mart. A single actor plays all the roles:

- Application Publisher,
- Application Developer,
- Mart Manager,
- Tool Developer.

An example of a similar Business Model can be given by online travel agencies like Expedia (www.expedia.com). These sites summarize different kinds of information from different sources to offer a one stop service to the customer. They aggregate information on flights, hotels, tourist attractions, cars and they sell them to the end customer. They develop the service, manage the relationship with the service Owners (e.g. Hotels), manage the run time and sell to the customers.

The characteristics of the Business Model described are summarized in the following table.

Table 2. Analysis of Service Deliver

SERVICE DELIVERER	
NECESSARY CONDICTIONS	Find a killer application for a market Find a market Build competence missing
OPPORTUNITIES	• Some market can be unsaturated (or without a big player) • Discover and exploit Long Tails
THREADS	• Data availability • How can we convince users to trust our application? • Long Tail can be too small (or with small profit margins) • Large specific investments • Market access
Possible ORGANIZATION CHART	SeCo Search Computing TD FNC, MKT FNC, AD FNC, RT FNC
COMPETENCES MISSING	• Learn to design killer applications SeCo-like • Learn to manage Service Owners and legal issues

In this business model the choice of the target market for the killer application is fundamental. Indeed, according to industry dynamics the analysis will completely change. For example, the presence of big players in a market can discourage the publication of new applications. Only through a deep analysis of the competitive environment it will be possible to understand the chance that the application will have to succeed.

Before starting with this configuration, it is therefore essential to find the context and the killer application to focus on. According to the market choice, it is necessary to do a deeper analysis of the threats and the present opportunities. It is also necessary to develop the missing competence.

That said, a potential SeCo spin-off would have a starting point focused on the development of a tool and demonstration application. The capabilities associated with these two activities are mainly technological. Skills for creating and managing a relationship with End Users and Service Owners should be built from scratch in terms of market capabilities (both for service concept and for service trading) and in terms of Marting Capabilities (find and engage Service Owners and manage the relationship: legal issues, Intellectual Properties, contracts...).

4 Second and Third Configurations: The Mart Provider and the Service Developer

The second and third configurations share the same approach: to let external actors manage the Market Capabilities maintaining a strong internal control on the Technological Capabilities and on the Marting ones. This approach may be declined into two different Business Models: Service Developer and Mart Provider. The main activity in both cases is the management of the mart, but in one case the development of applications on the mart is managed internally too, while in the other one it is delegated to external actors.

In these configurations, the main focus is on the creation and management of a rich mart (many different services included and connected) that gives the possibility to create different applications combining different services. The target customer can be the Application Publisher or Developer. The former will ask for a complex search application while the latter will ask for accessing directly to the mart and develop its own applications. Compared to the previous business model, in this case there is no need to focus on a single market: on the contrary, it is advisable to develop a large number of applications. Indeed, the service is sold to the end user by the Publisher and (even if the run time can still be managed internally) a part of the customers' fee will be appropriated by the Publisher. Therefore the value appropriation will be as in the previous configuration (fees by End Users and Service Owners), but also pay-per-use contracts with the customer or fixed fee for the mart utilization. Moreover developing a large number of applications will allow the exploitation of scale economies, needed to justify the investments for establishing and maintaining the mart.

Table 3. The Service Developer and the Marter Provider

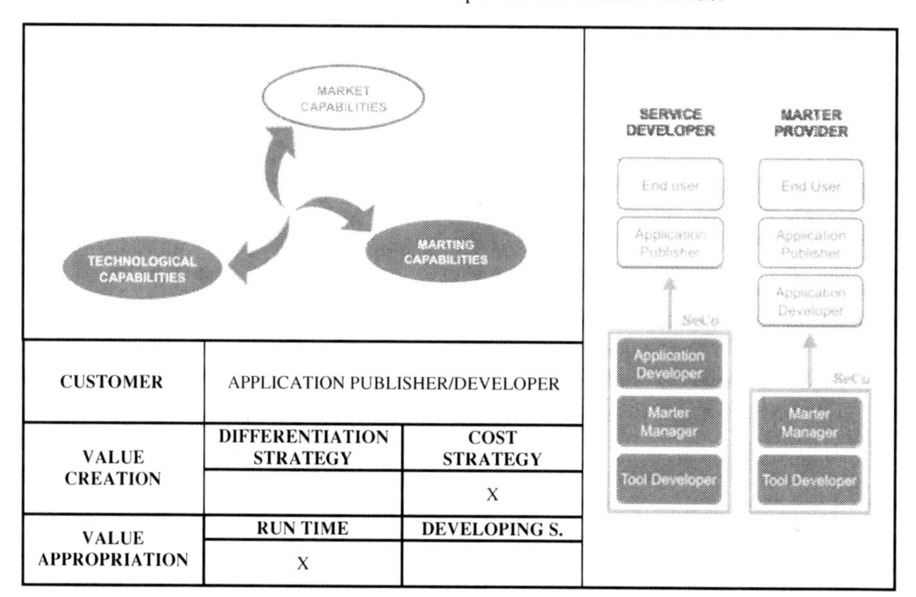

CUSTOMER	APPLICATION PUBLISHER/DEVELOPER	
VALUE CREATION	**DIFFERENTIATION STRATEGY**	**COST STRATEGY**
		X
VALUE APPROPRIATION	**RUN TIME**	**DEVELOPING S.**
	X	

In both Business Models the value creation is based on cost strategy. The Publisher (or the Developer) will be attracted by the possibility of creating a complex solution in a relatively easy way, without having neither to develop the core technologies nor to select and manage the Service Owners. Still the two business Models differ in terms of roles played.

In the first one, the Service Provider, the Application Publisher requires an application that can be done with the SeCo's tool. In this business model, the following roles are played internally:

- Tool Developer,
- Mart Manager,
- Application Developer.

To better understand the Business Model we can consider the case of Endemol (www.endemol.com) even though it is not a web-based service. Endemol is one of the leader firms in the creation of entertainment programs. This company develops standard television formats and sells them to broadcasting companies on a global scale. The broadcast companies manage the publishing of the shows but they buy format programs already developed. In other words Endemol develops standard applications and sells them to different Application Publishers.

The Service Provider Business model still asks for some market capabilities to be developed. If it is true that the Publishing (and therefore the contact with the end users) is managed by an external actor, still degree of market knowledge is needed to concept the new services, if the Application Publisher is not completely able to do it alone (as in the case of broadcast televisions). Nonetheless, focusing on the Mart, it is possible to find a Second Business model that almost does not require any market capability. The Mart Provider is focused on the mart and on Service Owners management. The covered roles are the following:

– Mart Manager,
– Tool developer.

In this Business Model, the role of the Application Developer is not even handled internally. This implies that the external developers must learn how to develop applications on the SeCo platform and mart. In other words this will increase significantly their cognitive barriers. Only when a community of these developers will be established, the Application Publishers will have someone to ask for having the development of complex research based services.

An example of this business model is Metaweb that developed Freebase (www.freebase.com), a service that permits to build a website using entities. Each entity is linked to different words that have the same meaning. In this way the developer of the site, choosing an entity, can link it not only to one source but to different service owners. In order to create a good network (critical mass), they had to create a community of developers committed to build the links between entities and words.

As mentioned above, involving these actors is not an easy task. The effort, both economic and cognitive, that the developer must bear in order to learn how to develop a new language is very high. It is necessary to find some way to reduce their costs and raise the benefits.

The characteristics of these two business models are summarized in the table below.

Table 4. Analysis of Service Developer and Marter Provider

SERVICE DEVELOPER	
NECESSARY CONDITIONS	Owning a large Service Mart together with agreement with Service/Data Owner Find an Application Publisher/ Application Developer
OPPORTUNITIES	• Exploit economy of scale (the business is serving multiple AP in the same domain) • Hard to copy (with the same cost)
THREADS	• Low margin for each application • Low profitability • Difficulty to find an elevate number of APs that repay investments • Large specific investments • Data availability
Possible ORGANIZATION CHARTS	SeCo Search Computing: TD FNC, AD FNC, RT FNC, BUSINESS CONSULT. SeCo Search Computing: TD FNC, RT FNC, BUSINESS CONSULT., DEVELOPER SUPPORT
COMPETENCE MISSING	• Learn to manage Service Owner and legal issues • Support to design killer application SeCo-like • (Support to develop the application)

These two Business Models are focused on managing the Mart, that is the main distinctive element of the SeCo value chain. To develop them, it will be essential to build the ability to manage different Service Owners. Relational, commercial, legal (as to conclude data-used contracts), skills will have to be developed. Moreover it is necessary to find the best way to interface with the data owners as to let them understand the potential value they may have once involved in the mart.

In both the Business Models there are some roles which are in external actors'hands and this leads to the problem of how to convince them to switch to another technology. They must be advised, helped and guided. How can this be done? In the Service Developer Business Model the main stakeholder to be involved would be the Application Publisher. These actors must be educated about what is possible to achieve with the SeCo technology and what kind of needs can be answered with it. In many cases it would be useful to help them in unveiling some needs that they are not able to express. The only way to do so is to build a staff Organizational unit for Business Consultancy. This unit would be focused on finding Application Providers with the "right problem to be solved". The basic idea is not far from the acquisition of Price Waterhouse Coopers operated by IBM. The technological company did it also because it felt that the level of its offering was not fully understood by the average of the potential customers. The consultancy company was needed to help the customers to understand their own needs and the potential benefits IBM could provide them.

The situation is even more difficult in the case of the Mart Provider Business Model. Here the external roles are two: Application Developer and Application Publisher. Unfortunately it is not possible to neglect the latter. Even though we can expect that the Application Developers will try by themselves to find customers with "the right problem to be solved", it would be too much risky to rely completely on them. Thus, to run the Business Model will request also in this case the establishment of a staff Organizational unit for Business Consultancy. Moreover, a second class of external stakeholders should join the cause to make the Business Model run: the Application Developers. These are software designers that will have to overcome their cognitive barriers to learn how to build software on the SeCo mart. This is a well known problem in the technological fields, and its possible solutions are well known as well. Building easy to use environments (e.g. Microsoft .NET) or even support kits (e.g. Apple SDK for iPhone and iPad OS) is a common answer to the problem. Also supporting services, tutorials or even resident engineers can be used. In other words, this Business Model does not require only the creation of a staff Organizational Unit for Business Consultancy but also another one for Developers Support.

5 Fourth Configuration: The Platform Leader

The last possibility analyzed is to leave all but core Technological Capabilities to external actors.

Table 5. The Platform Leader

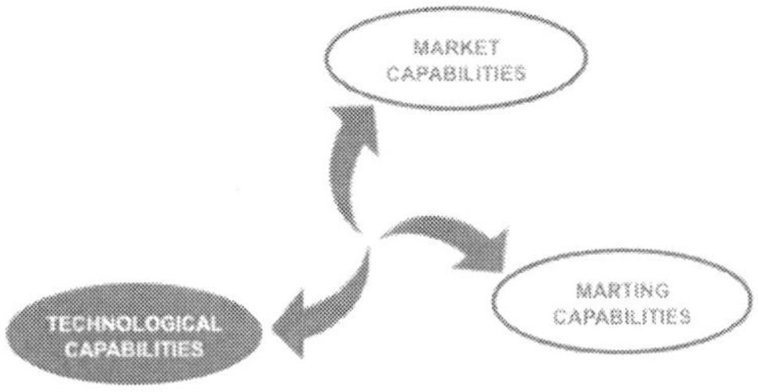

MARKET CAPABILITIES
MARTING CAPABILITIES
TECHNOLOGICAL CAPABILITIES

PLATFORM LEADER
End User
Application Publisher
Application Publisher
Marter Manager
SeCo
Tool Developer

CUSTOMER	APPLICATION DEVELOPER/MART MANAGER AND SERVICE OWNER	
VALUE CREATION	**DIFFERENTIATION STRATEGY**	**COST STRATEGY**
		X
VALUE APPROPRIATION	**RUN TIME**	**DEVELOPING STAGE**
		X

In this configuration, all the external actors converge on the SeCo technology platform to carry out their activities. The Application Publisher will ask for the development of a complex search service based on SeCo technologies. The Application Developer will write applications by picking and mixing the different services already existing in the Mart. In the meantime, the Mart will be designed, maintained and improved by the Mart Manager who will manage directly all the Service Owners. In this case the Tool Developer would be a Platform Leader.

Once again the value creation strategy will be based on cost reduction. The whole system must allow every actor to do its task efficiently. The Business Model will be effective only if a huge number of stakeholders will adopt the technology. Only with a huge number of actors involved, it is possible to create a network in which the network externalities are big enough to provide value to each member. Indeed, it is necessary that the tool should be used repeatedly to generate profits because, without handling the run time process, the Platform Leader can get the value only during the development stage. The developer may pay for the use of the tool but the publishers and the service owners run independent run times.

Given the large number of external actors and the poor control on the value chain, in this case more than in other ones, developing an efficient tool will not be enough. It is crucial to reduce the costs and increase the benefits of all the different stakeholders to move them toward the technological switch to SeCo.

This model is the one used by Yahoo! that developed a Development Kit to create web-based applications. Yahoo! Query Language (developer.yahoo.com/yql) lets the developer query, filter, and join data across Web services. Thanks to this tool the developer can create applications that are faster and with fewer lines of code.

The characteristics of the Platform Leader Business Model are summarized in the following table.

Table 6. Analysis of Platform Leader

PLATFORM LEADER	
NECESSARY CONDICTIONS	Find SeCo-like Application Developer Find Application Publisher
OPPORTUNITIES	• AP and AD want a technology that reduce the developing cost
THREADS	• Selling can be complex without a running prototype (trust) • How can we convince developers to use tools? • Big competitors • Hard to explain SeCo technology
Possible ORGANIZATION CHART	SeCo Search Computing TD FNC BUSINESS CONSULT. DEVELOPER SUPPORT RT CONSULT.
COMPETENCE MISSING	• Support to design killer application SeCo-like • Support to manage Service Owner • Support to developing the application

The Platform Leader Business Model is the hardest one to be implemented. It strongly depends on a large number of different stakeholders that must converge on identifying SeCo as a value added technology. A strong work to find this stakeholder, educate them, help and support them is the only key to increase the possibility of success of such a business model. In this case the staff organizational Units should be three. As in the previous cases one Organizational unit should be developed for Business Consultancy and one for Developers Support. Moreover in this case the Marter Manager must be supported too. It is particularly difficult to support this actor first of all because it needs both technical competences (to run the mart) and commercial/legal competences to manage the relationships with the Service Owners. Moreover, unlike the previous cases, it is very difficult to find already existing examples in tech based industries about how to support this role, because it is completely specific to the SeCo environment (and perhaps its main innovation and value).

6 Discussion and Recommendations

The first point this chapter starts from is that having a good (or even e great) technology is not enough to succeed on the market. The technological world is plenty of companies that had great success exploiting ideas and technologies introduced by others (e.g., Apple vs. Atari, MSFT and Intel vs. IBM, Sony vs. Nintendo). In many cases the winning technology is not the "best" one and in many cases the market winner is not the first mover. This chapter is not focused on understanding neither the market dynamics typical of technological industries nor their technology strategies. It only starts from the consideration that a great technology which is not used by any customer is useless.

For this reason the first two dimensions along which we compared alternative possible business models are related with the existence of a market. More specifically they are the Value Creation strategy and the Value Appropriation one. On the contrary the remaining two dimensions used are related with the costs to sustain to manage the competences and the building of the right organizational form able to support the Business Models.

Among the many roles that characterize the SeCo value chain (Fig.2) the Tool Developer is the hardest to be alienated: the core knowledge of the system is embedded in this role. Thus, the considered configurations may differ according to which other roles are managed internally or externally. One may feel that the involvement of external actors is a preferable alternative because it avoids to develop and manage competences internally. There are many cases of high-tech industries in which companies rely on third parties to deliver value to the final customer. The personal computer architecture is an example of this approach: a company develops the micro-processor, another one develops the Operating System and many others develop a multitude of applications. The case of the App Store by Apple is a similar recent example. The Cupertino's company is producing the tools (hardware and software) playing the role of Tool Developer and is managing the Store to sell the applications to the market playing the role of Application Publisher. Nonetheless the company relies on external Application Developers to create the apps.

Even if involving external actors may seem a panacea, there is a high risk related to this solution. Opening the system to external actors and allow them to run their business on a platform, using existing tools and architectures, does not imply that they will actually use them. First of all because they may lack some competences. A good example of this can be provided by the cell phone industry before the arrival of the smart phones. Apple competitors already opened their OSs to third parties and many actors tried to take advantage of it by by developing java apps or ringtones. Still, the diffusion of these softwares was poor if compared with the current diffusion of Apple apps. This is because these companies had not neither reputation, nor brand nor access to the market. If we consider the Apple case, the decision of managing directly the App Store and forcing it as the sole channel to market, allowed the third parties to leverage on their competences (concept and app development) providing them those competences that they hardly would develop (access to market). Another reason why third parties may not be willing to take advantage of an existing platform is because they need to invest time and money to "learn" it. This is another typical dynamic in multilayer platform based industries and companies know that they must pay a major effort in reducing the cognitive barriers and increasing the developers productivity (e.g. through APIs). In the Apple case this variable was managed by creating and distributing a Software Development Kit very easy to use as to "make it easy to develop for iPhone". In other words allowing external actors to build their businesses on a platform is not a sufficient condition to make them doing it. External actors must be involved and helped. Their cognitive barriers must be reduced and they must be provided with those capabilities and assets they may lack.

The decision of managing inside or outside some roles leads to reduce some costs but to increase some others. If a company wants to cover many roles in the value chain it will have high costs for the development of internal competences and low costs to support external actors, but of course the contrary is also true (Fig.4).

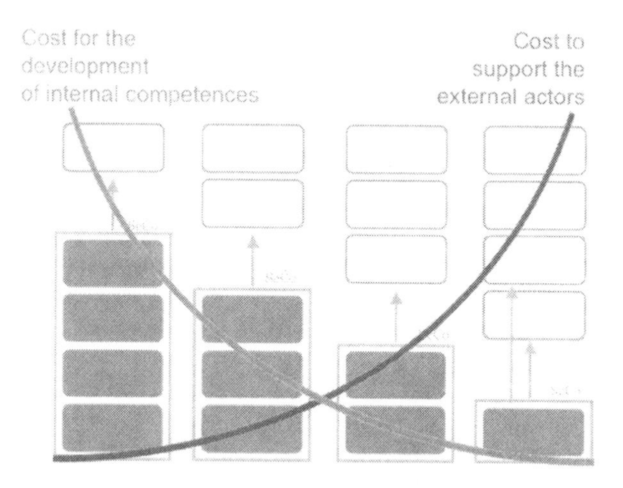

Fig. 4. Configurations development cost

The last point that should be taken into consideration to assess the Business Models is the starting point. The cost for developing any kind of Business Model is dependent on the starting quantity and quality of competences. At the current stage the SeCo Project encompasses mainly technological capabilities. Moving towards a Service Provider Business Model would imply the development of new internal competences, while moving towards a Platform Leadership Business Model would mean investing on the consultancy market.

The above considerations should be taken into account when planning the next steps of the SeCo Project. It seems that a unique or best way to proceed does not currently exist and all the different configurations identified have strengths and weaknesses to be leveraged and avoided. Still, we can say that moving towards one direction (e.g. the Service Deliverer configuration) should imply the internal development of brand new competences for the SeCo team (like the market or marter capabilities) and the further development of some technological capabilities that are currently not completely mastered by the team (e.g. the ones for running the Run Time). On the other side the development of configurations more related to become a service enabler (or even a platform leader) asks for the early involvement of the right partners, that means finding them, engaging them and managing the relationships with them. Even if an orientation for the project potential exploitation will be decided further along the schedule, in our opinion the very next steps of the project should take into account the implications of this decision in order to leave the different options really and not only nominally open till the end of the project.

References

1. Bhattacharya, S., Krishnan, V., Mahajan, V.: Managing New Product Definition in Highly Dynamic Environments. Management Science 44 (11 Part 2), S50–S64 (1998)
2. Cusumano, M.A., Selby, R.: Microsoft Secrets. Free Press, New York (1995)
3. Iansiti, M., MacCormack, A.: Developing Products on Internet Time. Harvard Business Review 75, 108–117 (1997)
4. MacCormack, A., Verganti, R., Iansiti, M.: Developing Products on Internet Time: The Anatomy of a Flexible Development Process. Management Science 47(1), 133–150 (2001)
5. Mizk, N., Jacobson, R.: Trading Off Between Value Creation and Value Appropriation: The Financial Implications of Shifts in Strategic Emphasis. Journal of Marketing 67, 63–76 (2003)
6. Porter, M.E.: Competitive Advantage: Creating and Sustaining Superior. Performance. The Free Press, New York (1985)
7. Teece, D.J.: Profiting from Technological Innovation: Implications for Integration, Collaboration, Licensing and Public Policy. Research Policy 15, 285–305 (1986)
8. Teece, D.J., Pisano, G.: The Dynamic Capabilities of Firms: an Introduction. Industrial and Corporate Change 3(3), 537–556 (1994)
9. Teece, D.J., Pisano, G., Shuen, A.: Dynamic Capabilities Management and Strategic. Strategic Management Journal 18(7), 509–533 (1997)
10. Verganti, R.: Planned Flexibility: Linking Anticipation and Reaction in Product Development Projects. J. Product Innovation Management 16(4), 363–376 (1999)

Author Index